MILITARY HELICOPTER DOCTRINES OF THE MAJOR POWERS, 1945–1992

Recent Titles in
Contributions in Military Studies

Hidden Ally: The French Resistance, Special Operations, and the Landings in Southern France, 1944
Arthur Layton Funk

Beyond Glasnost: Soviet Reform and Security Issues
David T. Twining, editor

For King and Kaiser! The Making of the Prussian Army Officer, 1860–1914
Steven E. Clemente

The U.S. Navy, the Mediterranean, and the Cold War, 1945–1947
Edward J. Sheehy

The Soviet Military and the Future
Stephen J. Blank and Jacob W. Kipp, editors

Soviet Military Doctrine from Lenin to Gorbachev, 1915–1991
Willard C. Frank and Philip S. Gillette, editors

Soviet Military Reform in the Twentieth Century: Three Case Studies
Raymond J. Swider, Jr.

The Laws of Land Warfare: A Guide to U.S. Army Manuals
Donald A. Wells

Raiders or Elite Infantry? The Changing Role of the U.S. Army Rangers from Dieppe to Grenada
David W. Hogan, Jr.

The U.S. Military
John E. Peters

Unconventional Conflicts in a New Security Era: Lessons from Malaya and Vietnam
Sam C. Sarkesian

"Mad Jack": The Biography of Captain John Percival, USN, 1779–1862
David F. Long

MILITARY HELICOPTER DOCTRINES OF THE MAJOR POWERS, 1945–1992

MAKING DECISIONS ABOUT AIR-LAND WARFARE

Matthew Allen

CONTRIBUTIONS IN MILITARY STUDIES, NUMBER 137

Greenwood Press
WESTPORT, CONNECTICUT • LONDON

Library of Congress Cataloging-in-Publication Data

Allen, Matthew.
Military helicopter doctrines of the major powers, 1945–1992 : making decisions about air-land warfare / Matthew Allen.
p. cm. — (Contributions in military studies, ISSN 0883-6884 ; no. 137)
Includes bibliographical references and index.
ISBN 0-313-28522-5 (alk. paper)
1. Close air support. 2. Military helicopters. I. Title. II. Series.
UG700.A58 1993
358.4′142—dc20 92-32230

British Library Cataloguing in Publication Data is available.

Library of Congress Catalog Card Number: 92-32230
ISBN: 0-313-28522-5
ISSN: 0883-6884

First published in 1993

Greenwood Press, 88 Post Road West, Westport, CT 06881
An imprint of Greenwood Publishing Group, Inc.

Printed in the United States of America

The paper used in this book complies with the Permanent Paper Standard issued by the National Information Standards Organization (Z39.48–1984).

10 9 8 7 6 5 4 3 2 1

To My Mother and Father

Contents

Illustrations

PHOTOGRAPHS

FIGURES

TABLES

Acknowledgments

My thanks go first to the Strategic and Defence Studies Centre at the Australian National University where I spent most of my time researching and writing this book; it also provided much-needed material support, including financial assistance for essential overseas research. Thanks to all the members of the Centre (Des Ball, Ross Babbage, Robert Glasser and Gary Waters in particular) and the ANU Department of International Relations for encouraging me in my endeavors. My overseas research trip, and subsequent correspondence, would also not have been possible without the hospitality and acts of kindness of many friends, relatives and strangers; and my thanks go especially to the serving and retired military officers who helped arrange most of the trip and who generously informed and entertained me.

As much as it was a journey of intellectual exploration, writing this book was a voyage of personal discovery—one that often took me into hazardous waters. My thanks to the friends (especially David Cullen, Chris Woodgate and Peter Rush) who helped me navigate these waters; and, Carolyn, your encouragement and reassurance were invaluable. Thanks, too, to my friends and colleagues at the Department of History of the University of Western Australia where I completed my writing. And, to my family, thank you most of all. Of course, all the help in the world cannot prevent errors and omissions, for which I take sole responsibility.

Abbreviations and Translations

GENERAL

AFV	Armored fighting vehicle
APC	Armored personnel carrier
ATGM	Anti-tank guided missile
FEBA	Forward edge of the battle area
FLOT	Forward line of own troops
ICBM	Intercontinental ballistic missile
MMS	Mast-mounted sight
NATO	North Atlantic Treaty Organization
NOE	Nap of the earth
SAM	Surface-to-air missile
S/VTOL	Short/vertical takeoff and landing
X-FLOT	Crossing forward line of own troops

UNITED STATES

AAFSS	Advanced Aerial Fire Support System
AAH	Advanced Attack Helicopter
ACCB	Air Cavalry Combat Brigade
AH	Attack helicopter
AHIP	Army Helicopter Improvement Program
ARCSA	Aviation Requirements for the Combat Structure of the Army

ARVN	Army of the Republic of Vietnam
CAB	Combat aviation brigade
CH	Cargo helicopter
HATAC	Helicopter air-to-air combat
JAAT	Joint Air Attack Team
LHX	Light Helicopter Experimental
OH	Observation helicopter
OSD	Office of the Secretary of Defense
Reforger	Return of Forces to Germany
ROAD	Reorganization Objectives Army Division
TOW	Tube-launched optically tracked wire-guided missile
TRADOC	Training and Doctrine Command
TRICAP	Triple capability
UH	Utility helicopter
USAAF	United States Army Air Forces
USAF	United States Air Force
USAREUR	United States Army, Europe

SOVIET UNION

BMD	*Boyevaya mashina, desantnaya* (airborne combat vehicle)
BMP	*Boyevaya mashina pekhoty* (infantry combat vehicle)
CIS	Commonwealth of Independent States
OKB	*Opytno konstruktorskoe biuro* (aircraft design bureau)
OMG	Operational mobile [maneuver] group
TVD	*Teatr voyennykh deystivy* (theater of strategic military action)
VDV	*Vozdushno-desantnyye voyska* (airbone troops)
VTA	*Voyenno-transportnaya aviatsiya* (air transport branch)
VVS	*Voyenno-vozdushnyye sily* (air forces)
Armyeyskaya aviatsiya	Army Aviation
Front	Unit equivalent to a Western army group
Frontovaya aviatsiya	Frontal Aviation

UNITED KINGDOM

2ATAF	2nd Allied Tactical Air Force

AAC	Army Air Corps
ATGW	Anti-tank guided weapon
BAOR	British Army of the Rhine
CGS	Chief of the General Staff
CINCAFCE	Commander-in-Chief, Allied Forces Central Europe
JEHU	Joint experimental helicopter unit
JWE	Joint Warfare Establishment
NORTHAG	NATO'S Northern Army Group
RAC	Royal Armoured Corps
RAF	Royal Air Force
SH	Support helicopter
SRT	Short-range transport

WEST GERMANY AND FRANCE

4ᵉ DAM	4ᵉ *division aéromobile* (4th Airmobile Division)
6ᵉ DLB	6ᵉ *division légère blindée* (6th Light Armored Division)
9ᵉ DIMa	9ᵉ *division d'infanterie de marine* (9th Marine Division)
ALAT	*Aviation légère de l'armée de terre* (light aviation of the army)
BSH	*Begleitschutz Hubschrauber* (escort/protection helicopter)
FAR	*Force d'action rapide*
GALCA	*Groupe aviation légère corps d'armée* (corps light aviation group)
GALDIV	*Grupe aviation légère division* (divisional light aviation group)
GHL	*Groupe d'hélicoptères légère* (light helicopter group)
HAC	*Hélicoptère antichar* (anti-tank helicopter)
HAP	*Hélicoptère d'appui et protection* (support and protection helicopter)
HOT	Haut supersonique optiquement réléguidé tiré d'un tube (high supersonic optically guided tube launched)
MBB	Messerschmitt-Bölkow-Blohm
PAH	*Panzerabwehr Hubschrauber* (anti-tank helicopter)
RHC	*Regiment d'hélicoptères de combat* (combat helicopter regiment)

RHCMS	*Regiment d'hélicoptères de commandement, de manoeuvre et de soutien* (command, maneuver and supply regiment)
Heeresflieger	Army Aviation

NOTES AND BIBLIOGRAPHY

AACJ	*Army Air Corps Journal*
AAH	*Armées d'Aujourd'Hui* (*Armies of Today*)
AD	*U.S. Army Aviation Digest*
AIK (*Aviatsiya i kosmonavtika*)	*Aviation and Cosmonautics*
AQDJ	*Army Quarterly and Defence Journal*
AWST	*Aviation Week and Space Technology*
BAR	*British Army Review*
CSI	Combat Studies Institute
GAO	Government Accounting Office
IDR	*International Defense Review*
IISS	International Institute for Strategic Studies
JPRS	Joint Publications Research Service
JAA	*Journal of Army Aviation*
JRUSI	*Journal of the Royal United Services Institution*
KZ	*Krasnaya zvezda* (*Red Star*)
PWLIWOPK	*Przeglad wojsk lotnicznych i wojsk obrony powietrznej kraju* (*Air Forces and Air Defense Forces Review*)
RUSI	Royal United Services Institution
SASO	Soviet Army Studies Office
SSRC	Soviet Studies Research Centre
UAC	*Report on USSR—Aviation and Cosmonautics*
UMA	*Report on USSR Military Afairs*
USACGSC	U.S. Army Command and General Staff College
V-IZ	*Voyenno-istoricheskiy zhurnal* (*Military History Journal*)
VM	*Voyennaya mysl* (*Military Thought*)
Kommunist vooruzhennykh sil	*Communist of the Armed Forces*

Pravda	*Truth*
Przeglad wojsk ladowych	*Ground Forces Review*
Tekhnika i vooruzheniye	*Arms and Technology*
Voyenny vestnik	*Military Herald*
Zarubezhnoye voyennoye obozreniye	*Foreign Military Review*
Zolnierz wolnosci	*Soldier of Freedom*

Introduction

> How can men attune their minds as clearly as possible to the constantly changing conditions and demands of war? How do military institutions adjust to new realities, what forces carry innovation forward, and what obstacles stand in its way?
>
> —*Peter Paret*[1]

After the Second World War, the application of nuclear energy and electronics to military activity, especially precision guidance technology, revolutionized warfare. Battlefields became vastly more lethal than ever before. Improved technology also saw aircraft flying higher, further and faster, while tanks and other ground vehicles became speedier and more reliable. As a result, the armed forces of the 1980s were much more mobile than those of 1945. However, and especially in comparison to the advances in firepower, most of the improvements in mobility were incremental, based on minor technical developments. Indeed, only one major innovation after 1945 brought with it a whole new form of military mobility: the helicopter. Yet, despite its significance, helicopters have only been subjected to limited study. This work is intended to correct this deficiency.[2]

Because the helicopter—for our purposes, a shorthand term that includes all the doctrines and technologies associated with helicopters—was a major innovation, the history of its development provides an excellent example of the processes by which military organizations change and adapt to new forms of warfare. There is considerable literature on decisionmaking and innovation in the post–1945 era, but little of it is directly relevant to this work. It concentrates on technology and weapons innovation; its main emphasis is on nuclear and strategic issues. In contrast, this book looks at operational and tactical change and is based foremost on Peter Paret's judgment that "the most important problem of innovation [is] not the development of new weapons, nor even their general adoption, but their intellectual mastery."[3] Thus it focuses on ideas and doctrines, the link between the actual use of helicopters and the weapons themselves.

While not ignoring events in the first two decades after the Second World War, the chapters that follow pay greater attention to more recent developments. A historian might well be on safer, steadier ground in the earlier period but would find fewer crucial developments, since helicopters were still on the periphery of military activity.[4] Despite the concentration on the last twenty years, this approach can reasonably be labeled contemporary history. There may be some argument that the events are simply too recent for the historical approach. However, on the whole, the last few years—what might be called the historical contemporary—were so important to the continuum of developments since the Second World War that it was impossible to exclude them, and the historical approach is suited to the general, inclusive picture painted here.

This work focuses on developments after the mid–1960s because it was in this period that military helicopters had a noticeable impact on thinking about major conflicts. Counter-insurgency campaigns may have been the predominant form of warfare since 1945 but, because of a mixture of sound strategic reasoning, assumptions about potential threats and military conservatism (some would say nostalgic inertia), it is the preparation for major conflicts that has provided the ultimate criteria by which the world's leading powers have judged the helicopter's effectiveness. Moreover, this book is particularly concerned with how change occurred and innovations were accepted. Counterinsurgency warfare was a relatively new military experience, and, as such, caused new ideas to be more easily integrated into existing plans. In contrast, major conventional warfare was surrounded by numerous preconceived notions, and the clash of innovation against this conservatism is a crucial theme in the following chapters. For similar reasons and because of space limitations, the employment of helicopters for amphibious warfare—an area in which there has been little doctrinal debate—is not discussed.

The monograph concentrates almost exclusively on the experiences and actions of the major contributors to the development of military helicopters. The first three chapters deal with the United States, the Soviet Union and the United Kingdom, while chapter 4 discusses, in less detail, West Germany and France. While this distribution on the one hand reflects the relative availability of source material—especially in the case of West Germany—on the other hand, it is a fair indication of the respective importance of each country's developments and the size of its helicopter deployments. Although the chapter on Britain might appear anomalous, its length is necessary in light of that country's special experience with the use and organization of helicopters. Each chapter discusses the development of helicopters from fragile, almost experimental light aircraft used mainly for artillery observation into sophisticated war machines capable of a multitude of tasks. It analyzes the reasons why helicopters were introduced, how they have developed and the way in which the changes were accepted and implemented. Given that intellectual and doctrinal innovation are their heart, these chapters give little detail about the many conflicts in which helicopters have been used, nor do they provide comprehensive technical details for helicopters or examine in depth their design, development and acquisition. There are already more than enough books covering these matters: what is needed is a fresh approach.

The chronological analysis in these four chapters is essential. It provides some of the narrative story of helicopter developments, which has, in large part, been absent until now, but it also prepares the way for the two succeeding chapters in which the last forty-five years of helicopter development are viewed as a single event and subjected to a broader analysis. Two different approaches are used in these chapters.

Chapter 5, *Helicopters and Air-Land Warfare*, summarizes the helicopter's development and its fundamental, technological characteristics. It argues that the helicopter's level of success and impact on the military were an inseparable part of the rise of a form of warfare in which the activities of two components, air and land, became closely integrated in pursuit of a land-oriented goal. Air-land warfare did not cause helicopters to develop, nor was it solely dependent on them. The helicopter's particular contribution was to make available to land forces the speed of movement by air, while retaining the capacity for direct and intimate interaction with ground troops and ground terrain. Some have argued that the helicopter is the new tank in that it will become the dominant arm of land combat, much as armored vehicles did in the 1930s and 1940s. They suggested that this process can be termed a "rotary-wing revolution."[5] Yet if there has been any revolution, it occurred not through competition between tanks and helicopters but because they complemented one another; "intellectual mastery" was obtained when the proper relationship between the two was established.

The last chapter is on decisionmaking in helicopter innovation. It identifies, classifies and examines the many determinants of doctrine: those factors that influenced and indeed comprised the decision-making process. It has been customary for scholars of this process to make prescriptions for more effective government policy and action. One result of this emphasis has been that ultimate explanations tended to be simple rather than complex. This view arose not necessarily because the processes were that elementary, but because it was much easier to make explicit or even implicit recommendations for changes in government policy if the basis for them was easily (if imperfectly) described. (In contrast, this work is not intended to provide suggestions for government or military policymakers.[6]) Some determinants were, at times, more important than others but overall the most crucial part of the process was their interaction. These factors were affecting one another and thus were influencing doctrine within the context established by their own interplay.

What is doctrine? In broad terms, doctrine is the process by which armed forces seek to impose control over war: it is their construction of how conflicts ideally ought to be fought. It is also a means of harnessing personnel and technology to produce combat power. Although it would be useful if writers on military affairs could achieve some consensus on the issue, unfortunately there are almost as many definitions of the scope and content of doctrine as there are authors working on the subject! Julian Lider, in *Military Theory*, gave one of the best summaries of these definitions. He concluded that doctrine can range from the level of grand strategy to tactics; it can be inchoate or finished, exist in the minds of soldiers or be officially published; and it can be either a set of loose principles or strict guidelines or both: clearly it is a word for all seasons.[7]

The difficulties are compounded when the term is applied to different countries. Soviet military doctrine was rigorous and essentially political in nature. It was:

> a state's system of views and instructions on the nature of war under specific historical conditions, the definition of the military tasks of the state and the armed forces and the principles of their development, as well as the means and forms of solving all of these tasks.[8]

The West, with its rather different culture and society, takes a much less formal view, with officials and independent analysts applying the word to a range of military decisions and ideas from strategic to tactical levels. Although generally scholars employ the word "doctrine" to mean military beliefs, principles and practices of a strategic nature, this usage does not reflect any inherent meaning in the word but rather the interests of the authors. For example, writers at the U.S. Army's Combat Studies Institute, with their own particular assumptions, have used "doctrine" in relation to low-level tactics.[9]

Despite this diversity, doctrine is an appropriate word to use because of the connotation that, as defined by the *Oxford Dictionary*, it is especially "that which is laid down as true . . . a belief, theoretical opinion; a dogma."[10] The element of compulsion to accept doctrine suitably suggests the hierarchical nature of the military. More deeply, the word implies that belief is an essential component of doctrine. This element of faith stems from the fact that doctrine is not objectively determined. While the process by which doctrine is constructed includes elements that might be termed objective (although "less subjective" is a more apt phrase), subjective interpretations of the worth of helicopters—based, for example, on similarly subjective perceptions of their use in past wars or peacetime trials—are at least as important. Moreover, political and organizational imperatives that often have little to do with the supposedly objective process of doctrine-making can play vital roles. There is no objective way of knowing that procedures or organizations determined by doctrine will work in combat; rather it is believed that this is the case. Often the belief will be justified, sometimes not. Either way, until the event occurs there is no way of telling.

To help gain a clearer understanding of doctrine, it can be broken down into some fundamental components, although it must be realized that in reality they remain innately connected. *Sources of Military Doctrine*, by Barry Posen, used the following questions to define doctrine: "*What* [military] means shall be employed? and *How* shall they be employed?"[11] Since the focus on helicopters answers much of the first question, this work concentrates mainly upon the elements Posen suggested for answering the second. These are the structure and employment of the forces and their relationship with other forces. Robert Doughty addressed this issue in more detail in *The Seeds of Disaster—The Development of French Army Doctrine 1919–1939*. Starting with the definition that doctrine was "the best thought available on what would usually work best on the battlefield," Doughty supposed that it was the basis for combat action, for training and for organization; included in all three was the idea that it brought some uniformity of effort to all parts of the army. While Doughty's analysis was influenced by the nature of doctrine in his own, contemporary U.S. Army, nevertheless it provides a sound footing from which any scholar seeking to examine doctrine can proceed.[12]

Although some analysts prefer to limit the meaning of the word "doctrine,"[13] the approach taken by Posen and Doughty is better because it more accurately reflects the very complex nature of doctrine and its components. This width of definition does not, however, extend to the actual weapons that were acquired to fulfill its requirements. Doctrine certainly guides weapons acquisition, but the latter process has its own momentum and dynamics, which are separate from this original guidance.[14] Hence, as most writers have agreed, ideas and weapons are best kept separate for the purposes of analysis. Thus helicopter doctrine is defined as those procedures and principles that were developed to guide the combat employment of helicopters, their organizations into units and the relationship of those organizations to other units, and guidance on the peacetime equivalent of combat—training.

As well as using the aspects of training, organization and employment principles to define doctrine, it is also assumed that doctrine can be divided into theories or doctrines in development, officially sanctioned doctrine and implemented doctrine. It is an important distinction to make since some ideas are not sanctioned by the military hierarchy, while others, although approved, are not in fact widely accepted and applied. Because desires for acceptance can influence which doctrines are actually proposed, and the realities of implementation can alter them, it is incorrect to treat the formal ideas of doctrine as separate from applied doctrine. Moreover, doctrines exist in many places at once: among thinkers, doctrine writers, decision-makers and combat soldiers. Thus doctrine is used as a shorthand term for a highly complicated subject in which certain elements can be distinguished, even though they are part of an indivisible whole at the same time. The term "doctrine-making" is used to define the process by which decisions are made about the development, consideration and acceptance of alternative doctrines.

For a number of years, but particularly since Graham Allison's excellent *Essence of Decision* was published,[15] studies of decisionmaking have looked beyond the so-called "rational actor" model for more subtle reasons and causes. It is now generally accepted that rational actors do not occupy the stage alone and that the stage itself plays a part. Works such as *Essence of Decision*, which one critic termed the "definitive statement" on the subject,[16] and successors that applied bureaucratic politics and organization theory to contemporary situations are certainly packed in this author's intellectual baggage.[17] However, their approach tended to concentrate on individual elements of decisionmaking within very specific cases and, as noted above, to concentrate on weapons acquisition. For a study of the development of doctrines relating to a class of weapons in various countries over a number of years, the more integrated style of historical works such as Robert Larson's *The British Army and the Theory of Armored Warfare, 1918–1940*, Doughty's *Seeds of Disaster* and Richard Holmes' *The Road to Sedan* is preferable.[18] Moreover, the technique of these works is better suited to analyzing a process in which the relationship between various causes can be crucial; indeed, as will be shown in chapter 6, the act of identifying individual determinants of doctrine can obscure the importance of their interaction.

Another concept that is used throughout the following chapters and that needs some preliminary definition is that of the helicopter advocate or supporter. It is axiomatic that an idea must have an originator, but of more importance to this study were the people who took up that idea and tried to advance it. On the whole, groups of these supporters were located in the structures within armed forces that were responsible for the development and provision of aviation support to the army: the British Army Air Corps, U.S. Army Aviation, the French ALAT and the West German *Heeresflieger*. Helicopter advocacy was by no means limited to these established groups but the desire, resources and capacity to achieve changes in helicopter doctrine were concentrated in them. While the shorthand term "advocacy group" is useful, it does not necessarily imply a unity of purpose, interest and effort. In reality there were competing ideas on the way in which helicopters ought to develop and, indeed, there were different advocacy groups within a particular country or even an individual organization. However, one dominant strand of advocacy usually emerged as the most important, thus justifying the descriptive simplification.

There is no attempt here to define exactly the composition of the advocacy groups, although their leading members can be identified and some general statements made about them. It might be a useful exercise to discover the details of participants in these groups, their specific interests, perceptions and influence both within the group and outside. Certainly, a study of this kind by Frederic Bergerson called *The Army Gets an Airforce*, which looked at the early development of American Army Aviation, was an important source for chapter 1.[19] However, such analysis would consume more space and time than this book allows and, moreover, it is unnecessary when the main focus is helicopter doctrine, rather than the supporters of such doctrine. Instead the work implicitly, though not uncritically, applies some of the characteristics of these groups that have been elucidated by organization theory and by works on bureaucratic politics such as Bergerson's.

Innovation is another important term used widely throughout the book. It should be understood to mean both small, incremental modifications such as the reorganization of the U.S. Army's helicopters from battalions into brigades in the early 1980s—as well as fundamental changes such as the development of anti-tank helicopters throughout the 1960s and 1970s. One reason the individual experiences of a number of countries have been examined in some detail is that the single innovation encapsulated by the term "military helicopter" can only be understood when it is realized that large innovations were often made up of an accretion of many lesser ideas—just as one major change also sparked a series of smaller ones.

Inevitably, any study of a military innovation cannot avoid making judgments about value and quality of the innovation. In an issue of *Air University Review* devoted to the question of military reform, Allan Millett defined the basic criterion for such judgments as whether or not a reform "increased military effectiveness."[20] However, establishing the criterion begs the question as to how one can determine whether effectiveness has increased or not. One method is to use "the test of battle," as Larson termed it in his study of armored warfare innovation.[21] For example, prior to 1940, France's concentration on firepower and defensive tactics was, in comparison to their doctrine before World War I, an apparently reasonable innovation based upon wartime experience and one that increased the effectiveness of its army. Historians, however, know this was not the case. They have been able to use the Battle of France to test whether Millett's criterion was met. Thankfully, no all-embracing test of battle can be applied to the military helicopter doctrines. While there have been some wars in which helicopters were used, these conflicts were not the ones for which the doctrines had been created.

The test of battle is, in any case, fraught with complications. Can historians discover if doctrines were at fault or simply their application, or if ineffectiveness was due to factors entirely separate from doctrine? An even greater methodological problem is that, in looking at reform in terms of increased military effectiveness, the assessment is not defined in the terms of those responsible for innovation. It is, instead, imposed with the benefit of hindsight. A more valid method is to pass judgment in terms that the innovators themselves would have used. They already believed in the combat worth of helicopters, otherwise they would not have supported them. This book is an attempt to trace the path of the new concepts presented by these people, and a more historically valid test of success was whether or not the innovations were accepted. Of course, judgments about military effectiveness must still be made, because ineffectiveness was one reason why new ideas might not have been accepted. However, the key question here is, once again, not whether the ideas were ineffective but whether it was reasonable for decision-makers to hold that belief. If it was unreasonable—and here it becomes impossible to avoid such judgments entirely—then this suggests that other factors such as organizational constraints or bureaucratic politics were of greater relevance to that particular decision.

Ultimately, innovations—what the dictionary calls "alteration[s] to the established"[22]—were successful when they themselves became part of the establishment. Successful innovation was the transformation of a novelty into something familiar, into a comfortable part of the culture and society concerned. This book is about military helicopters, focusing on doctrine to examine their development since the Second World War. Yet it is equally concerned with what they represent—innovation—and the process of change associated with them. Thus, this work tries to suggest some of the answers to the questions posed in 1966 by Peter Paret:

> How can men attune their minds as clearly as possible to the constantly changing conditions and demands of war? How do military institutions adjust to new realities, what forces carry innovation forward, and what obstacles stand in its way?[23]

NOTES

1. Peter Paret, *Innovation and Reform in Warfare* (Colorado: USAF Academy, 1966) 2.

2. The word "military" is used throughout the book as an antonym to "naval," that is, it implies land or land-oriented warfare.

3. Paret, *Innovation and Reform in Warfare*, 2.

4. The earlier period has already been covered to some extent by other authors, certainly more so than the later years. See, for example, Michael Hickey, *Out of the Sky* (London: Mills and Boon, 1979) and John R. Galvin, *Air Assault: The Development of Airmobile Warfare* (New York: Hawthorn Books, 1969).

5. Richard E. Simpkin, *Race to the Swift* (London: Brassey's, 1985) 117.

6. Matthew Evangelista, *Innovation and the Arms Race* (Ithaca, NY: Cornell University Press, 1988) 6–21 has an excellent analysis of this trend. Whether this book's rejection of policy prescription is itself a reason for the highly complex model of decisionmaking and innovation that emerges in chapter 6 is a moot point.

7. Julian Lider, *Military Theory* (Aldershot, Hants.: Gower, 1983) 308–10.

8. Major General S. N. Kozlov (ed.), *The Officer's Handbook* (trans.) (1971: Washington, DC: USAF, 1977) 62.

9. Robert A. Doughty, *The Evolution of U.S. Army Tactical Doctrine 1946–76* (Ft. Leavenworth, KS: Combat Studies Institute (CSI), 1979) and Timothy T. Lupfer, *The Dynamics of Doctrine: The Changes in German Tactical Doctrine during the First World War* (Ft. Leavenworth, KS: CSI, 1981).

10. *The Shorter Oxford English Dictionary*, vol. 1 (Oxford: Clarendon Press, 1985) 589.

11. Barry R. Posen, *The Sources of Military Doctrine* (Ithaca, NY: Cornell University Press, 1984) 13.

12. Robert A. Doughty, *The Seeds of Disaster—The Development of French Army Doctrine 1919–1939* (Hamden, CT: Archon Books, 1985) 6–8.

13. For example, Paret, *Innovation and Reform in Warfare*, 4.

14. See Gordon H. MacCormick, "The Dynamics of Doctrinal Change," *Orbis* 27.3 (1983): 267.

15. Graham T. Allison, *Essence of Decision* (Boston: Little, Brown, 1971).

16. Steven K. Krasner, "Are Bureaucracies Important? (or Allison Wonderland)," *Foreign Policy* 7 (1972): 160.

17. Some examples of the extension of Allison's and others' work can be found in R. G. Head, "Doctrinal Innovation and the A–7 Attack Aircraft Decisions," and K. N. Ciboski, "The Bureaucratic Connection: Explaining the Skybolt Decision," in J. E. Endicott and R. W. Stafford Jr., *American Defense Policy* (4th ed.) (Baltimore: Johns Hopkins University Press, 1977) 374–88 and 411–23.

18. Robert H. Larson, *The British Army and the Theory of Armored Warfare* 1918–1940 (Newark, NJ: University of Delaware Press, 1984); Doughty, *Seeds of Disaster*; and Richard Holmes, *The Road to Sedan* (London: Royal Historical Society, 1984).

19. Frederic A. Bergerson, *The Army Gets an Airforce* (Baltimore: Johns Hopkins University Press, 1980).

20. Allan R. Millett, "Military Reform in America," *Air University Review* 36.6 (1985): 3.
21. Larson, *The British Army and the Theory of Armored Warfare*, 234.
22. *The Shorter Oxford English Dictionary*, vol. 1, 1077.
23. Paret, *Innovation and Reform in Warfare*, 2.

MILITARY HELICOPTER DOCTRINES OF THE MAJOR POWERS, 1945–1992

1

"Above the Best"—Developments in the United States

> The helicopter is aerodynamically unsound. It is like lifting oneself by one's boot-straps. It is no good as an air vehicle.... No matter what the Army says, I know that it does not need any.
>
> *Unnamed Air Force General, 1950.*[1]

> I believe that [Army] aviation is the dominant combat arm of the future.
>
> *General G. K. Otis, U.S. Army, 1983.*[2]

After the Second World War, army helicopter developments in the United States were initially driven by the air force's reluctance to provide adequate tactical air support to the ground forces. Emboldened by the success of a fledgling Army Aviation force in Korea, these developments centered on the theory of "airmobility." Within this concept, some senior army officers proposed that army helicopters and fixed-wing aircraft could carry out many combat and support missions in close cooperation with infantry units, which would thereby become highly mobile. With more favorable political and financial circumstances in the early 1960s, the validity of this theory was accepted and implemented to some extent with the formation of an air assault division including some four hundred army aircraft.

While these developments had taken place mainly with conventional and nuclear warfare in mind, the war in Vietnam meant the air assault division and most other Army Aviation units were used for counterinsurgency missions until the early 1970s. Little theoretical progress occurred during these wartime years except in relation to anti-tank helicopters. While the Vietnam War benefited helicopters and led to a dramatic increase in the size and capability of Army Aviation, it altered the direction of further developments.

When the United States withdrew from Indochina the army underwent an extensive transformation to prepare it for mid- to high-intensity conflict, a process marked by the promulgation of a new operational doctrine known as Active Defense. Airmobility, the guiding theory of the 1950s and 1960s, became associated with the particular characteristics of the Vietnam War and with America's ultimate defeat. Questions were asked about the survivability and usefulness of the helicopter, and airmobility as defined in the 1950s was all but abandoned, except by those few units still preparing for a low-intensity conflict. Instead, there was an overwhelming concentration on the employment of helicopters for anti-tank combat. Even with this new approach, Army Aviation still struggled to overcome doubts and implement its helicopter doctrines. However, it had at least begun to establish itself as a credible part of an army preparing almost exclusively for a major conventional war.

While helicopters were by no means out of favor at the start of the 1980s, they became more significant in U.S. Army planning after the introduction in 1982 of a new doctrine called Airland Battle. It did not directly favor helicopters, but promoted the importance of maneuver warfare and created an atmosphere of change and innovation that helicopter advocates exploited. While the attack helicopter's importance increased even further, other activities—such as infantry air assault missions—were reconsidered and also given more approval. In concert with the helicopter's expanding role, Army Aviation became a more important and influential organization within the army, thereby setting the scene for further and more far-reaching developments in future years. Army Aviation proponents grew bolder and gained further evidence to convince doubters after the effective and crucial work of helicopters in operations Desert Shield and Desert Storm in the Persian Gulf.

This chapter examines these developments in more detail. It is divided into three sections. The first deals, in rather less detail, with the period from 1945 to the American withdrawal from Vietnam in 1971–72; much of this development has been covered in depth by other writers and it is not necessary to go over this ground again. The next two sections describe in much greater detail events in each of the following decades and emphasize the mix of factors that either led to or hindered innovation in helicopter doctrine. Although the many decisions made about military helicopters were expressed in terms of the army's overarching doctrines and usually owed much to current military planning, organizational and political factors were also important. Army Aviation played a significant role in gaining wider acceptance for helicopters. While individual doctrinal advances tended to be a response to a wide variety of specific issues, the general trend over the last forty-five years was that helicopter doctrines were developed to fulfill, within the army's land combat orientation, the need for aerial fighting power and to resolve the problems of combining air and ground forces of widely differing mobility.

THE FIRST TWENTY-FIVE YEARS

Early Innovations

By the end of the Second World War, the United States Army Air Force (USAAF) had acquired over four hundred Sikorsky R–4 and R–6 helicopters to perform a variety of light aviation tasks. However, they were a very minor adjunct to fixed-wing aircraft and played little part in actual combat. There were a few limited trials of armed helicopters, but of more significance was the use of a helicopter for casualty evacuation in Burma. This activity demonstrated that the helicopter's ability to hover made it better than fixed-wing machines for operations in close support of the army.[3]

After 1945 the USAAF became the United States Air Force (USAF), marking its complete separation from the army; at the dawn of the nuclear age, it concentrated almost exclusively on strategic aerial warfare.[4] The USAF nominally retained responsibility for close air support of ground forces, but the army became concerned that the air force was unwilling to provide it. In the recent past, during the Second World War, the USAAF seemed reluctant to conduct ground support missions. Then in 1948 the independent Tactical Air Command, set up in 1946 to meet the army's needs, was made subordinate to the strategic Continental Air Command.[5]

The army had begun to compensate for deficiencies in air force support in 1942 by establishing its own aviation units, although at that stage they were equipped only with light aircraft and primarily used for artillery spotting. To make up for the USAF's even greater disinterest in close air support after the war, the army continued to operate these light aircraft and began considering the acquisition of larger transport aircraft such as the De Havilland CV–2 Caribou; it also experimented with armed jet fighters. Most significantly it obtained in 1946–47 its first helicopters, fifty Bell OH–13 Sioux. Initially it was unsure how to use them but, as General James Gavin, an ardent helicopter proponent, later wrote, "It soon became apparent we had in our hands an air vehicle of great versatility."[6] Moreover, while the USAF fought vigorously against the expansion of Army Aviation,[7] it objected less to army ownership of helicopters than fixed-wing machines because helicopters did not appear to pose a threat to the air force's exclusive control of important airpower missions. Hence in the various army-USAF agreements of the 1950s that defined each party's aviation roles and missions, helicopters were never limited to the same extent as fixed-wing aircraft; indeed, by the end of the 1960s, Army Aviation was to consist mostly of rotary-wing aircraft.[8]

Despite attempts to restrict it, Army Aviation continued to expand, especially after its accomplishments in the Korean War. The helicopter was particularly useful in this conflict and, because it was not dependent upon airfields, showed itself to be more suitable than fixed-wing aircraft for many army missions. While the OH–13s evacuated casualties and carried small loads of emergency supplies over rugged mountain terrain, the bigger Sikorsky CH–19 Chickasaw helicopters of the U.S. Marine Corps (USMC) lifted impressively large loads. The army obtained its own CH–19s and, in the final stages of the war, carried out a small number of troop and supply movements with them. After the war, the army set a requirement for twelve transport helicopter battalions, each with sixty helicopters. As General John Tolson, official historian of Army Aviation in the Vietnam War, stated, "Though the twelve-battalion goal was overly optimistic at the time [because of financial restrictions], it dramatized how the impact of Korean experience had influenced Army planners."[9]

The period after the Korean War was one of great turmoil for the United States Army. Despite some successes, the army's performance in Korea had generally been disappointing. Moreover, the growing reliance on the threat of massive retaliation raised the question of what missions the Army could fulfill in the atomic era; it also meant a reduced budget for conventional forces as funds were used to acquire nuclear weapons. The USAF intensified its concentration and spending on strategic warfare at the expense of ground support; but even so the "watchwords were caution, prudence, thrift" for Army Aviation.[10] One senior proponent of helicopters recalled: "The army didn't oppose the tactical use [of helicopters], they just didn't have sufficient faith in its practicality to fund it."[11] Although the army had acquired just over 1,200 helicopters by 1955, many of them were obsolescent and this number was insufficient to meet its stated requirements. The army was only tentatively considering the design of new helicopters, and the USAF was determined that future army aircraft would continue to have limited capabilities. As General Matthew Ridgway concluded in 1956, financial and personnel restrictions ensured that the development of a strong Army Aviation force "in the near future [was] highly improbable."[12]

In stark contrast to the restrictions on equipment, there was no limit to the imagination of aviation advocates, and many new ideas about military helicopters emerged. More important, these officers began developing a general theory of airmobility, laying a foundation from which helicopters could start to play a much more direct role in battle. While

Photo 1.1
OH–13 Sioux

The Sioux was one of the first helicopters used by the U.S. Army and was also used by the Army Air Corps in the United Kingdom. (Bell Helicopter Textron. Used by permission.)

in other armies airmobility came to mean transport operations conducted by helicopter, in the mid–1950s in America the phrase meant much more. Coinciding with the army's growing interest in aviation, airmobility described the use of any form of aerial movement—parachutes, gliders, short take-off and landing assault aircraft and helicopters—to provide more mobility for infantry units. Reconnaissance, fire support, and logistic and other services for these units would also be provided by aircraft.

The airmobility proponents were mainly high-ranking paratroopers. These officers formed an "airborne club" and "dominated postwar military organization";[13] two of them, Generals Ridgway and Maxwell Taylor, became Chief of Staff of the army during the 1950s. Their interest in helicopters followed naturally from their airborne experience, and the link was strengthened when the OH–13s were first assigned to the 82d Airborne Division. These officers saw the helicopter as a way of providing tactical mobility to paratroops after they had landed, thus overcoming a major problem of airborne operations in the Second World War. Furthermore, the advent of nuclear weapons had rendered massed parachute drops all but impossible and helicopters, by allowing more dispersed and more rapid operations, offered a chance to preserve some form of aerial movement on future battlefields.[14]

The strongest advocate was General Gavin. In 1954, while Director of Army Doctrine and Combat Development, he published an article entitled "Cavalry . . . And I Don't Mean Horses!" Although it appeared in the popular *Harper's* magazine and was part of a concerted drive by senior officers to gain public support for the army, Gavin's article reflected the work of his directorate after the Korean War. Its central theme was that nuclear devastation would render speedy maneuvering by ground forces very difficult while, at the same time, mobility would be vital to exploit the effects of atomic weapons.[15] Gavin stressed the importance of cavalry actions such as reconnaissance, screening, exploitation and pursuit, which he believed would best be performed by units equipped with armed, troop-carrying and supply helicopters. The army's concentration on nuclear warfare ensured that there was little money for helicopters but also meant there was an atmosphere of innovation in which Gavin's ideas could prosper. Short on practicalities but long on vision, these theories were to underlie most of the work of the army's helicopter enthusiasts in the 1950s.

Although Gavin couched his ideas in terms of cavalry operations, it was the Infantry Branch, and the airborne troops especially, who took up the development of helicopters most enthusiastically. Another senior paratrooper, General Tolson, was appointed by Gavin to the School of Infantry at Fort Benning, where he promptly renamed the Airborne Department the Airborne-Army Aviation Department and produced the first field manual on helicopter operations.[16] When the infantry divisions were reorganized for atomic war into the Pentomic structure, which emphasized the use of decentralized combat groups, each was, on paper, given a unit of fifty-seven helicopters to provide troop mobility and logistic support for their widely dispersed subunits. Further, the Aviation Center at Fort Rucker (set up in 1954) expanded its activities and influence.[17] A Directorate of Army Aviation was created in the general staff, ensuring that a senior officer was available to promote new ideas and present an aviation perspective in the planning process.

Writing manuals and establishing centers did little immediately to give the army an aviation fighting force. However, these measures were vital to implant within the army as a whole a knowledge, a culture, of helicopters and aviation doctrine. For this purpose a name change was as important as weapons acquisition. Army Aviation proponents were well aware of the acculturation process. There were deliberate attempts to increase the number of senior officers and talented junior officers who were qualified pilots and who would thus have more knowledge of, and interest in, Army Aviation.[18]

But problems remained in meeting Army Aviation's equipment needs. The Army Aircraft Requirements Review Board of 1960 (known from its chair as the Rogers board) made relatively modest recommendations, calling for a new observation helicopter and, in light of financial constraints, deferring judgment on other aviation needs. While it also considered "the idea of the army having an aerial combat capability," no further action was taken because of the expected opposition that would arise from senior officers of both the army and the USAF. Gavin's ideas had already led to experiments with armed helicopters, but this work was limited because the Army was formally prohibited from having armed aircraft and the USAF opposed these developments. Therefore the work was at first carried out surreptitiously and did not obtain official army authorization until 1958; even then it was classified only as experimental.[19] The Rogers board therefore concluded only that more studies ought to be conducted by the army to assess the feasibility of "air fighting units."[20]

The Howze Board: Putting Gavin's Theories into Practice

The limited recommendations of the Rogers board were one of the last products of the financial restrictions of the 1950s. In 1959 the army started to move away from its reliance on nuclear weapons. Seeking to reorganize its divisions and strengthen them for conventional warfare, it began in 1961 to implement the Reorganization Objectives Army Division (ROAD) concept. In particular, ROAD stressed that the infantry needed more mobility and protection, and included the formation of the first mechanized infantry divisions in the U.S. Army. Each infantry division was nominally assigned more transport helicopters, and all divisions were given observation aircraft. The number of larger logistic carrying helicopters at corps and higher levels was also increased. However, as before, actual deployments of helicopters fell short of these authorized establishments.[21]

In 1961 the new administration of President John F. Kennedy brought fundamental changes to American defense policy. The strategy of massive retaliation was replaced by that of flexible response, which placed more emphasis on non-nuclear warfare. The development of helicopters benefited because more money was available for conventional weapons. More importantly, Robert S. MacNamara was appointed Secretary of Defense. Questioning the armed forces' competence to make their own decisions about what equipment to buy and how to use it, he sought to examine and reform America's defense policies and to make the civilian bureaucracy more powerful in managing them. In the process, the army's conservative attitude toward new ideas—such as airmobility—was countered to some degree by forward-looking civilians; there was considerable "background maneuvering"[22] as senior officers were bypassed in favor of their more modern subordinates and as new ideas proposed by these officers were used by civilian bureaucrats as issues over which they could attack entrenched army power in the Pentagon.

Basing his work on interviews of those involved, Frederic Bergerson characterized the politics of helicopter development in this period as a "bureaucratic insurgency," a kind of guerrilla war conducted by aviation supporters against established practice. While superficially it appeared that MacNamara was personally involved,[23] Bergerson argued that there was a relationship between an unnamed civilian systems analyst in the Office of the Secretary of Defense (OSD) and an army colonel, both of whom strongly favored helicopters and Army Aviation. They convinced MacNamara to reject the current cautious approach, drafting for him a key directive to the Army.[24] Calling for a "bold new look," it concluded:

> I shall be disappointed if the Army's reexamination merely produces logistically oriented recommendations to procure more of the same, rather than a plan of employment of fresh and perhaps unorthodox concepts which will give us a significant increase in mobility.[25]

Responding to an explicit suggestion in the directive, the army established the Army Tactical Mobility Requirements Board in May 1962 to determine how airmobility could be introduced. Its members had been appointed "in a most unusual departure from accepted practice" by MacNamara—or rather by the aviation proponents via MacNamara—and were favorably disposed toward their subject.[26] The board was chaired by General Hamilton Howze, who was a great supporter of helicopters. When Director of Army Aviation in the mid–1950s, he had been exposed to a number of airmobile concepts and had been involved in the Armair trial of a brigade-sized combat aviation unit.[27] Moreover, as a senior and respected cavalry officer, Howze's report was likely to carry more weight than one produced by a member of the airborne club.

The board began an intensive investigation of its subject. It conducted wide-ranging field trials and computer simulations of various airmobile formations and combat scenarios, studied British and French experience and visited South Vietnam, where helicopters were just beginning to be used. After three months the board's report was ready.[28] Its basic assumption was that aircraft, and helicopters in particular, were capable of fulfilling the army's five combat functions—reconnaissance, maneuver, firepower, command and control, and logistics. Its key recommendation was that there should be no fewer than five air assault divisions. They would be similar to infantry divisions but with their artillery and vehicles replaced by rocket-firing and transport aircraft and helicopters. These units would have equalled regular divisions in terms of firepower and exceeded them in mobility. The Howze board also proposed the formation of three air cavalry combat brigades, each including 144 anti-tank helicopters, and five air transport brigades with eighty fixed-wing and fifty rotary-wing aircraft each.

The report argued that the air assault formations ought eventually to replace two regular infantry divisions, the two airborne divisions and one of the new mechanized divisions. It suggested that in any future war the U.S. Army would probably be fighting alongside allied nations who could provide infantry divisions if necessary and that only America could bear the expense of the helicopter-based units. Thus the board compared the five air assault divisions not to thirteen other U.S. divisions, but to two hundred allied ones. Finally, the report argued that it was better to have air assault units in the regular army since, unlike infantry divisions, they could not be created rapidly in any general mobilization.[29]

Although the work of the Howze board carried much weight because of MacNamara's apparent involvement, the seniority of its members and the extent and detail of its recommendations, the suggested airmobile formations still existed only on paper, and the future of airmobility remained uncertain. If fully implemented, the report's recommendations would have seen Army Aviation more than double in size over five years, with the air assault divisions being the most expensive units in the army. Hence, airmobility threatened many vested interests. Some senior officers were fearful that the costly air assault divisions would mean less money for tanks and the new mechanized infantry units. Howze recalled:

> The Chief of the Armor Branch for example [was] not, as a matter of principle, against the funding of expensive Army aircraft development and purchase—

> but he [would] argue against such expenditures at the expense of urgently needed tank development and purchase.[30]

This anxiety was fueled by the exaggerated claims of some aviation enthusiasts that the helicopter would some day replace the armored vehicle.[31]

The USAF was also opposed to any expansion of Army Aviation, especially of the latter's fleet of large fixed-wing aircraft such as the Caribou and Grumman OV–1 Mohawk. The formations proposed by the board were anathema to the air force's strict doctrines on centralization of airpower—both for peacetime research, development and acquisition, and for wartime employment. The USAF was incensed with Army Aviation's proposal to convert Mohawk observation aircraft to a ground attack configuration in direct contravention of the existing roles and missions agreement. Indeed, the twenty-four heavily armed Mohawks in the air assault division caused greater political and bureaucratic problems between the services than its 404 helicopters,[32] further illustrating how the USAF worried far more about army fixed-wing aircraft than helicopters during this period.

MacNamara responded positively to the army's "bold new look" at airmobility but balked at the costs involved in implementing it. The 11th Air Assault Division (Test) and the 10th Air Transport Brigade were established in 1963 to explore the concept more fully before a final decision was made. A parachute officer, Brigadier General Harry Kinnard, was placed in command and given orders "to determine how far and how fast the Army can go, and should go, in embracing airmobility."[33] The air cavalry brigade was not formed because of resource limitations.

The army was, however, committed to a serious test, and Kinnard was allowed to purchase special equipment he needed and, despite the war in Vietnam, take priority in the assignment of pilots. After nearly two years of development the 11th Air Assault Division was tested for a month in maneuvers against the 82d Airborne Division. The conclusions drawn were that the air assault division had its limitations, especially when attacked by armor or when conducting prolonged combat, but that it had a great capacity for rapid reaction movement, could fight in many directions at once and was well suited to operations over large areas.

Moreover, the USAF's proposed alternative, a regular infantry division supported by more air force aircraft, was tested at about the same time and found to be nowhere near as capable because it lacked the close and continuous integration of ground and air assets. As much because of the "gut feelings" of the senior decision makers as the "hard data" that the maneuvers provided, the idea of an air assault division was approved.[34] However, as Tolson recalled, the actual retention of the division in the army's force structure was also due to the need for airmobile units in Vietnam. In 1965, the 11th Division merged with the 2d Infantry Division and together they became the 1st Cavalry Division (Air Assault).[35]

In previous years it had been thought that helicopters and fixed-wing aircraft were best employed in small supporting units, helping divisions and corps to fight more effectively. While not ignoring this task, Howze and Kinnard thought that

> Airmobility meant using Army aircraft whenever and however they improved our Army's ability to fight . . . previously Army aircraft were purely supplemental adjuncts to ground combat. Now we could develop . . . Army aircraft to perform all the functions of land combat.[36]

It was postulated that helicopters were best employed en masse in separate air assault and air attack units, permitting a style of combat based on the particular advantages and disadvantages of the fast, light airmobile forces and different from that used by armored units.[37]

Helicopter-based units, as distinct from units supported by helicopters, were far more capable because they were organized, trained and equipped for airmobility, and their commanders and troops had an airmobile frame of mind. By being concentrated, the airmobile forces gained unity of command and were able to operate over, and thus influence, a much wider area of the battlefield. In contrast to helicopters helping individual units of the army, airmobile units would assist the army as a whole.[38] Furthermore, by stressing the maneuver potential of the air assault division, helicopter-borne forces were placed on a more equal footing with armored and infantry divisions. The term "maneuver" implied much more than simple mobility—in fact it marked the difference between airmobile units and infantry divisions with air support. It meant an improvement in status for aviation because maneuver forces were the battle leaders, not just supporters. This change not only reflected the military value of helicopters, but also the desire of aviation advocates to increase their political power within the army.

Before proceeding to discuss the Vietnam War, it must be emphasized that the development of airmobility did not result from the emerging counterinsurgency in Indochina but, as in the 1950s, was due to the perceived requirements of a major conflict. Although in his study of the development of U.S. Army doctrine Doughty stated that the Howze board's tests assessed airmobile formations in combat against "an enemy force of irregulars,"[39] the cavalry brigade consisted mainly of anti-tank helicopters—hardly suitable against guerrillas—and the 11th Air Assault Division was designed for use in "higher scales of war," even involving nuclear weapons. "Vietnam was in the background" and only "one of many contingencies" for the emerging airmobile force.[40] These words were, however, to be forgotten once the Vietnam War had ended.

Vietnam

The U.S. Army concluded from the successful combination of Army of the Republic of Vietnam (ARVN) light infantry units with American helicopters in the early 1960s that airmobility was the best way in which regular military units could contribute to the counterinsurgency campaign in Vietnam. Movement by helicopters overcame Vietnam's difficult terrain and limited road system. It allowed a rapid response to enemy attacks and speedy redeployment of forces, compensating for a lack of numbers. With helicopters, the U.S. Army and the ARVN were better able to reply to the particular mobility advantages of an insurgent enemy operating in small, widely dispersed units. Thus it was hardly surprising that the 1st Cavalry Division (Air Assault) was one of the first major units deployed to Vietnam in the buildup after 1964. Moreover, although the number of helicopters deployed to Vietnam was for many months constrained by a shortage of new aircraft and pilots, General William Westmoreland concluded, "During 1966, airmobile operations came of age."[41]

For the rest of the war, utility helicopters were the single most important means of transport for American and South Vietnamese infantry operations, while larger cargo helicopters lifted artillery pieces and provided an essential supply service. Armed helicopters were used as "flying artillery" and escorts for transport helicopters. The Americans created another air assault division by reorganizing the 101st Airborne Division. Air cavalry squadrons, which combined observation, attack and utility helicopters with a reconnaissance infantry element, were introduced; they were so effective at finding and fixing the enemy that one was assigned to each division. Indeed, helicopters were so crucial to prosecution of the war that aviation units, along with the air assault divisions, were among the last American troops to be withdrawn from Vietnam.

The war had a profound effect upon the development of Army Aviation. First, its size expanded dramatically, its range of missions increased and it became almost entirely dependent on helicopters for combat missions. By the end of the 1950s the Army had 5,470 aircraft, of which 2,489 were helicopters. Ten years later it had 9,528 helicopters but only 2,297 fixed-wing aircraft.[42] Although the development of new and better performing helicopters such as the Bell UH–1 Iroquois (known as the Huey) and the Vertol CH–47 Chinook had been initiated before 1960, the Vietnam War ensured the rapid acquisition of vast numbers of these new machines. To staff its aircraft, Army Aviation required over 24,000 pilots as well as thousands more support personnel. Many senior officers—none more so than Chief of Staff Westmoreland—undertook pilot training and, by 1972, 36 generals and 230 colonels were drawing flight pay. The Vietnam War saw the confirmation of the culture of aviation that had begun in the 1950s. A network of officers in staff and command positions, a bank of combat experiences and readily available resources meant that Army Aviation was a force to be reckoned with—not just on the battlefield but also in the three-way struggle for money between and within the army, the Department of Defense and Congress.[43]

While the Huey and Chinook improved Army Aviation's transport capacity, the most significant expansion of its capabilities came with the acquisition of the Bell AH–1 Cobra attack helicopter. The initial development of armed helicopters had, of course, preceded the Vietnam War. However, it was not until the United States actually began conducting airmobile operations that the need for such aircraft was fully recognized. Even so, utility helicopters equipped with an improvized variety of machine guns and rockets had to suffice for some years. The USAF maintained that armed helicopters encroached on its close air support mission and thus they were only permitted to fire defensively, just before the transport helicopters began to land. Although the armed UH–1 was relatively effective, it was slower than the unarmed version it was accompanying, while its large size and lack of armor also made it vulnerable, especially as enemy air defenses intensified.[44] Both equipment and doctrine were thus severely limited.

In the mid–1960s, however, the army was given the opportunity to rectify this problem. In 1964–65, with a directive from MacNamara to review its future aircraft needs, the army began to consider acquiring a fast, well-armed, armored helicopter specifically to provide fire support during air assaults.[45] Then in 1966 yet another agreement on air support of ground forces was signed. In return for the army giving up its right to operate fixed-wing aircraft such as the Caribou, the USAF relinquished "all claims for helicopters and follow-on rotary-wing aircraft which are designed and operated for intratheater movement, fire support, supply and resupply of Army forces."[46] The agreement ensured the final supremacy of rotary-wing over fixed-wing within Army Aviation. As well as protecting Army Aviation's growing fleet of transport helicopters, it also allowed the army to acquire a purpose-built attack helicopter.

The army promptly expanded its experimental program for a specialized escort helicopter to cover the full range of aerial fire support missions. While this project (like many U.S. military aerospace developments) became more and more complicated and was delayed for technical reasons, the need in Vietnam for a new armed helicopter was increasing rapidly. By a mixture of good fortune and good management, the Bell Helicopter Company was privately testing a relatively simple attack helicopter developed from the UH–1. The Cobra, as it was known, did not extend the bounds of technology but did solve the problems of the armed Huey. It was much faster and had a larger weapons load; more agile, smaller and with some armor, it was also less vulnerable. Cobras were first sent to Vietnam after only one year's development and they immediately became an essential part of the Americans' war effort.[47] Attempts continued to develop a more complex helicopter, termed the Advanced Aerial Fire Support System (AAFSS), and the Lockhead AH–56 Cheyenne was eventually chosen. While repeated technical failures and cost overruns meant that the Cheyenne never entered production, the Vietnam War finally confirmed the army's use of armed helicopters and provided them with the equipment to match.

Indeed the army's general perception of helicopters and their use in combat was much more favorable in 1972 than in 1960 because the Vietnam War gave airmobility "the ultimate legitimacy" of "proven combat effectiveness."[48] However, this legitimacy was gained only at the expense of the link between conventional war-fighting and airmobility that had driven American developments before 1965. Further theoretical progress in this direction had been limited by the ever-increasing demands from the front line for more helicopters and airmobile units. Moreover, the counterinsurgency character of the war meant that airmobility lost the breadth envisaged in early 1960s when it had been thought that airmobile divisions would be employed by corps and armies in concert with large armored striking forces. Although the 1st Cavalry Division was sometimes used as a whole in major offensives such as the relief of Khe Sanh, the vision of airmobility narrowed considerably in Vietnam, and battalion- or even company-sized actions were far more important. As "the Army's attention remained riveted on the infantry-intensive war in Vietnam,"[49] airmobility was linked with that conflict's small-unit, counterinsurgency actions and, undoubtedly, with some of the defeat and despondency that followed the United States' withdrawals.[50] As one aviator put it: "There were many reasons for this shift in emphasis, but I believe the fundamental one was that 'there must be no more Vietnams.' "[51]

THE 1970s: ACTIVE DEFENSE

The Struggle for Credibility

After the withdrawal from Vietnam, the army went through a painful process of reconstruction as it came to grips with the legacies of the war: poor morale, substandard leadership, inadequate training and the failure of conscription. In turning away from "the Vietnam trauma," the army had to begin again to face the threat of a major war against the

Photo 1.2
AH–1S Cobra

The U.S. Army's first dedicated attack helicopter, shown here in its post-Vietnam anti-tank configuration. (Bell Helicopter Textron. Used by permission.)

Soviet Union and its allies after many years in which its thinking about and organization for conventional warfare had been stagnant.[52] Although the need to attend to these problems was clear enough by 1972, the immediacy and scope of the changes required was reinforced by the intense conventional combat of the Arab-Israeli War of October 1973. This war had a major impact on the United States' military thinking.[53] It demonstrated the increased lethality of modern warfare, the need for better training, the likelihood of vast ammunition expenditures and the importance of anti-tank guided missiles (ATGM). The ensuing redirection was led by General William DePuy, the first commander of the newly created centralized Training and Doctrine Command (TRADOC). It resulted in a fundamental new doctrine called Active Defense, which was to guide army planning throughout the 1970s.[54]

Army Aviation's senior commanders were under no illusion about the need to adapt quickly to the "high threat environment" of "mid-intensity warfare." Brigadier General William Maddox, the director of Army Aviation and a key figure in the reorientation, summed it up: "Those of us who *remain* in Army Aviation must continue to strive and to meet new goals. We must consider new thoughts and we must adjust to new conditions. . . . Our thinking and our techniques are subject to great change."[55] This "great change" aimed to bring Army Aviation out of the Vietnam era and to develop tactics, training and equipment necessary for the next war. More importantly in the short term, it sought to establish that Army Aviation could operate in mid-intensity conflict and be an important part of the doctrine of Active Defense, thereby countering the perception that helicopters were suitable only for counterinsurgency wars.[56]

Although Tolson asserted that, because of the Vietnam War, America had "airminded, airmobile commanders . . . [who] will never be satisfied without a '*Three Dimension Army*,' " many in the army were in fact too ready to forget the painful memories of Vietnam in their haste to adapt to mid-intensity conflict. Helicopters seemed to symbolize the failure in Vietnam—even though they had probably prevented a more resounding defeat. General Maddox recalled that "our chief obstacle was credibility. Very few believed that the helicopter could survive and could perform an essential role in a heavy combat environment."[57]

Aviation faced a number of problems in trying to argue that the helicopter could operate on the mid-intensity battlefield. Because of the relatively weak air defense threat that helicopters faced in the Vietnam War, there were still questions about helicopter vulnerability—particularly when compared to armored vehicles. As one aviator confessed, "There exists an inherent misgiving about the reliability and survivability of any piece of equipment marked with the self-confessing 'NO STEP.' There, thinks the man who will fight protected by armor plate, is a toy."[58]

The arguments of a few extreme helicopter enthusiasts, who still suggested that heavily armored helicopters could fight it out with enemy air defenses, only emphasized the apparent failure of aviation doctrine to face reality.[59]

Even if helicopters could survive, many saw them as peripheral to a battlefield that was dominated by armored vehicles. Furthermore, the continued development of attack and utility helicopters to replace the aging Cobras and Hueys had always been dependent upon the army agreeing to forego some other capability.[60] Thus, at a time of general financial cutbacks, there was opposition from those parts of the army seeking priority in the allocation of resources at the expense of Army Aviation.[61] While the entire army was reduced after 1971, Army Aviation was more at risk because it had gained the most during the war: the only major reorganization of combat formations in the aftermath of the Vietnam War was the reduction from two to one in the number of air assault divisions in favor of another armored unit.

Realizing the dangers in this situation Army Aviation's senior officers urged their colleagues to forget "last year's war stories," "find new ways that aviation can contribute to the Army's mission" and "get out and sell aviation."[62] While believing in the military value of helicopters, Army Aviation was also not prepared to let its organizational influence simply disappear. It wanted to retain the status and access to resources it had won during the Vietnam War but could only do so if the value of its helicopters in mid-intensity conflict was recognized.

Selling Aviation

The need to "sell aviation" was particularly important because of Army Aviation's place in an army structure based on combat branches. A branch was the proponent for its particular aspect of land combat. It determined the doctrine, wrote the training manuals, inducted and educated personnel at its own school, and generally promoted itself throughout the army and the wider defense community. Aviation had originally been part of the Artillery Branch. As it grew during the 1950s and began to perform many more combat functions, responsibility for it was shared among the existing branches. At this time, there had been considerable debate among Army Aviation proponents about whether or not an individual branch was preferable. However, senior Army Aviation officers decided that, due to its political weakness, Army Aviation should not be established as an insignificant separate section. Because each branch concentrated on its own specialty, Army Aviation was accorded a lower priority when shared among them but, in this way, it could prosper by association. It was thought that the separate Directorate of Army Aviation in the Pentagon would provide an adequate focus for Army Aviation developments.[63]

During the Vietnam War, Army Aviation expanded enormously and took on many of the characteristics of a branch. For example, because of the shortage of pilots, officers were allowed to keep serving with Army Aviation rather than continue their careers in their own branch as was normally required; as a result, their loyalty was often to Army Aviation instead of their own branch. Indeed, the huge involvement of helicopters in the conflict gave Army Aviation considerable power within the army—probably equal to that of an established branch—and there was less pressure than in the 1950s for Army Aviation to be formally recognized as such.

However, with the end of the Vietnam commitment, there was a decrease in the size and status of Army Aviation; it no longer resembled a branch of its own.[64] The decline was exacerbated by the decision, as part of a general restructuring of the army, to disestablish the Directorate of Army Aviation and reduce Fort Rucker's role in developing new Army Aviation equipment and doctrine. Aviator officers now had to pursue specialization in their own branch; it usually proved impossible to maintain their flying skills as well, and the lack of a distinct career path in aviation discouraged many officers from maintaining their close links with helicopters.[65]

The importance of convincing nonaviators about the value of the helicopter was reinforced by the increased emphasis in American military doctrine on the combined arms team concept after Vietnam and the 1973 war. Although the two maneuver arms—infantry and armor—retained their preeminence, it was recognized that they would be much less effective unless they operated as part of a team with supporting arms, such as artillery, engineers and signals.[66] Army Aviation had to belong to the combined arms team or else its place on the future battlefield would have been called into question. Moreover, practical aviation employment was largely governed by combined arms commanders, who were not aviators.

Army Aviation's quest for membership in the combined arms team symbolized its fight to gain acceptance from the other combat arms and establish itself as a credible part of an army that was thinking almost exclusively about mid- to high-intensity warfare. This struggle was reflected in the conflicting approaches of those writing about the combined arms team. Some stated that Army Aviation was already a member of the team, thereby keeping up an optimistic front; at the same time, others would exhort aviators to make the integration of Army Aviation into the combined arms team a reality. Aviators were urged to make it clear that Army Aviation was trying to fit into the combined arms ground battle: "Once [ground] commanders realize we know the score . . . we will have the opportunity to employ our wares as an accepted member of the combined arms team."[67] Yet these commanders were not necessarily willing or able to accept Army Aviation in this way. After the Reforger (Return of Forces to Germany) exercise in 1974, the first in which Army Aviation was employed, it was concluded that "we must institute a training program to tell and show the ground commander what Army aviation can do. Too many still think of Army aviation in the Vietnam environment."[68]

The respective reiteration that Army Aviation was "a member of the combined arms team" disguised the fact that helicopters did not combine easily with armored and mechanized forces. Despite the rhetoric of combined arms, even as late as the mid–1980s ground commanders were not effectively integrating helicopters into the "ground scheme of maneuver." Many combined arms team commanders saw helicopters as they had originally been conceived—a form of close air support to be called upon only in emergencies—and there was "reluctance among other members of the combined arms team to accept or recognize Army Aviation's combat capabilities."[69]

General Donn Starry, a leading helicopter proponent, concluded that the problem was with infrequent combined training since helicopters and ground units "only get together when they go to the field for maneuvers. That isn't good enough."[70] In addition, while tank, infantry and even artillery soldiers could identify with and understand one another's specific doctrines and techniques, helicopters were very different, thus exacerbating the problems of infrequent combined training. Attack helicopter units were not widely deployed for many years, further hampering training, nor was Army Aviation and its requirements considered in the original design of the National Training Center.[71] Unit commanders, almost always with little recent aviation experience, were often "unaware of or disregard[ed] aviation training requirements."[72] These conditions produced a circular dilemma in which Army Aviation found it difficult to be accepted into the combined arms team because of training problems that were themselves the result of Army Aviation's lack of acceptance.

Survivability

Army Aviation was well aware that the threat to its helicopters in a mid-intensity conflict would be very different from that in the Vietnam War.[73] The importance accorded to survivability can be seen in Army Aviation's principal doctrine manual, FM 90–1 *Employment of Army Aviation Units in a High Threat Environment*, which tried to ensure that Army Aviation would survive on the future battlefield.[74]

Although the Vietnam War had been characterized by a relatively benign air environment, there were occasions when U.S. aviators had to face heavy enemy air defenses. In particular, during Operation Lamson 719 in 1971, they encountered a very high concentration of enemy anti-aircraft guns that had been arranged specifically to counter an air assault operation. Critics felt Lamson 719 had demonstrated that helicopters were vulnerable, but supporters thought it proved they were capable of operating in high-threat environnments. As an examination of the testimony given to various congressional Armed Services Committees shows, the ultimate conclusions drawn from Lamson 719 were in fact based on initial assumptions about the helicopter's value and no objective judgment was possible.[75] The lessons learned depended on the skill in advocacy and argument of the various proponents.

Nevertheless, aviators did not ignore the threat. As the use of Soviet air defense weapons, particularly the ZSU–23–4 *Shilka*, in the October 1973 war had demonstrated, helicopters would need to avoid or neutralize enemy air defenses if they were to be effective.[76] In response, Army Aviation took General Tolson's advice that "now, we must go back to many of our original [1950s] practices," which had been designed specifically to allow helicopters to operate in and to survive a major war.[77] The key was to try and avoid detection in the first place and then to avoid being hit; the main means of achieving this goal was to fly at nap of the earth (NOE) altitudes below the line of any buildings or vegetation, thus remaining hidden. Nighttime operations also offered a way in which helicopters might avoid detection and thus reduce their vulnerability.[78]

While NOE flying was reintroduced toward the end of the war in Vietnam to counter the threat of the SA–7 Grail surface-to-air missile, for most of the time it had been more effective to avoid the enemy's anti-aircraft machine guns and fly above 600 meters. Moreover, because of the minimal enemy threat, many pilots thought that helicopters were relatively immune to enemy anti-aircraft fire. Most of them had not experienced the more serious threats of 1971–72 and believed the greatest danger lay in inadequate flight safety.[79] Thus, as Army Aviation sought to refocus its attention on the high-threat environment, it was both very necessary and very difficult to overcome the habits acquired by many hours of flying in Vietnam. There was confusion over exactly how high above the ground a helicopter should be flown, and training had to be radically altered to accommodate NOE operations after years in which they had actually been prohibited for safety reasons. Numerous articles in *U.S. Army Aviation Digest* warned of the "most hostile environment" of future battlefields. Indeed, Tolson called on aviators to "unlearn" the lessons of Vietnam, once again suggesting that recent helicopter developments were marked not by continuity from the Vietnam War but by a disjunction from it.[80]

Following the advice of FM 90–1 that survivability also required tactics and proper equipment, as well as training, combined arms was seen as a way of reducing vulnerability since helicopters could benefit from tank and artillery attacks on enemy air defenses.[81] It was also recognized that helicopters had to be physically more survivable.[82] They needed to have reduced visual, infrared, radar and audible signatures, and increased agility and better resistance to damage. To allow them to operate at very low altitudes they had to be more maneuverable and powerful, while night vision equipment was needed to enable helicopters to exploit the cover of darkness.[83] Existing helicopters were modified and the design of the new attack and utility helicopters, which had begun in the mid–1960s, was reassessed to ensure that these characteristics were included in the final product, although one negative result was that costs increased and production was delayed.[84]

The Anti-tank Mission

Simply being able to survive in the high-threat environment was not sufficient to secure for Army Aviation a place in the combined arms team: it also had to demonstrate that helicopters could fight effectively in a mid-intensity conflict. Given the fundamental importance of tanks within U.S. doctrine, combat effectiveness was basically measured by a weapon's ability to destroy enemy armored vehicles: as one aviator stated rather jauntily, "The most popular game on the battlefield for NATO forces will be to kill tanks."[85] Although it was by no means the only task for Army Aviation, many senior aviators, such as General Maddox, concentrated on the anti-tank mission.[86] They not only agreed that attacking enemy armor was the most important battlefield task but, given the emphasis placed on anti-tank combat by the rest of the army, realized that Army Aviation would be likely to receive more resources if they gave it priority. The concept of anti-tank helicopters also appealed strongly to the army's senior leaders, particularly because they believed that "the Russians are overweight in tanks. If you can stop their tanks, you can blunt the attack."[87] Their support helped protect helicopters during the difficult period of readjustment after Vietnam, countering Army Aviation's political weakness within the army and helping it to overcome some of the doubts about its contribution to mid-intensity conflict.

The army valued attack helicopters so highly because of yet another dispute between it and the USAF. Despite the improvement in USAF support during the Vietnam War, the army still feared that the USAF could not be relied on.[88] The move away from counterinsurgency suggested that the USAF would return to its previous concentration on air supremacy and operational and strategic strike missions. Even if close air support assets were acquired (and at that stage the specific details were uncertain), they would remain centralized and subordinate to the air force's primary tactical mission of winning air superiority. In essence the army believed that "close air support could be highly effective or zero at any given moment!"[89] The army wanted to control its own aerial fire support and nothing less.

The whole issue was subjected in 1971 to close scrutiny by a special subcommittee of the Senate Committee on Armed Services. To overcome the obstacle that, as then defined, close air support was an accepted air force role, the army argued that helicopters were in fact providing "direct aerial fires." However, the differences were more than just semantic. Army fire support was more timely, more responsive and more accurate because the helicopter was able to operate close to the front lines and could fire in a hover or at slow speeds; attack helicopters were integrated into the army's combat organization, ensuring very good coordination with ground troops. Yet the army supported the retention of USAF close air support aircraft whose heavier firepower and longer range complemented army helicopters.[90]

The USAF position was more extreme. Its spokesperson before the subcommittee, General William Momyer, rejected attack helicopters as vulnerable and labeled them "low-performing"—reflecting his orientation toward jet aircraft. Moreover, he argued that because they were flown by the army they were a dispersion of air power; and that these helicopters duplicated close air support aircraft and hence were a waste of money.[91] However, the army prevailed. Its argument was expressed more tactfully and was politically more astute because, unlike the air force position, it did not assert exclusive control over aerial fire support. Ultimately, the extent of the USAF's opposition to armed helicopters simply strengthened the army's resolve to make its own preparations for close air support.[92]

Why was aerial, as opposed to ground-based, support so attractive? First, since attack helicopters flew, they were able to move their weapons—the ultimate source of the fire support—unrestricted by terrain and thus much more quickly than ground forces. The doctrine of Active Defense assumed that the army would be outnumbered and would have to redeploy its forces very rapidly to deal with successive enemy attacks. The helicopter was therefore seen as a "tool which will permit us to stretch the capabilities of our outnumbered forces."[93]

The attack helicopter was also appealing because it offered a way of better exploiting the firepower of ATGMs.[94] Active Defense stressed the importance of more accurate, more powerful and longer ranged weapons systems, particularly the ATGM. However, their range was often negated by intervisibility problems at ground level, since hills, trees, buildings and the like could prevent firing over long distances (Figure 1.1). Anti-tank helicopters, which could fire missiles from a few feet

Figure 1.1
U.S. Army Diagram of Intervisibility Problem

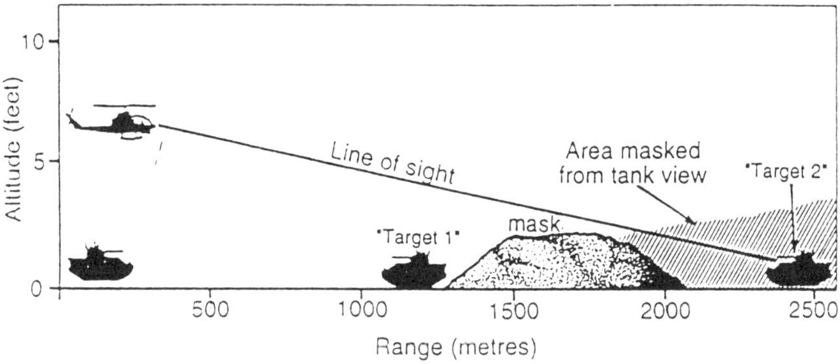

Source: FM 100–5 *Operations* (1976), 2(22).

above the ground, helped considerably to alleviate this problem, and allowed the capability of ATGMs to be more fully realized.[95] Because they were so mobile, helicopters also ensured ATGMs could be concentrated rapidly in response to enemy threats. The link between ATGMs and helicopters was so strong that the chief of staff, General Fred Weyand, declared in 1976 that, even though only ATGMs and not anti-tank helicopters had been used in October 1973, this war had demonstrated the importance of such helicopters.[96]

Support for the use of attack helicopters for anti-tank combat also came from Weyand's predecessor, General Creighton Abrams. Abrams included the Advanced Attack Helicopter (AAH) in his package of new equipment acquisitions, known as the Big Five, and argued that it was the anti-tank counterpart of the proposed M–1 tank and of equal importance.[97] General George Blanchard, the senior American commander in Europe was another firm ally.[98] Attack helicopters also had the approval of General Starry, who commanded the School of Armor from 1973 until 1976, was a corps commander in Europe and then became the head of TRADOC (the key doctrinal development agency). Starry, who had become convinced of the value of helicopters while studying air cavalry units in Vietnam, had considerable input into the 1976 edition of FM 100–5 *Operations* and helped to link the attack helicopter with the Active Defense doctrine.[99] Moreover, his support was particularly important because Armor was the proponent branch with responsibility for attack helicopters. These senior officers and many of their colleagues had been given flight training in the 1950s as part of Army Aviation's campaign to demonstrate its usefulness: this investment was beginning to pay off.[100]

Final acceptance of the attack helicopter hinged on one of the few lessons of the Vietnam War that was directly applicable to the conduct of a mid-intensity conflict: that helicopters were indeed an effective anti-tank weapon. The idea of army aircraft engaging enemy tanks had first arisen in the Korean War but had taken second priority behind the development of airmobility in the 1950s and early 1960s.[101] Although a small number of early model UH–1s, which had been armed with the SS–11 ATGM for tests following the Howze board, had been deployed to Vietnam in the mid–1960s, there were no armored vehicles for them to engage. In the latter stages of the war, however, there was a pressing need for anti-tank helicopters. For example, during Lamson 719, over sixty-six enemy armored vehicles were sighted but only six were destroyed because the Cobras lacked an adequate anti-armor weapon. Coincidentally, trials were being conducted in America with UH–1s armed with the new Tube-launched Optically-tracked Wire-guided missile (TOW), and two were deployed to Vietnam for combat testing. During North Vietnam's spring 1972 offensive, these helicopters performed very well and, in many situations, were much more effective against tanks than fixed-wing close air support aircraft. In eight months they destroyed twenty-seven tanks and sixty-one other targets without suffering any losses.[102] Even though the threat to helicopters in Vietnam was considerably weaker than it would have been in mid-intensity warfare, the combat experience of these anti-tank helicopters "sold the AAH both in the Pentagon and on Capitol Hill."[103] It was a good example of the way in which a specific incident of success in Vietnam was isolated from the more generally gloomy picture and used to good effect. Indeed, the contrast between the success of anti-tank helicopters and other failures made this example dramatically convincing.

A more controlled test, the Joint Attack Helicopter Instrumented Evaluation, was then conducted at Ansbach, West Germany by the United States, Canada and West Germany in June 1972 (Table 1.1). Two combinations of scout and attack helicopters were opposed to a force of tanks and air defense weapons and covered by fighter aircraft in three different scenarios—in each case the helicopters were defending. All sides used laser designators to simulate the effects of their attacks. While only an approximation of wartime conditions, the Ansbach trial was one of the most realistic experiments ever conducted on the place of helicopters in conventional warfare. The results, shown in the following table, were expressed in terms of tank or helicopter casualties. Of most interest were the exchange ratios, in which between twelve and seventeen armored fighting vehicles (AFVs) were destroyed for each helicopter. Because of the numerical inferiority of NATO's armored forces, helicopters were

Table 1.1
Joint Attack Helicopter Instrumented Evaluation

Hits and Exchange Ratios

Defenders	2 Observation and 1 Attack Helicopter				2 Attack Helicopters Only				
Scenarios	Defence	Delay	Break-Through	Total	Defence	Delay	Break-Through	Total	Overall
Hits									
Enemy AFV	12	19	36	67	33	21	73	127	194
AH	2	0	0	2	1	2	6	9	11
OH	1	1	2	4	na	na	na	na	4
Exchange									
AFV:AH	6:1	0	0	33:1	33:1	10:1	12:1	14:1	17:1
AFV:Total	4:1	19:1	18:1	11:1	33:1	10:1	12:1	14:1	12:1

Source: Howze, "The Case for the Helicopter," 17–20.

thus a highly attractive, cost-effective weapon system. The trial answered questions about survivability, confirming that NOE flying, when combined with the standoff range of ATGMs, greatly reduced the vulnerability of helicopters.[104] It also demonstrated that the attack helicopter was more effective and less at risk when used against armored vehicles that were moving, since these targets could be more easily acquired and engaged from maximum range.

A most important tactic emerged from Ansbach and later trials, which combined NOE flying and "engaging [the enemy] from maximum effective range, with minimum exposure time to threat air defense weapons."[105] Called the pop-up maneuver, it involved an attack helicopter hiding behind vegetation or buildings, rising up to acquire a target, firing its weapons and then resuming cover. The procedure would take about thirty seconds at most and it was thought (and rightly so) that enemy air defense weapons would not be able to acquire, track and engage the helicopters inside this time. The basic attack helicopter unit was a company, with three platoons each of four scout and seven attack helicopters, which fought either together or in rotation, one platoon at a time while the others refueled and rearmed, thereby maintaining continuous pressure on the enemy. The attack helicopters were equipped with TOW ATGMs, rockets and cannon to provide a diverse attack capability.

Official doctrine held that attack helicopters could either be kept back as a reserve against a breakthrough by enemy units or integrated with other forces and employed "early [in combat] as far forward in the battle area as possible," thus attempting to prevent any breakthrough.[106] However, the conceptual link between attack helicopters and close air support, which had initially driven army interest in them, remained very strong. Views were still shaped by memories of Vietnam when armed helicopters had often been, to the ground soldier, nearly indistinguishable from close air support aircraft. Moreover, anti-tank helicopters were widely thought to be at their most effective when attacking armored units that had advanced beyond the cover of their air defense units—in other words, after they had broken through.[107] Therefore, despite the doctrinal emphasis on using helicopters early in the battle, normal practice was to hold them in reserve and then employ their speed and mobility to react quickly to an unexpected situation. They came to be regarded mainly as the last ditch reserve—"the finger in the dike."[108]

The army's interest in anti-tank helicopters also saw the creation of an experimental Air Cavalry Combat Brigade (ACCB), which had first been suggested by the Howze board. Having survived the failure of the TRICAP test (see below) with which it was associated, it was permanently established in 1975 as the 6th Cavalry Brigade (Air Combat), nominally comprising one air cavalry squadron, two attack helicopter battalions and supporting elements. The brigade's mission was specific: "To destroy or disrupt enemy armor or mechanized forces by aerial firepower,"[109] revealing again the perception that the tank was the most dangerous enemy on the modern battlefield. Once again it was to be a reserve, used for counterpenetration missions at the corps or even army level. It was expected that, within Europe, this unit could "shoot for V Corps in the morning and VII Corps in the afternoon. . . . [It was the army commander's] 'Fire Brigade.' "[110] Yet, despite some interest in the brigade among aviators, almost all army helicopters were deployed at divisional level in smaller units. It was symptomatic of the move away from the aviation concepts of the early 1960s that 6th Cavalry was never fully equipped and remained the only corps-level air combat unit until the 1980s.

As well as forming the ACCB, Army Aviation reorganized its other units to emphasize anti-tank firepower. Following the third Aviation Requirements for the Combat Structure of the Army (ARCSA III) study in 1976,[111] each mechanized and armored division received two attack and one utility helicopter company and each infantry division got two of the latter and one of the former. Yet, although ARCSA III contained an administrative concentration of aviation assets, there was no corresponding centralization of operational control. Since the emphasis in the doctrine of Active Defense was on fighting battles at the battalion and brigade level, helicopter companies were to be assigned to brigades and actually fight in battalion combined arms teams. Attack helicopter platoons were even considered as possible reinforcements for individual companies.[112]

It was politically more appropriate to integrate attack helicopters at "the lowest level," as General Maddox put it, since this organization helped the army to emphasize the differences between USAF close air support and helicopter fire support, and thus win approval for armed helicopters.[113] This close integration of attack helicopters and ground units, which was very different from the organization that had been suggested by Howze in the early 1960s, also reflected Vietnam War experiences when small infantry and armor units had gained greatly from very close, easily available aerial fire support. Clearly only some of the war's memories were too painful.

Low-level integration also fitted the strategy that aviation commanders pursued of "augmenting the capability of the Army,"[114] which made aviation less threatening to the army's most powerful arms, infantry and armor. Since these were the maneuver arms, which controlled the conduct of battle, they also had the most influence over doctrinal developments in peacetime. Therefore, Army Aviation relied greatly on their acceptance of helicopters and did not challenge their dominance. Occasional suggestions that attack helicopter companies and air cavalry troops were "combat maneuver units"[115] were not widely accepted and certainly not implemented.[116] Yet the existence of the ACCB—no matter how peripheral—reminded more eager aviators that, perhaps, the future would see both large helicopter units within the U.S. Army and maneuver status for Army Aviation.

Coincidentally, just before the Ansbach trial, the U.S. Army had finally canceled the Cheyenne project because of its expense and initiated a new design process for the AAH.[117] This competition was won by the Hughes (later McDonnell Douglas) AH–64 Apache.[118] The AAH was at first intended to be "much more austere" than the Cheyenne, thereby avoiding the cost and technical problems that had plagued the earlier program. However, the Apache became more complex with the addition of an advanced target acquisition and night vision system in the mid–1970s and, like the Cheyenne, it suffered numerous delays and near cancellations. Its acquisition was also held up by an army decision to buy the M–1 Abrams tank more quickly than had first been envisaged. Combat units received the first Apaches in 1986, some ten years after the doctrine for which it was designed had first been approved.[119]

The Apache's survivability was called into question a number of times during its development. Yet the major reason for these questions and the accompanying threats of cancellation, which emanated from the Department of Defense, was the cost of the Apache and not its effectiveness. Although the army remained committed to the AH–64, between

Photo 1.3
AH–64A Apache

The most advanced helicopter in the world. It is seen here with the prototype *Longbow* radar, which will further enhance its capacity to destroy enemy targets at long range. (McDonnell Douglas Corp. Used by permission.)

1977 and 1979 a series of tests was carried out to determine whether fewer helicopters and more Fairchild A–10 Thunderbolt close air support aircraft—or vice versa—ought to be acquired.[120] What emerged from these tests was that close cooperation between the Apache and the A–10 would in fact result in much better survival rates and more powerful combat effects for both. The Joint Air Attack Team (JAAT) was born.[121] JAAT's military value cannot be ignored, but its political importance was at least as great. JAAT ensured that both the army and the USAF fire support aircraft would be acquired: by cooperating instead of competing, the two services were able to outmaneuver the Department of Defense.

Anti-tank doctrine drove other technical developments. A new scout helicopter was needed, and eventually—and to save money—the Army Helicopter Improvement Program (AHIP) was initiated. The program sought to make major modifications to the existing observation helicopter, the Bell OH–58A Kiowa.[122] In order to limit the chances of detection, these helicopters were equipped with a mast-mounted sight (MMS)—a sensor package mounted on a mast projecting above the rotor blades. This device was designed to allow the AHIP to perform its mission while keeping all but the MMS out of sight of the enemy and thus making the aircraft much less detectable.[123] After suffering financial and technical delays, mainly because of the complex and expensive MMS, the OH–58D was first deployed in 1988. The development of a new ATGM, specifically designed for helicopters and known as Hellfire, was also driven by the specific tactics of the anti-tank helicopters. With a longer range and shorter flight time, and with the capacity to be guided from another helicopter or ground observer, Hellfire enhanced the attack helicopter's survivability by allowing it to fire at a much greater standoff distance and ensuring that it did not have to come out of NOE for very long when firing.[124]

Other Missions in the High-threat Environment

While anti-tank action prospered, air assault—the movement of infantry units and their supporting arms by helicopter—went from being the most important to the least important mission after the Vietnam War. While it was recognized that future counterinsurgency campaigns would require air assaults, and likely intervention forces were all well equipped with helicopters, the army's main attention was on mechanized, mid-intensity conflict.[125] Air assault operations did not seem relevant and there was "[a]n underflow of skepticism regarding the ability of assault forces to survive."[126] It was understandable that, at a time when the army was stressing the need to use long-range ATGMs, NOE, and ambush tactics to ensure the survivability of its attack helicopters, assault

Photo 1.4
OH–58D AHIP

The vastly improved version of the OH-58A. Its mast-mounted sight enables it to hide from view and still observe and target enemy vehicles (its armament is nonstandard). (Bell Helicopter Textron. Used by permission.)

operations "deep behind enemy lines" were regarded as impractical and "would not be contemplated in Europe."[127]

At first, there was some willingness to try to find a way in which air assault could play a part in mid-intensity conflicts. In 1971–72 the 1st Cavalry Division (Air Assault) was reorganized into a test formation called TRICAP—short for "triple capability"—with an armored brigade, an air assault infantry brigade and an ACCB. TRICAP came out of the last stages of the Vietnam War when airmobile infantry, with air cavalry and attack helicopter support, operated in close conjunction with armored units.[128] However, TRICAP was a failure and in 1974 the 1st Cavalry Division was converted to an armored unit; as noted above, the ACCB survived as a separate entity. According to General Kinnard, who had helped set up the test, "there were too many divergent goals, too many cooks, too many addenda tests and a general watering down of the principles I had laid down."[129] It seems likely that the combination of attack helicopters, airmobile infantry and tanks, all with widely differing capabilities, was too difficult to reconcile in a single unit.[130]

However, the most important reason for abandoning TRICAP had less to do with helicopters than with growing doubts about the values of unarmored infantry on the high-threat battlefield. In comparison to mechanized troops, they were more vulnerable to artillery fire and lacked mobility once they had landed. While TRICAP began by combining tanks, airmobile infantry and anti-tank helicopters, in the end all that survived were an armored division and the anti-tank ACCB. Even this brigade progressively rid itself of light infantry and air assault helicopters with the presumption that doing so reduced its vulnerability. Originally paired with the airmobile brigade, it soon had just its own airmobile infantry battalion, which was then reduced to a company. By 1975 it had no infantry whatsoever.[131] The general concentration on tank and anti-tank combat at the expense of air assault was exacerbated by the fact that the TRICAP tests—in contrast to those of the 11th Air Assault Division in the 1960s—were arranged by the Armor Branch, which naturally had far more interest in anti-tank helicopters than in infantry.

More satisfactory trials were conducted by the 101st Division, the remaining air assault unit. The commander of the Infantry Branch in the early to mid–1970s also remained committed to adapting air assault to the high-threat environment.[132] Furthermore, as part of the Big Five modernization package, the United States continued with the 1960s project to produce a new utility transport helicopter that would eliminate many of the Huey's problems, selecting the Sikorsky UH–60 Blackhawk, which was first fielded in 1979.[133] But as the Infantry Branch increasingly concentrated on developing its mechanized combat power, air assault became less and less important. The idea of helicopters became more associated with anti-tank combat; for example, the outstanding contribution of the 101st Division to Reforger '76 was reported to have come from its attack helicopter company.[134]

Army doctrine reflected the tension, inherent in the adjustments made after the Vietnam War, between abandoning what had been learned about air assault and the difficulty of applying it to a high-threat environment. An excellent example was the unusual choice of words in Army Aviation's own manual, which stated that "air assault operations in the high threat environment present *interesting* challenges [emphasis added]."[135] While the manual did not rule out air assaults, it hedged their execution with so many constraints that, reading between the lines, it was obvious they were not being seriously considered. Other doctrine manuals, such as FM 100–5 *Operations*, treated air assault similarly, making general comments about the high mobility of airmobile infantry and their use in "restricted" terrain but not giving the same level of specific advice as for attack helicopter operations.[136] A new version of a detailed air assault manual to replace the 1971 edition did not appear until 1980, well after the publication of other manuals. Nor were attack helicopters readily available to provide the necessary support for air assault because they had been devoted to anti-tank combat. Finally, even though it was argued that all infantry units were capable of being airmobile, mechanized infantry units rarely had the helicopter support needed to make air assault training possible.[137]

Although utility helicopters were included in divisions as part of ARCSA III, they were dedicated to the "essential" resupply mission. The intense fighting in the October 1973 war had shown that if the army was to fight a mechanized, conventional war it would have "tremendous logistical requirements."[138] Because of their mobility, flexibility and independence from terrain obstacles, helicopters were seen as an important asset for logistical support, especially urgent resupply missions. Both utility and CH–47 cargo helicopter squadrons were to be used for this task. Reflecting the importance attached to the mission, the Blackhawk was designed with a much greater lifting capacity than the UH–1, and the existing Chinook fleet was modernized. Command and control, artillery and air strike observation missions were also performed by helicopters, with each brigade and division having a small number of helicopters for these purposes.[139]

Army Aviation was also used for reconnaissance missions. The early development of armed helicopters had focused on this task, and divisional cavalry squadrons began receiving air cavalry troops in 1962. In Vietnam, all divisions had an air cavalry squadron and, at first, senior officers returned from the war believing that air cavalry was needed in conventional warfare as well because the "mobility differential" between helicopters and ground vehicles made the former better suited to reconnaissance.[140] However, despite the arguments by supporters of air cavalry against too great a concentration on attack helicopter units, anti-tank action soon became paramount.[141] Air cavalry units were assigned fewer helicopters and lost their special reconnaissance infantry platoons, which limited their effectiveness.[142] Many officers probably thought that, if the U.S. Army ever had to defeat a massive Soviet armored offensive in the narrow confines of northwest Europe, finding the enemy would be the least of their worries. Thus, although less severely affected in comparison to air assault, air cavalry also lost much of its Vietnam-era importance.

Summing up the 1970s

Perhaps the most significant of the many changes that Army Aviation underwent in the 1970s was the emergence of a narrower approach to Army Aviation's battlefield employment as the wide-ranging doctrines of previous years lost their appeal. During the initial development of the Active Defense doctrine, there was concern that the new doctrine "signalled a retreat from air mobility and too narrow a focus on mounted or mechanized warfare"[143] and, perhaps to allay this concern, FM 100–5 *Operations* asserted that "the 'airmobile' concept is the most dramatic organizational advance in the U.S. Army."[144] However, this sweeping statement did not accord with actual practice in the 1970s, either before or after the publication of the Active Defense doctrine. The comments of three senior officers exemplify this decline. General William Richardson, Commander of the Combined Arms Center, stated that the army had "little understanding of airmobility in its total context"; Kinnard warned that the lack of integrated aviation support for attack helicopters meant that "we will never develop [their] full potential"; and Howze thought that the army had "quite inexplicably drawn back from the idea of tactical airmobility."[145]

Airmobility, as developed in the 1950s and 1960s, was based on air assault with armed, reconnaissance and utility helicopters to support the infantry. This infantry-based concept appeared outdated by the 1970s when the army was facing the prospect of intense, mechanized combat on battlefields increasingly dominated by firepower. Airmobility was linked inextricably with air assault—indeed the latter became known as airmobile operations for some years—and hence its decline was in some respects linked to that of air assault. Even the concept behind airmobility—the use of integrated aviation units in which all combat functions were performed by helicopters or helicopter-based units and which could be employed as separate, operational-level formations—was not broadly applied. With the exception of the experimental 6th Cavalry Brigade (Air Combat), aviation assets were not arranged in large self-supporting units capable of coordinating with the army as a whole. Instead, there was a concentration on small-unit, tactical employment of aviation assets, mainly attack helicopters.

Yet this outcome was not solely due to the link between airmobility and air assault, because another broad theory, an extension of the air cavalry mission and based on concepts more appropriate for the mid-intensity threat of the 1970s, also failed to gain acceptance. Proponents of an air cavalry approach looked to the Vietnam War, during which attack helicopters had been used purely for fire support missions, while air cavalry had been employed as a maneuver force.[146] Unlike airmobility, air cavalry favored armed helicopter action over infantry air assaults and stressed close integration of both helicopter and ground forces at the tactical level. Yet it was similar because it provided a unifying, all-emcompassing approach to Army Aviation employment and sought to establish helicopters as a significant and distinct part of the army.

The broad concepts of airmobility and air cavalry were rejected as underlying principles for the employment of Army Aviation. Instead, the attack helicopter dedicated to anti-tank missions—the part of Army Aviation most clearly relevant to mechanized, mid-intensity conflict—became the focus of attention and the main symbol of what Army Aviation was all about. Kinnard commented that Army Aviation tended "to speak and think of attack helicopters as if they are the only helicopters worth mentioning."[147] Airmobility as Gavin or Howze would have understood it was dead: the new voice for Army Aviation, General Maddox, now equated it with anti-tank action.[148] This approach stemmed both from the need to address changed military circumstances and to be seen to be addressing them by the rest of the army.

Whatever the change, Army Aviation survived the difficulties of the withdrawal from Vietnam and, by the start of the 1980s, had become relatively well established, with the anti-tank helicopter as its most valuable contribution to the doctrine of Active Defense. It had overcome doubts about relevance and survivability and, having gained the confidence of senior army officers, had fought off most attempts to reduce or cancel its equipment acquisition programs. Yet while this relatively solid foundation was a necessary precondition for new developments, it was not of itself sufficient to cause them. The progress that was to come in the 1980s in applying helicopters to conventional warfare was not inevitable and required the impetus of the upheaval associated with a new army doctrine: Airland Battle.

1980s: AIRLAND BATTLE

A New Beginning for Aviation

In contrast with the previous decade, the 1980s saw an increase in Army Aviation's role in the army and more recognition of its contribution. The attack helicopter retained its primary position, with its capabilities and missions continuing to expand, and other helicopter missions regained some of the prestige they had lost after the Vietnam War. This process was helped by the eventual deployment of more capable helicopters, such as the Apache, and from the army's increased attention to potential conflicts outside Europe. In these areas, greater distances, more dispersed units and less threatening air defenses all favored helicopter operations. But the main concentration remained in Europe, as did the impetus for a new doctrine, Airland Battle, which replaced Active Defense. Although it had its roots in theories developed in the late 1970s, Airland Battle's impact was first felt in 1982 when it was published in a new edition of FM 100–5 *Operations*.[149]

In the same year, Army Aviation underwent an exhaustive series of reviews designed to recommend changes to doctrine, organization and equipment to rectify deficiencies. By 1984, a new manual had been published, with the noticeably more positive title of *Combat Aviation Operations*, establishing Army Aviation's basic doctrines for Airland Battle.[150] Finally, a new unit organization, first developed as part of the "Army '86" studies in the late 1970s and confirmed in the "Army of Excellence" review of 1983, began to be implemented. Instead of the unwieldy aviation battalions of ARCSA III, each corps' and division's assets were reorganized into combat aviation brigades (CABs) with individual companies becoming battalions; this improved commanders' control over their helicopters, eased maintenance requirements and meant "better use of limited resources."[151]

Although Airland Battle was by no means the revolutionary change from the Active Defense doctrine that some people first thought, there were a number of significant differences. Airland Battle was based on offensive instead of defensive action. Furthermore, where Active Defense had favored firepower and attrition-based "target servicing," Airland Battle promoted combat in depth, mobility and agility, maneuver, seizing the initiative, and the importance of morale, leadership and the will to win. It also placed less emphasis on battalion and brigade battles and more on corps action at what was termed the operational level of warfare.[152]

Helicopter proponents eagerly seized upon words such as "agility" and "mobility," and suggested that, since helicopters were very agile and mobile, the success of Airland Battle depended on them. It was even claimed that because helicopters were part ground vehicle and part aircraft, they were the essential connecting force between the two components of Airland Battle.[153] In other words, air and land would come together in the shape of the helicopter. Such assertions reflected the desire of Army Aviation to create a place for itself in Airland Battle and should not be taken too literally. Nor were helicopters the cause of the army's doctrinal reassessment. Despite his support for Army Aviation, when Starry used the term "Airland Battle," "air" referred primarily to the USAF and its role in interdicting the enemy's rear areas, and not to Army Aviation.[154] Nor did other authors of the new edition of FM 100–5 *Operations* give helicopters any special support or credit them with a particularly important role in Airland Battle.[155] In fact, the new manual was less openly supportive than its Active Defense predecessor. But written support in the 1970s had not been translated into practice and, with Airland Battle, the actual words in the manual were peripheral. What Airland Battle brought was an atmosphere of change and a new approach to warfare, both of which was very favorable for those advocating greater employment of helicopters: they lost little time in exploiting these conditions.

Airland Battle introduced two specific but related changes of considerable importance to the employment of attack helicopters. First, there was a move away from a narrow defensive approach. The aim was now "the earliest possible initiative of offensive action."[156] Although in the late 1970s Army Aviation proponents had already convinced themselves that helicopters were capable of offensive combat, they had not been able to change the generally held view that attack helicopters were too vulnerable except when used in defensive ambushes; they had also been restricted by the Active Defense doctrine which rarely favored offensive operations.[157] It had already been recognized by these advocates that helicopters were not suitable for attacks against prepared defenses and, if they were to be used offensively, should instead strike less well-protected units in the enemy rear. While Airland Battle placed much importance on the "Main Battle" at the front-line, it also stressed—for the first time in recent army doctrine—the value of "Deep Battle," in which army units would operate as much as 150 kilometers into the enemy's rear area, neutralizing opposing reserves, preventing them from reaching the main battle and destroying command and logistic services.[158]

At the same time that helicopter enthusiasts began contemplating offensive operations, they also started to reconsider the close, company-level integration of Army Aviation and ground forces. They thought that, at this level of command, the mobility and flexibility of helicopters were underemployed and the helicopters' firepower used piecemeal. Instead, they argued that divisional helicopter assets should never be assigned lower than brigade level and that attack helicopter battalions should not be broken up into their constituent companies.[159] This view was in fact confirmed in later doctrinal statements and helped to stimulate interest in deep offensive missions since larger attack helicopters units were better suited to such attacks.

The concept of deep battle was particularly appealing to Army Aviation proponents because helicopters appeared to offer the mobility necessary to strike deep into the enemy rear and avoided the problem of concentrated air defenses near the front line.[160] Although in 1981 General Carl McNair, Commander of the Aviation Center, named air force and artillery units as the key deep attack assets, he also suggested that "future Army Aviation systems will further enhance the ability of the combined arms team to conduct deep attack"; in 1982 FM 100–5 *Operations* clearly stated that Army Aviation would be used to extend the battle into the enemy's depths. With continuing delays to the army's missile and associated targeting systems, deep attack by maneuver grew more important and, by 1984, Army Aviation was being described as "a major means for conducting the deep attack."[161]

Offensive deep battle was a key part of Airland Battle and one of the most significant changes from the doctrine of Active Defense. Army Aviation's participation in this form of combat raised the organization's status within the army and helped gain acceptance for a wider range of helicopter missions. Army Aviation proponents were not backward in exploiting their assumed new importance, as can be seen in comments by McNair at a time when the future funding of the AH–64 Apache was in doubt. He claimed that, without the Apache, "we would have to give serious thought to altering our entire Airland Battle Strategy and the concept of the extended battlefield. This would be a real setback."[162] McNair later wrote that "[army] aviation is essential if we are to execute the doctrine of the Airland Battle." While there was a good deal of truth in the comment, Airland Battle was equally essential if Army Aviation was to increase its status, power and resources within the army.[163]

Developing Doctrines for Deep Attack

In the 1970s, it had been thought that deep attacks were too risky and, given the needs and advantages of the defensive battle, that helicopters would be better used in positions behind the friendly side of the front.[164] However, with the new emphasis on deep attack, this negative approach to crossing the forward line of own troops (X-FLOT) changed. While previously the difficulties of overcoming enemy air defenses had been stressed, in 1981 a threat analysis appeared in *U.S. Army Aviation Digest* entitled "The Leak in the Soviet Air Defense Umbrella." In particular it presented a more balanced view of the ZSU–23–4 anti-aircraft gun, which had been considered the most serious threat in the 1970s. It pointed out that there would only be a small number of ZSU–23–4s in any one place, their ammunition supply was limited and their radar system was fragile and often not used for fear of detection by radar warning receivers.[165]

The ZSU–23–4 had not changed dramatically overnight: it had always been like this. Rather, under the impetus of Airland Battle's support for deep attacks, Army Aviation assessed its threat differently. Previously the attitude had been that survival was paramount and the range of missions was strictly subject to the tactics for countering enemy air defenses; now it was thought that "every weapon and system has weaknesses which can be exploited";[166] the mission now took priority, and survival procedures had to develop to allow that mission to occur.

Existing procedures for X-FLOT missions were inadequate, and the 1980s were dominated by efforts to develop the techniques required for successful penetration of enemy front lines.[167] Accurate intelligence gathering was to be used to identify areas with few enemy air defenses. Helicopters were to use their speed and mobility when crossing the FLOT. Night vision equipment (which became widespread in the army in the late 1980s) allowed helicopters to use the cover of darkness and other periods of low visibility to aid them in penetrating the front line. As with any attack, artillery and possibly close air support and electronic warfare were to suppress enemy air defenses.[168] Underpinning all these developments was a belief that, on the fluid and nonlinear battlefields of any future conflict, front lines might in fact be avoided altogether and penetrations made in gaps between enemy formations.

There were, however, a number of problems with implementing these doctrines. At first it was very hard to demonstrate X-FLOT missions convincingly because of the limitations and vulnerabilities of the AH–1—especially in terms of night vision, navigation and endurance. Practical operating procedures, as opposed to general theories, could not be developed without actually using the Apache. Only after the AH–64 was deployed in 1986 could Army Aviation "seriously examine the prospects of the [X-FLOT] mission and realistically address the doctrinal and tactical implications."[169] Other problems with the concept of deep operations concerned the difficulty of resupplying helicopter units once they had crossed the FLOT, the need for accurate, prompt intelligence to guide the deep attack and the need for close coordination with the air component of Airland Battle. Not surprisingly, deep attack was described as "new and evolutionary."[170]

But despite doubts (mainly from outside the army) about the capacity of the Apache to undertake these missions, Army Aviation persevered.[171] The results of recent trials and exercises, including Reforger in 1987, indicated that Army Aviation was beginning to master the X-FLOT employment of helicopters.[172] Although a minor skirmish compared to what was to come in the Persian Gulf, Operation Just Cause in Panama showed that Army Aviation could successfully deploy and operate its Apaches.[173] The Gulf War was, however, to provide Army Aviation with its best ammunition yet to reply to critics of its innovative doctrine.

Attack helicopter units of the 82d Airborne Division were the first heavy anti-tank forces in Saudi Arabia and were the backbone of the initial response of what became known as Operation Desert Shield. The vast distances that they patrolled, checking for Iraqi incursions beyond the Kuwait-Saudi border could not have been covered as effectively by ground-based forces. Intensive maintenance and technical improvization meant that at no stage were helicopters unduly affected by the extreme climatic conditions. Readiness figures of 80 to 90 percent were used by Army Aviation as an effective rejoinder to criticism that the Apache was prone to breakdowns.[174] When the counteroffensive Desert Storm began, Apaches fired the first shots of the air war, knocking out key early warning radars behind Iraqi lines. The flight to the objective was over 750 kilometers, and Black Hawks successfully penetrated with the Apaches, carrying additional fuel.[175] During the ground offensive, only one Apache was actually shot down—out of hundreds deployed. There were a few helicopter losses but almost all occurred during training flights and not as a result of enemy action. Survival equipment, especially items designed to affect infrared and radar guidance, seemed to work superbly.[176] In attack, Apaches proved—like most other U.S. Army combat systems—to be lethal against Iraqi tanks and other vehicles.

Undoubtedly, the Gulf War provided Army Aviation with a "show case" in which to display its qualities, as its current commander exulted soon after Desert Storm was finished.[177] Nevertheless, particularly severe problems remained in division-level missions since the logistics, intelligence and fire support necessary for deep attack were only readily available for corps aviation units. Moreover, deep battle was mainly seen as a corps responsibility, with divisional commanders likely to use Army Aviation in their main area of concern, the close battle at the front line. Some Army Aviation sources referred to divisional attack helicopter battalions as maneuver units capable of carrying out deep attacks.[178] However, by the late 1980s other doctrinal statements—including one by army Chief of Staff General John Wickham—differentiated between corps and divisional CABs, suggesting that, because of the pressures on divisions in the main battle, only the former would regularly conduct deep attacks.[179] It is still too early to judge whether the Gulf War has altered this separation of functions.

As well as the Gulf War, 1991 was an important year for Army Aviation with the announcement of the winner of a competition to build a new light helicopter. The history of this development, first known as the Light Helicopter Experimental (LHX), began in the early 1980s. Although it was intended that the Blackhawk and Apache would be in service well into the twenty-first century, they were large and expensive aircraft. Needing to maintain a mix of light and heavy aircraft, Army Aviation had to replace its existing light helicopters—the UH–1, Hughes OH–6 Cayuse, OH–58 and AH–1—since these machines were already becoming obsolete. Their successor was conceived in light of the basic requirements of Army Aviation's mission in Airland Battle—deep attack, helicopter air-to-air combat (HATAC) and X-FLOT operations—and was designed to "meet the threat challenge, solve the fleet obsolescence problem and win the Airland Battle of the future.[!]"[180] It was to improve further on current standards of survivability, reliability, availability and maintainability; have sophisticated night vision equipment; and, by virtue of its small size, be easily transported by aircraft over long distances.

The LHX faced numerous problems in the 1980s with Army Aviation trying to work out exactly what characteristics the new helicopter ought to have and what tasks it would perform, as well as attempting to keep the cost within reasonable limits and convincing senior army and civilian officials to continue funding its development. In the hope of gaining weight and long-term personnel cost benefits, the helicopter industry pursued the possibility of a highly sophisticated single-pilot helicopter, while a small number of aviators and designers from Bell pressed for a very fast helicopter—using Bell's advanced tilt-rotor system—which would be more capable of performing air-to-air combat.[181] However, after many months of debate and confusion, it was determined that the LHX would be only an evolutionary step forward from the Apache since there was no military justification for the expensive and unproven technologies that had been suggested.[182]

The results of this development were prototypes from two teams: Boeing-Sikorsky and Bell-McDonnell Douglas. The winner, Boeing-Sikorsky, and the official name for the helicopter—the RAH–66 Comanche—were announced in April 1991. The helicopter is, in some ways, a smaller version of the Apache in terms of its armament and equipment. But, of great importance in light of future force personnel reductions, the full replacement of existing light helicopters by the Comanche will mean a 60 percent reduction in Army Aviation staffing. It will also be far cheaper and easier to maintain and have further improvements in survivability, flight characteristics and sensor systems.[183]

Despite escalating costs and problems with defining its missions, the army remained committed to the eventual acquisition of over five thousand LHXs for much of the 1980s. Showing their political power within the army, helicopter advocates were for some time able to maintain funding for the LHX at the expense of production of existing helicopters. Even when the LHX program was reduced to two thousand aircraft in 1988, the money was still committed to Army Aviation in the form of increased acquisition of Apaches and AHIPs.[184] By fiscal year 1991, the cost had become $30 billion for the 1,292 helicopters planned. Thus the Comanche acquisition process demonstrated the army's continuing and even increasing faith in Army Aviation in the era of Airland Battle.

Photo 1.5
RAH–66 Comanche

A mock-up of the next U.S. Army helicopter that will perform light reconnaissance and observation tasks; it will come into service in the next century. (Compliments of Boeing.)

In 1991 Comanche had become the "Army's number one modernization priority."[185]

Air Assault

Airland Battle also played a major part in the reemergence of a more favorable view of air assault. Instead of being limited to low-intensity conflicts, air assault in the 1980s was applied directly to mid- to high-intensity warfare—a change directly attributable to the concepts introduced by Airland Battle. For example, the new version of FM 90–4 *Air Assault Operations* first released in 1985, compared a number of ideas from the Active Defense doctrine and its successor. It pointed out that, unlike the former doctrine, Airland Battle stressed "the human dimension of battle" (i.e., attacking the enemy's morale, not equipment), "offensive actions," "deep battle" (including "maneuver attack to [the] depth of enemy second echelon forces"), "maneuver warfare" and "accept[ing] reasoned risks." All these new ideas, it was asserted, favored the movement of personnel in an attack against the enemy's rear areas: in other words, air assault.[186] Since it depended on crossing the front line, air assault also benefited from the more favorable assessment of using helicopters in X-FLOT missions.

The light infantry that air assault helicopters would carry was also undergoing a revival. FM 100–5 *Operations* (1982) gave it more prominence, and in the early 1980s four light infantry divisions were established.[187] In 1984 General Galvin, commander of U.S. VII Corps in Europe, wrote that light infantry was well suited to supplementing the heavy divisions already deployed in Europe because of the large areas of terrain that did not favor mechanized operations. He also argued that the threat of enemy air assaults could only be matched by "a friendly airborne or air assault force" in which the helicopter provided the mobility that the infantry lacked.[188]

In the 1970s, air assault doctrine had clearly stated that close coordination between the infantry and the helicopters was essential; thus, limited Army Aviation resources and other training priorities meant such operations by mechanized infantry were unlikely. In the mid- to late 1980s the change was dramatic, and "mechanized infantry units of the heavy division" were told that "they *must be proficient* in the conduct of air assault operations [emphasis added]."[189]

Potential problems with the concept of air assault had been emphasized previously; now limitations were minimized as they were far outweighed by the advantages of air assault.[190] It was, perhaps, more than just coincidence that the army Chief of Staff at this time was General Wickham, who had commanded the 101st Division and had been an army commander in Korea, where air assault operations were more accepted than in Europe. The recent successes in Panama and the Gulf have provided air assault advocates with potent evidence with which to continue these developments. In the former, Panamanian soldiers, "cut-off by lightning quick, NVG [Night Vision Goggles]-equipped air assaults . . . surrendered instead of fighting,"[191] demonstrating the attack on enemy morale that Airland Battle had suggested. In the far more intense environment of Desert Storm, the 101st Air Assault Division moved hundreds of kilometers into the Iraqi rear to cut off units retreating from Kuwait.[192] Nevertheless, as with much of the recent history of Army Aviation, change in theory was not necessarily change in practice: while the trained troops of the air assault units were successful, "the bottom line is that mechanized infantrymen need to spend time working with Aviation units" to make sure official doctrine was implemented.[193]

Helicopter Air-to-Air Combat

While many threats to helicopters were assessed more optimistically in the 1980s, the opposite was true for the danger posed by enemy helicopters. The rapid growth in Soviet armed helicopters after 1974 had not gone unnoticed in the United States, but in the early to mid–1970s any discussion of helicopter air-to-air combat (HATAC) added to fears that Army Aviation would, like the USAF, conduct aerial combat missions instead of assisting the land battle.[194] The military requirement for HATAC strengthened in the 1980s with the deployment each year of more Soviet helicopters. U.S. analysts claimed in the mid–1980s that the Soviets were developing a fighter helicopter called Hokum specifically for HATAC, and the presumed threat that it posed was used repeatedly by pro-HATAC aviators.[195] Given the lack of clear evidence for this view, the assumption that Hokum's role was air-to-air combat probably stemmed more from the Americans' own interest in the issue. A more realistic cause for concern was that Army Aviation's growing interest in X-FLOT missions meant helicopters would have to operate outside the range of friendly air defense, and thus it became imperative to give helicopters their own means of defense against air attack.[196]

Although General McNair wrote in 1981 that Army Aviation had a "threat problem to solve with little time or resources to waste,"[197] and the first doctrine manual was written and HATAC trials were held, it took until 1986–87 to overcome the army's general reluctance to give Army Aviation an air-to-air capability.[198] General Ellis Parker wrote in 1987, "We have been able in the last 6 months to make more progress toward the Army air combat capability than was made during the preceding 10 years."[199] The future of HATAC capability appears assured, however, as this requirement has been built into the RH–66 Comanche.

Aviation Seeks a Maneuver Role

As well as changing missions, Airland Battle brought with it the impetus for upgrading the role that Army Aviation played in the army. In the 1970s senior aviators had accepted Army Aviation's status as a supporting arm. With the advent of Airland Battle their successors sought a more active role, stating that Army Aviation could "conduct prompt and sustained combat operations.... Gone is the nuance of Army Aviation being simply a supporter, a transporter or an observer."[200] The emphasis on maneuver warfare in the new doctrine allowed proponents to claim that helicopters were now more important since they could provide "the rapid mobility that is so inherent to winning on the modern battlefield."[201] While Army Aviation would continue to perform its combat support and combat service support tasks, it sought recognition that its attack helicopters, air cavalry and even air assault helicopters were maneuver forces just like infantry and armor. Army Aviation was not challenging the supremacy of the tank as the key battlefield vehicle but was certainly claiming a close second place, and thus increased power and influence in peacetime doctrine-making.[202]

Although official doctrine superficially reflected Army Aviation's desire for maneuver status,[203] actual acceptance and implementation of the idea by the rest of the army were not as easily gained, particularly because helicopters were unable to occupy terrain in the same manner as infantry or armor. Although advocates argued that helicopters could dominate ground by firepower, it was not an idea that was easily proved or accepted—especially for soldiers who had served all their lives in an armor-oriented army. In 1987 a ground battalion commander asked Chief of Staff General Wickham: "Much confusion seems to reign over [aviation]... should it maneuver or be a provider [of support]? What are your views on this issue?" Wickham replied that "aviation provides the fourth dimension to the combined arms team on the battlefield, and aviation, therefore,... provides the capability to maneuver."[204]

Yet the same battalion commander continued: "We just don't know what to do about this [issue] and I think that people are confused."[205] Indeed, in the previous year an Army Aviation conference concluded that many officers said helicopters were a maneuver force but did not use them as such; training exercises confirmed that few understood Army Aviation was a maneuver force to be integrated into the battle from the start. Ground commanders instead tended to use Army Aviation as an emergency fire support unit when their infantry and armor were about to be overwhelmed.[206] Once again, gaps were emerging between doctrine and its implementation.

For the change from support to maneuver to take root, a window of opportunity in the existing framework of the army's doctrinal house was

Figure 1.2
Comparison of U.S. Combat Aviation Brigades

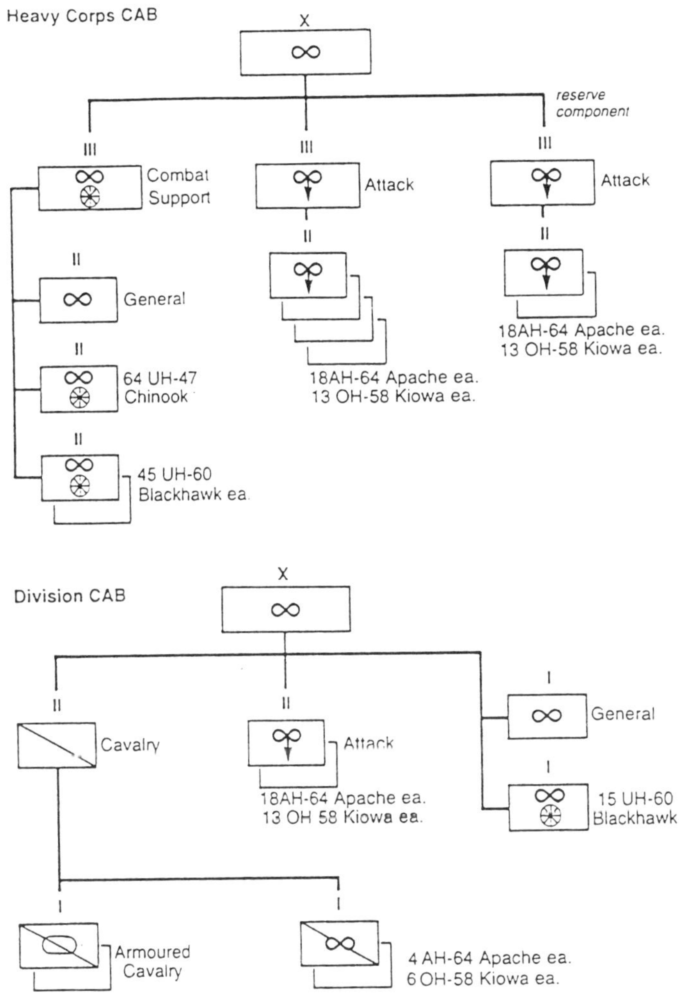

Comparison of Total Helicopter Assets:

	Corps	Division
AH-64 Apache:	108	44
CH-47 Chinook:	64	-
OH-58 Kiowa:	108	50
UH-60 Blackhawk	108	22
Other Aircraft	39	9
Total	427	125

Source: FM 90–1; FM 1–100.

required (Figure 1.2). Because the experimental ACCB had already established itself as a maneuver unit, it was accepted that the very similar corps CAB could also maneuver.[207] However, the development of a similar style of organization for divisions provided that opportunity. But it also depended on the question of whether the new structure was merely an administrative change or whether it actually altered the unit's role. Therefore, most of the confusion centered on the divisional CAB. While Army Aviation advocates stated that it was "certainly capable of operating as a maneuver unit," battalions and not brigades were the basic unit of maneuver. What was really meant was that the brigade was a maneuver headquarters: "The CAB provides the division commander with a fourth planning brigade headquarters . . . [which can accept] control of armor and infantry, while supported by divisional combat support and combat service support elements."[208] Yet, while it provided command assets, centralized control and maintenance, as well as a structure similar to other units, at the same time it had many functions other than commanding maneuver battalions. It had to maneuver—both in the close and rear battle—and provide combat support and combat service support, all from the one headquarters.[209] Thus, there were serious doubts about its command and control capacity.

Army Aviation Gains Branch Status

The controversy about having a maneuver or support role reflected one part of Army Aviation's fight for increased status in the 1980s. But whatever its role, the measure of Army Aviation's acceptance was its place within the combined arms team and the actual integration of helicopters with other weapons systems. Once again, since the combined arms team commanders controlled the way in which army doctrine was put into practice, their acceptance was critical to the implementation of Army Aviation doctrine.[210] Yet, in the early 1980s, a review had found that, of the 160 areas of army operations concerning combined arms, only eight were causing problems—and of these the only major area was Army Aviation. No wonder aviators thought that "despite its growing importance . . . Army Aviation is still at times mistreated, misused and misunderstood by ground commanders at all levels."[211] The fundamental question was, "How can aviation be declared a full member of the combined arms team when, in fact, it is not organized, trained, managed and recognized as a full member?"[212]

The 1982 Army Aviation Systems Program Review, which considered this question, rejected the method of the 1960s and 1970s, when the integration of Army Aviation had been sought by spreading responsibility for it among the other branches. It argued that this approach had in fact retarded combined arms effectiveness and concluded that Army

Aviation could not be a full member of the team unless, like the other members, it was given branch status. Without a branch's own center and school, which served as a focus for the development of doctrine and its implementation through organization and training, Army Aviation was "fractionalized" and had no "institutional base" from which to address "training voids." There was no clear proponent of Army Aviation ideas and doctrine. Furthermore, the technical complexity of Army Aviation and the many changes it was experiencing made it impossible for other branches to assume responsibility for it.[213]

A major justification for an aviation branch was the need for training that produced aviators who understood combined arms rather than infantry or armor officers who understood aviation, but there were also subtle reasons concerned with how officers identified with Army Aviation and their loyalty to it. Indeed, since Army Aviation had become a combat arm and an officer speciality in 1979, many of the training needs were already beginning to be addressed, although of course branch status made this process more effective.[214] The transformation of Army Aviation into a branch was at least as important in confirming the growing equality between it and other aspects of the army—an equality already implied by combat arm status and heightened by the doctrinal advances encouraged by Airland Battle. Finally, the introduction of CABs made Army Aviation organizationally similar to the other branches. It allowed for an orderly progression through the officer ranks and, while increasing the ratio of officers to helicopters dramatically, also increased the number of officers in the army as a whole who were aviation conscious. Most of all, it aligned the "Army Aviation command structure with that of the *other* combat arms [emphasis added]."[215] (See Figure 1.3.)

The concern—almost an obsession—about status and equality was clearly expressed in many sources. In 1979, "the hawk, a magnificent bird of prey" was chosen as a symbol of Army Aviation's combat arm status.[216] Aviators wanted recognition that helicopters could be used "as decisively in the future as have Artillery, Armor or Infantry units in the past," and indeed, after it had become a branch, a senior officer, General Parker, wrote that Army Aviation was "*blessed with the full rights* as are the other combat maneuver arms [emphasis added]."[217] Words such as "coequal" and "full-fledged" (how appropriate for an aerial force!) appeared constantly in descriptions of Army Aviation's membership in the combined arms team. Branch status not only assisted in attaining this equality but signified throughout the army that Army Aviation had regained the prominence that it had previously had only during the Vietnam War. While some helicopter enthusiasts in other branches feared a loss of control and input into decisionmaking, a new breed of officers whose loyalties lay only with Army Aviation was emerging. Finally, it was argued that the helicopter was a principal

Figure 1.3
Aviation Officers Required: 1970s and 1980s

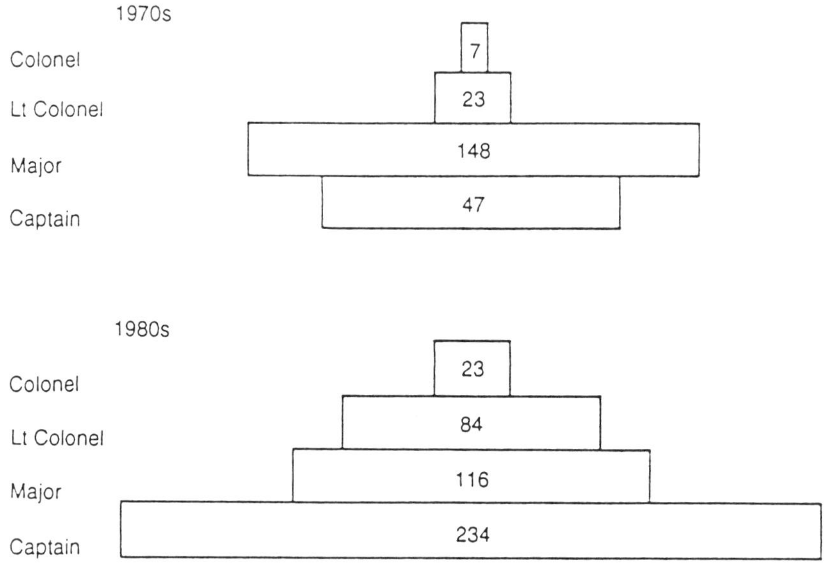

Source: Merryman, "Air Cavalry Attack Brigade," 5.

weapon system in much the same way as a tank, infantry combat vehicle or artillery piece. Since all other principal weapons systems were identified with a specific branch, the helicopter could only develop effectively within a branch of its own.[218]

As if reflecting the sort of patronizing attitudes that aviators thought they detected in other branches, there was some concern that branch status meant Army Aviation would "drift away" from the army, much as the Army Air Forces had done when they became the USAF, and that the opposite to closer integration would occur.[219] Critics saw aviators as reluctant soldiers and frustrated pilots who only barely kept in touch with the presumed realities of land warfare and who knew little about soldiering and needed an education from the infantry, armor or other branch to provide knowledge of land warfare. As one aviator recalled, "Those outside Aviation have a tendency to see us ... displaying a lot of flair and flamboyance without much substance."[220] Prejudices against the air force and some Army Aviation practices in Vietnam were mainly responsible for these perceptions. However, aviators resented any implication of similarity with the USAF. They assumed that Army Aviation was just as much a part of ground warfare as any other branch and argued that branch status would establish that Army Aviation was in fact a valued member of the army. Nevertheless, senior officers were aware that these criticisms had to be addressed—both for political and practical reasons. They emphasized that "[Army Aviation] must not be a separate, elite organization."[221] New training schemes sought to give aviators "a comprehensive understanding of the basic operations and, specifically, how ground commanders plan and conduct them" so that they were clearly seen to be "fighters first, fliers second."[222]

In some ways, the branch was a victory for image over substance in its first years. As a brief review in 1991 concluded, the branch "is still struggling for solid recognition and identity" and that, at first, aviators tended to assume that just because they now had branch status all the problems that had led to its establishment would simply vanish.[223] And so, for much of the 1980s, helicopters continued to be left out of the "scheme of maneuver." Perhaps recognizing that many aviators and nonaviators were not capable of being trained to integrate Army Aviation properly, senior commanders stressed that one solution was to use the best aviators as liaison officers in divisional and brigade headquarters. They would have to be the "sharpest people," with as much experience as possible, and thus would be "credible" and "good representatives of Army Aviation."[224] The Army Aviation commander, General Parker, pointed out that infantry and artillery had always coordinated their activities very well because of the strength of their liaison officers, and concluded: "I cannot overemphasize the absolute importance of liaison officers." Once again there were political considerations. Reviving the "bureaucratic insurgency" of Army Aviation advocates in the Pentagon in the 1950s and 1960s, Parker even went so far as to suggest that these officers "infiltrate [the ground headquarters] to the point they think you are theirs. Take control . . . [but] make them think you work for them."[225] As if echoing the internal warfare in which Parker imagined his Army Aviation officers were engaged, the branch's approach within the Army in the 1990s will be guided by "patience and cunning" so it will "grow old and prosper."[226]

Summing up the 1980s

Army Aviation began a new era when Airland Battle was introduced. Although it had already begun to recover from the doubt and uncertainty of the early 1970s concerning helicopter operations in mid-intensity conflict, it had only a limited—although important—role envisaged under the doctrine of Active Defense. The fundamental change in Airland Battle was its reemphasis of maneuver warfare over the firepower-oriented, positional approach of the 1970s.[227] While Active Defense did not disregard the helicopter's mobility, weapons were regarded as more important. Airland Battle, on the other hand, was not concerned with increasing the total weight of firepower as much as ensuring that it was delivered at the most appropriate location. In other words, as long as a relatively powerful force was used, the crucial determinant of success was effective maneuver before battle. Since the helicopter was the most mobile army weapons system available, it naturally benefited from the new approach.

The less dominant position given to firepower and the increased recognition of psychological factors in the Airland Battle doctrine meant that air assault became more popular again. It could, by delivering light infantry into sensitive rear areas, lead to results disproportionate to their low combat power. Yet the movement of troops by air never regained its predominance of the 1950s and 1960s and, with the arrival of the Apache, the focus on attack helicopter operations intensified.[228]

In part, this outcome resulted from the obvious combat power of the Apache and its fearsome appearance: using this helicopter as a symbol of what Army Aviation could achieve was politically sensible and psychologically irresistible. The Comanche, when it is deployed, will continue this trend. Senior officers, who had been involved in aviation since the 1950s, warned that the idea of a total force, integrating air assault and attack helicopters, was still vital: "We must continue to think of Army Aviation and airmobility as being all-inclusive of the three dimensions and other battlefield functions of combat . . . we may overlook the enormous potential of a fully rounded airmobile force."[229] But, whereas before these warnings had been directed at nonaviators seeking to limit Army Aviation's existence and reduce it to a supporting role, now they were intended for the new breed of aviators who regarded the Apache and attack helicopter operations as their raison d'etre. With the growing prospect of U.S. involvement in low-intensity warfare, the army might do well to remember the effectiveness of airmobility in Vietnam.

Yet, the concentration on attack helicopters in the 1980s was very different from the emphasis on anti-tank helicopters in the 1970s because Army Aviation at last began to sever the restraining, narrowing link between attack helicopters and close air support. The difference can best be seen in the following comments, intended to educate students at the Command and General Staff College, and worth quoting at length:

> Close air support is fire support, not maneuver. . . . Control of [support] assets is retained by the parent organization, integration is accomplished by ALOs and FACs [liaison officers and forward air controllers] . . . and stable association of supporting air with supported ground formations below army-tactical air force level is deliberately forfeited in return for mission versatility. In sharp contrast . . . the more stable the air combat-ground relationship, the greater the contribution of air combat assets to tactical success. . . . In short, the model for air maneuver should be the ground maneuver battalion, not the close air support squadron.[230]

While the actual implementation of this concept will require a major training effort, for both aviators and ground officers, at least by the mid- to late 1980s this view had been confirmed in Army Aviation doctrine and was actively supported by senior army commanders. Whereas in the 1960s and 1970s, helicopter proponents sought a maneuver concept based on air assault forces, in the 1990s it is attack helicopters that are the maneuver force.

Maneuver status implied control over the way in which combat missions were carried out, and air assault had required the sort of partnership between aviation and infantry that had been proposed in the 1950s and 1960s by the helicopter-promoting airborne club. Attack helicopters, on the other hand, resided entirely within the Aviation Branch. Hence, even as the issue of whether an Aviation Branch aided or hindered combined arms training was being debated, the whole notion that Army Aviation had to be an integral part of the ground maneuver forces began to be questioned. In 1984 FM 1–100 *Combat Aviation Operations* had suggested that:

> The advantage aviation offers to the [combined arms] team will be wasted on missions *which do not require aviation's speed and agility*. This outcome must be avoided, even if it means that attack helicopters and air cavalry units must sometimes be used as *pure aviation task forces* [emphasis added].[231]

There were obvious disparities in helicopter and mechanized movement rates and thus, in mixed forces, Army Aviation had to slow down to ground tempo. There were differences in operating techniques and problems with command and control; also, ground forces would usually be engaging the enemy at the point where air defense coverage was heaviest, thus increasing the risk to helicopters. Pure helicopter missions were also very attractive to helicopter advocates because they removed any hint that Army Aviation was a supporting and not a maneuver arm. Helicopter units would not try to do everything but rather would replace the ground-based maneuver component of the combined arms team. As with the airmobility concept of the 1950s, it would still be combined arms, but at helicopter tempo.[232]

Although there were "many arguments" over this issue,[233] and they have yet to be fully aired, let alone resolved, Airland Battle doctrine did not concentrate exclusively on deep attack and stated that deep operations were only to be conducted if they would have an effect on the main battle of close combat. Thus there was a need for operations at both air and ground tempo, with the former to be used mainly in deep attack and the latter in close combat. This strengthened the distinction between the corps CAB and its divisional counterpart, with the former being seen as an independent force maneuvering at aerial speed and the latter being integrated with ground forces and operating at their tempo.[234]

If the integration of helicopters into the ground scheme of maneuver continues to be uncertain because of the inability of the many nonaviator combined arms commanders to implement Army Aviation doctrine, then pressure within Army Aviation for more independent operations will grow. The more helicopter proponents become confident of their status within the army and of the value of their contribution to its activities, the more they will want control of how Army Aviation concepts are put into practice. If so, fears of another USAAF episode, with Army Aviation becoming obsessed with its doctrine and seeking some form of institutional independence from the army, will again be high. Assurances from some senior aviators that the predominant operation for the CAB will be in conjunction with the ground maneuver forces are at odds with comments by others that compare deep helicopter attacks to air force battlefield air interdiction.[235] However, the development of both corps and divisional CABs suggests there will be sufficient flexibility to allow both independent and closely integrated operations, thus satisfying the requirements and beliefs of both the Aviation Branch and the rest of the army.

Helicopters had gained widespread acceptance in the U.S. Army by 1992. Aviation had both an extensive fleet of helicopters and doctrines that recognized its importance. Overall, there was a much more mature understanding of the contribution that helicopters might make to the army. Nevertheless, development has not stopped. As the former commander of the U.S. Army, Europe, General Glen Otis, stated:

> The tactics and techniques in use [with helicopters] today are in their infancy compared to what they could do. Emphasizing technological developments, particularly in the realm of aviation, will give us an advantage on the battlefield not available today.[236]

CONCLUSION

The history of U.S. military helicopters must start with the theoretical development of airmobility in the 1950s, which was an intellectual advance equal to that of early tank proponents, for it was based upon the notion of a future war radically different from that which had just been fought. Yet, like armored warfare, airmobility in fact owed much to the war that had preceded its appearance. The airborne club of paratroop officers who were at the forefront of these developments did not create a new military theory simply from thin air. Instead, they linked the helicopter, the potential of which had been revealed in the Korean War, with their existing military views. Driven by their belief in the value of airborne operations but tempered by wartime experiences and the expected difficulties of using paratroops on an atomic battlefield, they produced the concept of airmobility.

The term "sky cavalry" captured the essence of the new theory. The definitive characteristic of cavalry was its speed about the battlefield but, more importantly, its speed relative to foot soldiers. Airborne officers such as Gavin, Taylor and Ridgway valued air transport because it was a much faster way of deploying troops to a vital position than ground movement. However, as had been demonstrated in the Second World War, this strategic or operational speed advantage was lost when the troops landed because mechanized ground forces had better tactical mobility. What distinguished airmobility from these earlier airborne operations was the employment of helicopters. The helicopter could interact easily with terrain and provide continuous mobility to the troops it was transporting. Thus the proposed airmobile formations had a movement advantage over ground units even at tactical level. Airmobility was also characterized by the insistence that, to gain the most from helicopters, they had to be concentrated into large units and not be separated out among the rest of the army; they were to perform all five functions of combat towards a common purpose within these units. This was an all-encompassing approach to helicopter doctrine. The separate units ensured that helicopters would not be commanded by officers untrained in their use and still thinking in ground-based terms.

The exact shape of airmobility doctrine was not just determined by military requirements, however. The army became locked into a long-running battle with the USAF over the issue of air support to ground forces and, unhappy with the air force, proceeded to develop its own aviation organization. Yet official acceptance of airmobility was constrained by agreements that gave near-exclusive responsibility for air support to the air force. Implementation was restricted because money and personnel were not available in sufficient quantities to satisfy both the requirements of the more established parts of the army and the fledgling Army Aviation force. Only the success of the "bureaucratic insurgents" gave Army Aviation access to increased financial resources in the early 1960s. Even then, final and absolute acceptance was secured only because of the needs of the Vietnam War—exactly the sort of conflict for which airmobility had not been intended. When airmobility was used in battle, mainly against Vietnamese insurgents, the speed of helicopters proved vital in countering a highly dispersed enemy operating in small groups in heavy terrain. It also demonstrated its value against conventional forces, albeit with poor air defenses, when deployed against North Vietnamese regular army units.

Yet, despite the apparent confirmation of all that had gone before, there was a dramatic change in the U.S. Army's approach to helicopters following the Vietnam War. While not solely responsible for the transformation, the war played a major part in separating the infantry-based airmobility doctrine of the 1950s and 1960s from the needs of conventional warfare that had originally inspired Gavin, Howze, Kinnard and the others. Vietnam was the first occasion on which most army officers had been exposed to helicopters and they gained valuable insight into the new weapon's qualities. But in the process airmobility became identified with the small-scale, low-intensity, minor air-defense-threat battles that characterized much of the Vietnam War; it also became associated to some extent with America's eventual defeat. Airmobility of this type did not die out, but with the army concentrating predominantly on a mid-intensity conflict, it became peripheral to major doctrinal developments in the 1970s.

Nevertheless, the historical conjunction of the passing of infantry-based airmobility with the U.S. withdrawal from Vietnam should not obscure the fact that the military environment of the 1970s was very different from that of the 1950s and 1960s, and that this also contributed to airmobility's decline. As demonstrated in the October 1973 Arab-Israeli War, aircraft of all types were now exposed a much greater array of lethal weapons, and survivability was a crucial issue. Equally, the light infantry central to airmobility was now subject to much greater threats, especially from artillery. In the 1970s air assault divisions had to be compared favorably with mechanized infantry and not just with "leg" infantry if they were to be accepted. It might be thought that, because airmobility had first been developed for use on an atomic battlefield, these conventional threats would have been easily dismissed. However, with hindsight it can be seen that, as with some other military theories that arose just after the Second World War, the devastating impact of nuclear weapons had not been fully understood by Gavin and his contemporaries, and their theories were not necessarily convincing.

The involvement in Vietnam had distracted the U.S. Army from changes in conventional warfare. It could well be argued that, but for the conflict, Army Aviation proponents would have adapted their theory of airmobility to suit the changing circumstances. It may well have been that the mechanized infantry air assault capability until recently being pursued by the former Soviet Union would have emerged fifteen years earlier in the United States but for the Vietnam War. The problem with this particular game of historical "What if?" is that, without the war, Army Aviation would not have obtained the resources or personnel to implement any radical doctrine of airmobility. Nor would its political power within the army have grown in such a way as to help it gain access to those resources. For all that helicopters were associated with the defeat in Vietnam, nonaviators in the U.S. Army gained a great deal of practical experience with them and, despite reservations, these officers were more favorably disposed to them after the war than they had been before it. Finally, without the Vietnam War, it is likely that Army Aviation would have continued to concentrate equally on fixed- and rotary-wing aircraft, leading to more, and more determined, USAF opposition than its predominantly helicopter-based force actually had to face.

These differing aspects of the Vietnam War and helicopter development resonate with the twin difficulties that Army Aviation proponents faced—and continue to face. They imagined themselves, quite properly, to be innovators—at the cutting edge of modern land warfare. This self-image required a certain degree of indifference towards other, more traditional combat arms. To suggest the need for their new approach, innovators also stressed its distinctive, separate nature from existing methods. Yet until the innovation becomes dominant (and is itself challenged), such innovation must somehow dovetail into existing practice if it is to continue to receive the support of the bulk of the army. Therefore, in the 1970s Army Aviation concentrated upon the combined arms team, eschewing radical innovation in return for acceptance. In a similar vein, the Vietnam War provided the possibility for expansion and increased innovation; yet satisfying the daily combat requirements, adjusting to enemy initiatives, meant that the opportunity could not be grasped in its totality.

When the Vietnam War ended, theoretical development recommenced and, while the changed battlefield environment ruled out airmobility, it did not necessarily rule out every similar all-encompassing theory. Air cavalry was one such concept, developed by senior officers in the Armor Branch and based on their Vietnam experience. However, the most powerful aviation advocates such as Maddox and Merryman rejected broad ideas and the large units they entailed, concentrating instead on making Army Aviation just another member of the combined arms team at the lowest organizational level. As they fought to establish the credibility of helicopters in mid-intensity conflict and to avoid too great a decline in numbers after the Vietnam War, these aviators wished to prevent Army Aviation from being seen as too distinct from the rest of the army.

By clearly establishing that the helicopter was a part of ground combat,[237] this strategy helped to defeat the air force's claim to be the exclusive provider of close air support since one of the most convincing differences between close air support and helicopter fire support was that the latter was directly available to ground commanders. To prevent opposition from within the army, Army Aviation had to proclaim loudly that it was not trying to usurp the leading role of the armor and infantry branches but was merely assisting them. While Starry and other air cavalry proponents wanted helicopters to have a maneuver role, senior aviators declined to compete with armored vehicles for attention and downplayed the relative importance of helicopters. Instead of an unequal competition within the army, they matched themselves against the USAF, playing upon the army's old fears about possible failures of air force support and thus ensuring they had the full approval of the army.

With the emphasis on combined arms it was not surprising that the independent Air Cavalry Combat Brigade was never given the same appreciation as the divisional anti-tank companies, thus reversing the view of the airmobility advocates who preferred to concentrate helicopters at as high a level as possible. Equally, after Vietnam, helicopters did not perform the "five functions of combat" in a concerted and self-supporting way but in isolated packages. For example, attack helicopters were rarely supported by cargo helicopters for resupply or by airmobile infantry carried in utility helicopters. Instead, these activities were performed in cooperation with and for the benefit of ground-based forces. The exception was the last remaining air assault division, which mainly concentrated on low-intensity warfare.

There were, however, valid military reasons for the 'penny-packet' approach, which spread helicopters throughout the army in small units. Active Defense was based on the notion of concentrating all available firepower into a connected series of combined arms firefights at brigade and battalion level with the aim of inflicting an unsustainable rate of attrition on the enemy. Helicopter tactics reflected this doctrine. Although the great mobility of helicopters was restricted by this approach because the area in which the helicopters would operate was reduced, it also meant that firepower was more readily available. Given the army's hierarchical organization, it was not surprising that the constituent doctrines, such as those directly relating to helicopters, followed the lead of the main doctrine. It was genuinely believed that, given the conditions established by the doctrine of Active Defense, the low-level, combined arms approach was best. Thus the real problem for the most innovative aviation proponents was that they had to change not just the rest of the army's thinking about helicopters but also its whole approach to land warfare. Helicopter advocates could not achieve this result and had to bide their time until Airland Battle emerged.

Although other tasks such as logistics movement and reconnaissance did not disappear completely, and even retained considerable importance, attack helicopters became the focus of Army Aviation activity. When using NOE flight and pop-up tactics, they were capable of overcoming the anti-aircraft threat and were very acceptable given Active Defense's orientation on firepower. Yet while they were apparently easier to combine with ground forces, attack helicopters did not achieve the simple integration into the combined arms team that had been expected. They were often perceived as a form of close air support—an ironic reminder of the way in which army leaders first became convinced of the need for armed helicopters—and were treated as a last-ditch reserve, to be used when other forces were unable to help. In the 1970s Army Aviation doctrine reflected the political and military circumstances within the U.S. Army and received official sanction and considerable resources, but was very difficult to implement. It failed because of the inherent tensions caused by trying to use new equipment and new ideas within established units and patterns of behavior.

To understand helicopter developments in the next decade, it is useful to imagine that airmobility (divisional, all-encompassing and stressing the unusual nature of Army Aviation) and the approach after the Vietnam War (battalion-based, specific and promoting membership of the ground combined arms team) were two extremes on the scale of possible helicopter doctrines. In the 1980s a balance was found between these extremes and helicopter doctrine became rather more flexible as the effects of the overall doctrinal change to Airland Battle filtered down to individual combat arms.

Army Aviation, with branch status and a new-found importance as a means of performing deep attack maneuvers and air assaults in Airland Battle, became rather more obviously an aerial arm of the army—at first rekindling some of the passions that had attended the USAF's separation from the army during and just after the Second World War. Although there was growing support for employing aviation forces in independent operations, Army Aviation continued to pursue the goal of combined arms integration with mechanized ground units and, after initial difficulties, it began to achieve its aim more effectively than before. This distinction is best seen in the appearance of CABs for both divisions and corps, and the assumption that they had rather different purposes. The latter was weaker and intended to fight rather like the helicopter units of the 1970s; the former gave to corps commanders many of the options for aviation operations associated with airmobility in the 1950s. Whether these distinctions continue, or whether the divisional CAB becomes more independent, remains to be seen.

There were specific, short-term reasons and imperatives for the decisions made in the 1980s about doctrines to allow helicopters to operate in the era of Airland Battle and about the final creation of a branch structure for Army Aviation, making it an equal member of the army along with older combat arms such as armor and artillery. But the decade can also be seen as the resolution of a number of fundamental problems associated with the original introduction of the helicopter in the 1940s and the army's continuing attempts to satisfy its needs for integrated air support. The organization of assets into divisional and corps combat aviation brigades gave the U.S. Army the chance to use helicopters flexibly—either in small supporting units at brigade level, in bigger units at divisional level or in large formations at corps. This organization addressed the diverse requirements at these various levels and recognized that there was more than one way of employing the helicopter's primary characteristic—its mobility. In tactical combat, at battalion and brigade level, attack helicopters would be used more frequently and over shorter distances, fighting at the tempo of the ground forces with which they were cooperating. Corps commanders, however, needed to be able to dispatch strong forces over hundreds of kilometers—possibly into enemy territory—using helicopters at a higher relative speed and more independently, without tying them to the lower mobility of ground forces.

The different organizations, with different approaches, resolved (at least temporarily) the competing aims of satisfying the existing, established army view about helicopters and its need for their support, while providing the room for more innovative approaches to be developed and practiced. This differentiation reflected both political and military considerations but was only possible because of the massive defense buildup of the United States in the 1980s. In the most assuredly smaller army of the next few decades Army Aviation will have to fight continually to maintain its access to resources. But whatever the actual number of helicopters (and the U.S. Army will continue to be the best equipped in the world), it will be the relative status of the divisional and corps CABs that signals whether or not Army Aviation has achieved complete dominance in obtaining for its units the equipment and doctrine it believes they need. One is tempted to ponder the possible meanings of the Branch motto, "Above the Best"—will Army Aviation simply roam the skies above the best army—or will it, in the other meaning of "above," consider itself superior to the other arms and fly beyond them into new military theoretical territory?

NOTES

1. James M. Gavin, *War and Peace in the Space Age* (London: Hutchinson, 1959) 114.

2. General Glen K. Otis (Commander, U.S. Army, Europe [USAREUR]), Memorandum to General William Richardson, USAREUR (unpublished: U.S. Army, 1983) 1.

3. General sources for early helicopter developments in the United States are Galvin, *Air Assault*; and Robert P. Weinert, *A History of Army Aviation 1950–1962, Phase I: 1950–1954* (Ft. Monroe, VA: CONARC, 1971).

4. Richard I. Wolf, *Basic Documents on Roles and Missions* (Washington, DC: USAF, 1987) 74.

5. Richard G. Davis, *The 31 Initiatives—A Study in Air Force-Army Cooperation* (Washington, DC: USAF, 1987) 8–9.

6. Gavin, *War and Peace in the Space Age*, 112.

7. General (Ret.) Robert R. Williams, correspondence (1990). Williams was one of the most influential advocates for helicopters in the U.S. Army.

8. Wolf, *Basic Documents on Roles and Missions*, 237–46 and 291–302; and Bergerson, *The Army Gets an Airforce*, 46–53.

9. General John J. Tolson, *Vietnam Studies: Airmobility 1961–1971* (Washington, DC: Department of the Army, 1973) 4.

10. Gavin, *War and Peace in the Space Age*, 113; Galvin, *Air Assault*, 275; and Maxwell D. Taylor, *The Uncertain Trumpet* (London: Stevens and Sons, 1960) 11–18. For the Korean War, see Andrew J. Bacevich, *The Pentomic Era* (Washington, DC: National Defense University Press, 1986).

11. Williams, correspondence, and General Harry W. O. Kinnard, correspondence (1990).

12. Matthew B. Ridgway, *Soldier: The Memoirs of Matthew B. Ridgway* (Westport, CT: Greenwood, 1974) 314.

13. Bergerson, *The Army Gets an Airforce*, 34, 101.

14. Gavin, *War and Peace in the Space Age*, 112–14, 257, and Ridgway, *Soldier*, 311–12.

15. The article was also printed in the U.S. Army's *Armor* journal: General James M. Gavin, "Cavalry . . . and I Don't Mean Horses!" *Armor* 63.5–6 (1954): 18–22.

16. Tolson, *Airmobility*, 6.

17. William E. Butterworth, *Flying Army* (New York: Doubleday, 1971) 83–95; and Bergerson, *The Army Gets an Airforce*, 104, 107.

18. Williams, correspondence; and General (Ret.) Hamilton H. Howze, correspondence (1990).

19. Bergerson, *The Army Gets an Airforce*, 72–73, 76; and Wolf, *Basic Documents on Roles and Missions*, 317–24.

20. Bergerson, *The Army Gets an Airforce*, 108; and Tolson (who was also a member of the board), *Airmobility*, 9.

21. Doughty, *The Evolution of U.S. Army Tactical Doctrine 1946–76*, 19–25.

22. Tolson, *Airmobility*, 19.

23. Dr. Alan Enthoven in United States Congress, Senate Committee on Government Operations, Subcommittee on National Security and International Operations, *An Inquiry into Planning-Programming-Budgeting*, 91st Congress, 2nd Session (Washington, DC: USGPO, 1970) 281; but compare with the rather different implication in Alan C. Enthoven and K. W. Smith, *How Much is Enough?* (New York: Harper and Row, 1971) 100.

24. Bergerson, *The Army Gets an Airforce*, 110–11.

25. Senate, *An Inquiry into Planning-Programming-Budgeting*, 279–81.

26. Tolson, *Airmobility*, 19.

27. Robert P. Weinert, *A History of Army Aviation 1950–1962, Phase II: 1955–1962* (Ft. Monroe, VA: TRADOC, 1976) 131–39.

28. Galvin, *Air Assault*, 276–77; confirmed in Howze, correspondence.

29. General (Ret.) Hamilton H. Howze, "Speculations on the Future," *U.S. Army Aviation Digest* [hereafter *AD*] 20.6 (1974): 6–7; and *Final Report, Army Tactical Mobility Requirements Board*, in Davis, *The 31 Initiatives*, 17.

30. Howze, correspondence.

31. Galvin, *Air Assault*, 277–78.

32. Tolson, *Airmobility*, 10–13; and Bergerson, *The Army Gets an Airforce*, 112, 118.

33. United States Congress, House of Representatives, Committee on Appropriations, Subcommittee, *Hearings on Department of Defense Appropriations for Fiscal Year 1964*, Part 1, 88th Congress, 1st Session (Washington, DC: USGPO, 1963) 415; and Lt. General (Ret.) Harry W. O. Kinnard, "Airmobility Revisited Part I," *AD* 26.6 (1980): 3.

34. Lt. General (Ret.) Harry W. O. Kinnard, "Airmobility Revisited Part II," *AD* 26.7 (1980): 14.

35. Tolson, *Airmobility*, 61.

36. Kinnard, "Airmobility Revisited Part I," 3.

37. Tolson, *Airmobility*, 56–57.

38. Howze, "Speculations on the Future," 6–7.

39. Doughty, *The Evolution of U.S. Army Tactical Doctrine 1946–1976*, 28.

40. Tolson, *Airmobility*, 23, 50, 56; and General Robert R. Williams in United States Congress, Senate Committee on Armed Services, Special Subcommittee on Close Air Support, *Hearings on Close Air Support*, 92nd Congress, 1st Session (Washington, DC: USGPO, 1971) 72.

41. General Westmoreland quoted in Tolson, *Airmobility*, 125.

42. U.S. Army Aviation Systems Command figures in Davis, *The 31 Initiatives*, 23.

43. Bergerson, *The Army Gets an Airforce*, 121–22; and Major General (Ret.) William J. Maddox Jr., "The Hinge," *AD* 25.6 (1979): 2.

44. Tolson, *Airmobility*, 29–33.

45. United States Congress, Senate Committee on Armed Services, *Hearings on Department of Defense Appropriations for Fiscal Year 1972* Part 5, 92nd Congress, 1st Session (Washington, DC: USGPO, 1971) 4168ff.

46. Wolf, *Basic Documents on Roles and Missions*, 382.

47. Tolson, *Airmobility*, 146.

48. Bergerson, *The Army Gets an Airforce*, 117, and Colonel (Ret.) John W. Oswalt, correspondence (1990).

49. Doughty, *The Evolution of U.S.Army Tactical Doctrine 1946–1976*, 40.

50. See for example Major General William J. Maddox Jr., "Where We Are—Where Are We Going," *AD* 20.11 (1974): 21. Maddox was the senior Aviation officer after the Vietnam War.

51. Kinnard, correspondence.

52. General Donn A. Starry, "To Change an Army," *Military Review* 63.3 (1983): 24; and "A Tactical Evolution—FM 100–5," *Military Review* 58.8 (1978): 3–4. Starry, then head of the Armor School, had a major influence on the development of Active Defense.

53. General Creighton W. Abrams in United States Congress, Senate Committee on Armed Services, *Hearings on Department of Defense Appropriations for Fiscal Year 1975* Part 1, 93rd Congess, 2nd Session (Washington, DC: USGPO, 1974) 684; and Starry, "To Change an Army," 24.

54. FM 100–5 *Operations* (Washington, DC: Department of the Army, 1976).

55. Brigadier General William J. Maddox Jr., "Looking into the Future," *AD* 18.3 (1972): 2. The term "high-intensity" was reserved for nuclear war.

56. Brigadier General James H. Merryman, "Bringing Army Aviation through the 70s... Into the 80s," *AD* 20.6 (1974): 4; and Major General (Ret.)

James H. Merryman, correspondence (1990). Merryman succeeded Maddox as the senior Army Aviation officer in 1977.

57. W. Maddox, "The Hinge," 3.

58. Captain Woodrow W. Turnbow, "Attack Helicopter Survivability," *Armor* 62.9–10 (1973): 22; and Major General William J. Maddox, Jr., "Army Aviation Priorities," *AD* 20.12 (1974): 29.

59. General William W. Momyer (Commander, USAF Tactical Air Command) in Senate, *Hearings on Close Air Support*, 173–87; and Williams, correspondence.

60. Charles M. Clifford, *1970 Defense Budget and Defense Program for Fiscal Years 1970–1974* (Washington, DC: USGPO, 1969) 81.

61. Harold Brown, *Department of Defense Annual Report, Fiscal Year 1982* (Washington, DC: USGPO, 1981) 11.

62. W. Maddox, "Looking into the Future," 2; and Lt. Colonel William Fraker, correspondence, *AD* 20.12 (1974): 7.

63. Lt. Colonel Charles M. Putnam, "Proponency for Aerial Fighting Vehicles," *Armor* 74.5–6 (1970): 7–10.

64. Brigadier General (Ret.) John C. Bahnsen, "A New Army Air Corps or a Full Combat Arms Team Member?" *Armed Forces Journal Intenational*, October 1986: 70; and W. Maddox, "The Hinge," 3.

65. Maddox, "Where We Are—Where Are We Going," 3; and Major Frank T. Peterlin, "Combat and Training Developments," *AD* 19.8 (1973): 4, 19—21.

66. FM 100–5 *Operations* (1976), 3(10). Combined arms was, of course, by no means new but the particular emphasis placed on it was a significant change from past practice. See Brigadier General Charles E. Canedy in Senate, *Hearings on Department of Defense Appropriations for Fiscal Year 1975*, Part 9, 4680–81.

67. Merryman, "Bringing Army Aviation through the 70s . . . Into the 80s," 4; and Major General John C. Smith, "AVNEC Report," *AD* 24.7 (1978): 2.

68. CW3 Miguel S. Lopez, "Terrain Flying in Bavaria . . . Reforger '74," *AD* 21.2 (1975): 21.

69. "On Army Aviation," an interview with General Donn A. Starry, *AD* 24.11 (1978): 5.

70. Ibid., 5.

71. Major General John C. Smith in ibid., 5.

72. Betty Goodson, "A Resolution Conference at the Army Aviation Center," *AD* 27.3 (1981): 38.

73. Maddox in United States Congress, Senate Committee on Armed Services, *Hearings on Department of Defense Appropriations for Fiscal Year 1974*, Part 7, 93rd Congress, 1st Session (Washington, DC: USGPO, 1973) 4762–63.

74. FM 90–1 *Employment of Army Aviation in a High Threat Environment* (Washington, DC: Department of the Army, 1976) 2(24).

75. Captain J. E. Fulbrook, "Lamson 719: III Reflections and Values," *AD* 32.8 (1986): 11; and Colonel John Molinelli in United States Congress, House of Representatives Committee on Armed Services, *Hearings on Military Posture for Fiscal Year 1972*, Part 1, 92nd Congress, 1st Session (Washington, DC: USGPO, 1971) 3117–31.

76. Lt. Colonel D. R. Hull, "The Way of the Future: Low Level," *AD* 18.10 (1972): 3.

77. Lt. General John J. Tolson, "The New Challenge," *AD* 18.11 (1972): 17.

78. Maddox in United States Congress, Senate Committee on Armed Services, *Hearings on Department of Defense Appropriations for Fiscal Year 1976*, Part 8, 94th Congress, 1st Session (Washington, DC: USGPO, 1975) 4421–22.

79. Tolson, *Airmobility*, 198–99.

80. Major General Robert M. Shoemaker, "Tactics Outlook," *AD* 20.8 (1974): 1; and Tolson, "The New Challenge," 17.

81. FM 90–1, 3(2).

82. Ibid., 2 (22)–2 (23), 3 (7); and Maddox, "Army Aviation Priorities," 27.

83. Various sources were used: G. Costello, "Blackhawk," *AD* 24.2 (1978): 34–39. "On the Attack Helicopter," an interview with Brigadier General Edward M. Browne (Program Manager, Advanced Attack Helicopter), *AD* 25.1 (1979): 2–7; and Donald E. Fink, "YAH–64 Design Leans upon Viet Experience," *Aviation Week and Space Technology* [hereafter *AWST*], 18 August 1975: 42–47.

84. The modifications took many years to be implemented fully because of a lack of money.

85. Major Frank T. Peterlin, "Offensive Attack Helicopter Operations," *AD* 24.7 (1978): 42.

86. For example, W. Maddox, "Where We Are—Where Are We Going," 22. See also United States Congress, House of Representatives Committee on Armed Services, *Hearings on Military Posture for Fiscal Year 1983*, Part 2, 97th Congress, 2nd Session (Washington, DC: USGPO, 1982) 253.

87. Senate, *Hearings on Department of Defense Appropriations for Fiscal Year 1972*, Part 5, 4149.

88. Davis, *The 31 Initiatives*, 21.

89. Statement of Secretary of the Army S. R. Resor in Senate, *Hearings on Department of Defense Appropriations for Fiscal Year 1972*, Part 1, 713.

90. Packard in Senate, *Hearings on Close Air Support*, 50–51, and Williams and Maddox in ibid., 72–172.

91. General William W. Momyer in ibid., 173–87, 188, 195.

92. Bergerson, *The Army Gets an Airforce*, 121–42.

93. Major General William J. Maddox, Jr., "That First Battle," *AD* 21.10 1975, 3; and FM 100–5 *Operations* (1976), 1 (2), 2 (21)–2 (22) and 5 (5)–5 (6).

94. As explained by Resor in Senate, *Hearings on Department of Defense Appropriations for Fiscal Year 1972*, Part 1, 709.

95. Chief of Staff General Edward G. Meyer in United States Congress, House of Representatives Committee on Armed Services, *Hearings on Military Posture for Fiscal Year 1981*, Part 1, 96th Congress, 2nd Session (Washington, DC: USGPO, 1980) 781.

96. Chief of Staff General Fred C. Weyand in Senate, *Hearings on Department of Defense Appropriations for Fiscal Year 1976*, Part 2, 378.

97. Statement of Chief of Staff General Creighton W. Abrams in Senate, *Hearings on Department of Defense Appropriations for Fiscal Year 1974*, Part 2, 494–96.

98. Brigadier General Canedy in United States Congress, Senate Committee on Armed Services, *Hearings on Department of Defense Appropriations for Fiscal Year 1978*, Part 6, 95th Congress, 1st Session (Washington, DC: USGPO, 1977) 4043.

99. For example, see Major General Donn A. Starry, "Aviation—Part of the Armor Family," *AD* 20.6 (1974): 24–28.

100. Williams, correspondence.

101. Oswalt, correspondence.

102. Maddox in Senate, *Hearings on Department of Defense Appropriations for Fiscal Year 1974*, Part 7, 4764–65.

103. Maddox, "The Hinge," 4.

104. Quoted in General (Ret.) Hamilton H. Howze, "The Case for the Helicopter," *Army* 29.3 (1979): 19.

105. Colonel John C. Bahnsen, "Attack Helicopter Fundamental Tactics," *AD* 25.4 (1979): 6; and FM 90–1, 3(16)–3(21), 3(54)–3(56).

106. FM 90–1, 3(4).

107. FM 71–100 *Armored and Mechanized Division Operations* (Washington, DC: Department of the Army, 1978) 5(4).

108. Lt. General D. R. Keith in House of Representatives, *Hearings on Military Posture for Fiscal Year 1981*, Part 2, 603.

109. FM 90–1, 3(22).

110. Colonel Charles E. Canedy, "From ACCB to the 6th Cavalry Brigade (Air Combat)," *AD* 21.5 (1975): 3–4, and FM 17–47 *Air Cavalry Combat Brigade* (Washington, DC: Department of the Army, 1982) 3(2)–3 (9).

111. Major G. R. Hall et al., "ARCSA III," *AD* 23.7 (1977): 2–3, 17–19; and Canedy in Senate, *Hearings on Department of Defense Appropriations for Fiscal Year 1975*, Part 9, 4676.

112. FM 100–5 *Operations* (1976), 3(5)–3(10); and Starry "Aviation—Part of the Armor Family," 26–27.

113. Maddox in Senate, *Hearings on Close Air Support*, 90.

114. FM 90–1, 1(1), 3(10)–3(48). Note also the continued reference to a "support" role by senior aviators at this time.

115. Starry, "Aviation—Part of the Armor Family," 26–27.

116. The idea that attack helicopters were a maneuver force was described in 1978 as being "a significant departure from the current usage." Major D. Carlin, "The Antitank Battalion," *AD* 24.12 (1978): 9.

117. United States Congress, Senate Committee on Armed Services, *Hearings on Department of Defense Appropriations for Fiscal Year 1973*, Part 6, 92nd Congress, 2nd Session (Washington, DC: USGPO, 1972) 3660ff.

118. Major General Edward M. Browne in United States Congress, House of Representatives Committee on Armed Services, *Hearings on Military Posture for Fiscal Year 1982*, Part 2, 97th Congress, 1st Session (Washington, DC: USGPO, 1981) 406–411, 421, 424; and Parker in House of Representatives, *Hearings on Military Posture for Fiscal Year 1983*, Part 2, 321–61.

119. United States Congress, Senate Committee on Armed Services, *Hearings on Department of Defense Appropriations for Fiscal Year 1981*, Part 4, 96th Congress, 2nd Session (Washington, DC: USGPO, 1980) 2088. See also "Budget, Technical Items Delay Attack Helicopter," *AWST*, 2 July 1979: 17–18.

120. Canedy in Senate, *Hearings on Department of Defense Appropriations for Fiscal Year 1978*, Part 6, 4051–58.

121. The best source on JAAT is Major W. J. Filippini and Captain D. S. Prewitt, "JAAT—A Present Concept," *AD* 28.11 (1982): 2–7.

122. David H. Rumsfeld, *Department of Defense Annual Report, Fiscal Year 1978* (Washington, DC: USGPO, 1977) 166. See also Comptroller General's Report to Congress, *Army Helicopter Improvement Program's Future May Depend Upon Success in Controlling Cost*, GAO MASAD–83–2 (Washington, DC: USGPO, 1983) 1–8.

123. Robert R. Ropelewski, "Mast-Mounted Sight, New Rotor Boost Effectiveness of OH–58D," *AWST*, 8 September 1986: 66.

124. Rumsfeld, *Department of Defense Annual Report, Fiscal Year 1978*, 167–68.

125. FM 100–5 *Operations* (1976), 14(1)–14(27).

126. Captain David W. Henk, "The Threat to Air Assault Operations," *AD* 22.2 (1976): 6; and Brigadier General (Ret.) John C. Bahnsen, correspondence (1990).

127. Resor in Senate, *Hearings on Department of Defense Appropriations for Fiscal Year 1972*, Part 1, 700.

128. Packard in Senate, *Hearings on Close Air Support*, 55.

129. Kinnard, "Airmobility Revisited Part II," 14.

130. Major General William B. Fulton, "TRICAP—The Challenge of the Seventies," *Armor* 81.5–6 (1972): 22.

131. Canedy, "From ACCB to the 6th Cavalry Brigade (Air Combat)," 1–4, and "New Helicopter Combat Roles Planned, *AWST*, 29 September 1975: 57. For earlier configurations, Colonel David R. Redmond, "The Role of Helicopters in Conventional Warfare," *AD* 18.1 (1972): 9; and Lt. Colonel Jimmy D. Weeks, "ACCB: Tank finder and killer," *Armor* 82.3–4 (1973): 33.

132. Major J. R. Mills, "Air Assault Update," *AD* 21.9 (1975): 6–7, 28–29; and Major General T. M. Tarpley, "Dear Fellow Infantrymen," *Infantry* 64.3–4 (1974): 3.

133. Major General Peter G. Olenchuk in United States Congress, House of Representatives Committee on Armed Services, *Hearings on Military Posture and Procurement for Fiscal Year 1976*, Part 1, 94th Congress, 1st Session (Washington, DC: USGPO, 1975) 1129.

134. For just one example, Colonel L. J. Baughman and Major R. E. Jones, Jr., "101st in Reforger 76," *AD* 22.12 (1976): 2–3, 18–22.

135. FM 90–1, 1(5).

136. FM 100–5 *Operations* (1976).

137. Major General W. J. Livsey, "Commandant's Notes," *Infantry* 68.7–8 (1978): 5–6; and FM 90–4 *Airmobile Operations* (Washington, DC: Department of the Army, 1980) i.

138. FM 100–5 *Operations* (1976), 5(6)–5(7); FM 90–1, 3(26)–3(28); and FM 71–100, 6(12)–6(13).

139. FM 90–1, 3(34)–3(39), 3(43)–3(53). For the Chinook see Donald E. Fink, "CH–47D Offers Upgraded Performance," *AWST*, 18 July 1983: 54–65.

140. General Donn A. Starry, *Armored Combat in Vietnam* (Poole, Dorset: Blandford, 1981) 221.

141. Brigadier General D. K. Doyle and Major William V. Chiaramonte, "We'll Have No Problem Finding the Enemy?" *AD* 24.1 (1978): 30–36; and Brigadier General D. K. Doyle and Major William V. Chiaramonte, "Scouts Out," *Armor* 87.9–10 (1978): 30.

142. Bahnsen, "A New Army Air Corps or a Full Combat Arms Team Member?" 70; and Merryman, correspondence.

143. General William E. DePuy, Letter to General Fred C. Weyand (Chief of Staff), 18 February 1976, reproduced in John L. Romjue, *From Active Defense to Airland Battle: the Development of Army Doctrine 1973–1982* (Ft. Monroe, VA: TRADOC, 1986) 84.

144. FM 100–5 *Operations* (1976), 2(30).

145. Lt. General William R. Richardson, "Airmobility in the 1980s," *AD* 27.8 (1981): 3; Kinnard, "Airmobility revisited Part I," 5; and Howze, "The Case for the Helicopter," 16.

146. General William E. Depuy, "Keynote Address to the Armor Conference 1977," *Armor* 68.7–8 (1977): 34.

147. Kinnard, "Airmobility Revisited Part I," 5. For senior officers' implicit rejection, see Merryman, "Bringing Army Aviation through the 70s . . . Into the 80s," 2; and Maddox, "Where We Are—Where Are We Going," 22.

148. Maddox in Senate, *Hearings on Department of Defense Appropriations for Fiscal Year 1976*, Part 8, 4423.

149. FM 100–5 *Operations* (Washington, DC: Department of the Army, 1982). See also Major General Carl H. McNair Jr., editorial, *AD* 28.12 (1982): 1.

150. Major M. Snyder and Captain R. Hancock, "Aviation Employment Conference," *AD* 30.1 (1984): 8; and FM 1–100 *Combat Aviation Operations* (Washington, DC: Department of the Army, 1984).

151. Major A. R. Stewart, "ACAB—Air Cavalry Attack Brigade," *AD* 26.7 (1980): 6.

152. FM 100–5 *Operations* (1982).

153. Major Charles B. Cook, "Shaping the Airland-Echeloned Battlefield with Army Aviation," *AD* 29.10 (1983): 9.

154. General Donn A. Starry, "Extending the Battlefield," *Military Review* 61.3 (1981): 47.

155. Lt. Colonel John S. Doerfel, "The Operational Art of the Airland Battle," *Military Review* 62.5 (1982): 6, 8; and Lt. Colonel Hubert Wass de Czege and Lt. Colonel L. D. Holder, "The New FM 100–5," *Military Review* 62.7 (1982): 57.

156. Major General Carl H. McNair, Jr., and Captain John Reinsprecht, "Army Aviation Forces in the Airland Battle," *AD* 27.7 (1981): 8.

157. Major J. F. Fields, "AVNEC Part II—How to Fight: Offense," *AD* 24.7 (1978): 4.

158. FM 100–5 *Operations* (1982), 7(13)ff.

159. Lt. Colonel E. H. Grayson, Jr., and Lt. Colonel G. R. Allen, "Attack Helicopters for a Successful Defense," *AD* 25.11 (1979): 12, 15. For the application of this thinking in the 1980s, see FM 1–100, 27.

160. Lt. General Crosbie E. Saint and Colonel William H. Yates, "Attack Helicopter Operations in the Airland Battle: Close Operations," *Military Review* 68.6 (1988): 8.

161. McNair and Reinsprecht, "Army Aviation Forces in the Airland Battle," 10; FM 1–100, 37; and FM 100–5 *Operations* (1982), 1(3), 2(8).

162. Deborah G. Meyer, Interview with Major General Carl H. McNair, Jr., *Armed Forces Journal International*, May 1982: 49.

163. Major General Carl H. McNair, Jr., and Captain John Reinsprecht, "Army Aviation Systems Program Review '82," *AD* 28.6 (1982): 41; and Brigadier General Ellis D. Parker in United States Congress, House of Representatives

Committee on Armed Services, *Hearings on Defense Department Authorization and Oversight for Fiscal Year 1984*, Part 3, 98th Congress, 1st Session (Washington, DC: USGPO, 1983) 498.

164. Lt. Colonel W. C. Page, Jr., "Roller Coaster Operations Across the FEBA," *AD* 26.3 (1980): 34–38.

165. Major Frank E. Babiasz, "The Leak in the Soviet Air Defense Umbrella," *AD* 27.11 (1981): 35–37.

166. FM 17–50 *Attack Helicopter Operations* (Washington, DC: Department of the Army, 1984) 2(6). Compare it with FM 90–1, chapter 2.

167. Major General Jack V. Mackmull and Captain John Reinsprecht, "Army Aviation Systems Program Review '82—Concepts, Doctrine and Tactics Panel," *AD* 28.7 (1982): 12.

168. "White Paper—Employment of Combat Aviation" in *Dynamics of Battle* (Ft. Leavenworth, KS: USACGSC, 1987) 18, and "Employment of the Divisional Combat Aviation Brigade" (unpublished: U.S. Army, 1987) 2–3.

169. Lt. Colonel Keith R. McGinty and Captain G. A. Brockman, "Cross-flot with the AH–64 Apache," *AD* 33.1 (1987): 2. For problems, see Lt. General Jack V. Mackmull, "Army Aviation in the Real World," *Defence Helicopter World* 3.9–11 (1984): 7–9.

170. FM 1–111 *Aviation Brigade* (Washington, DC: Department of the Army, 1986) C(1).

171. Comptroller General's Report to Congress, *Issues Concerning the Army's Light Helicopter Family Program*, GAO MSIAD–86–121 (Washington, DC: USGPO, 1986) 11.

172. See, for example, Robin Adshead, "Exercise 'Certain Strike,'" *Armed Forces* 6.12 (1987): 570.

173. Major General Rudy Ostovich III, "Quality of the Aviation Force," *AD* 36.7–8 (1990): 1.

174. David A. Fulgham, "U.S. Army Aviation to Assume Frontline Role in Saudi Arabia," *AWST*, 15 October 1990: 80–81.

175. "Defence Hoverview," *Defence Helicopter* 10.4–5 (1991): 44.

176. Frank Colucci, "Surviving the Storm," *Defence Helicopter*, 10.10–11 (1991): 7–10; and "Defence Hoverview," *Defence Helicopter* 10.4–5 (1991): 44.

177. Major General Rudy Ostovich III, "Aviation's Contribution to War in the Gulf," *AD* 37.3–4 (1991): 1.

178. FM 17–50 (1984), J(4); and Major Charles L. Barry, "Planning Aviation Cross-FLOT Operations," *Military Review* 64.1 (1984): 42. See also Lt. General Robert W. RisCassi (Commander, Combined Arms Center), "Army Aviation in the 1980s," *AD* 32.1 (1986): 5.

179. "Employment of the Divisional Combat Aviation Brigade," 1–3. See also FM 1–111, C(1). Note the different treatment of the two brigades in FM 1–100, 23.

180. Major General Carl H. McNair, Jr., "LHX—the Conceptual Formulation of Why, What and When," *AD* 28.12 (1982): 5; and Comptroller General's Report, *Issues Concerning the Army's Light Helicopter Family Program*, passim.

181. "LHX Studies Include Tilt-Rotor Option," *AWST*, 16 June 1986: 29.

182. "Army Opts for Conventional Rotorcraft Technology for LHX," *AWST*, 11 March 1985: 27.

183. Stanley W. Kandebo, "Boeing Sikorsky LH Technologies Will Increase Army Combat Capabilities," *AWST*, 15 April 1991: 20.

184. Benjamin F. Schemmer, "Army Aviation to Modernize, but with 2,000 Fewer Planes," *Armed Forces Journal International*, April 1988: 84–85.

185. Major General Rudy Ostovich III, "1991: A Year for the Record Books," *AD* 37.3–4 (1991): 1; and David F. Bond, "Cost, Supportability Key to Boeing Sikorsky LH Award," *AWST*, 15 April 1991: 19, quoting Stephen A. Conver, Assistant Secretary, Department of the Army.

186. FC 90–4 *Air Assault Operations* (Washington, DC: Department of the Army, 1985) 1(2)–1(10).

187. Compare FM 100–5 *Operations* (1976), 4(7), 5(8)–5(9) and FM 100–5 *Operations* (1982), 7(4)–7(5).

188. Lt. General John R. Galvin, "Heavy-Light Forces and the NATO Mission," *Infantry* 74.7–8 (1984): 12, 14; and FM 100–5 *Operations* (Washington, DC: Department of the Army, 1986) 41.

189. FM 90–4 *Air Assault Operations* (Washington, DC: Department of the Army, 1987) 1(2).

190. Compare FM 90–1, 3(84), and its 1980s successor, FM 1–100, 29–30.

191. Ostovich, "1991: A Year for the Record Books," 1.

192. "Operation Desert Storm," *AWST*, 4 March 1991: 21.

193. Colonel E. H. Grayson, Jr., "The Past: What Might Have Been. The Future: What Can Be," *AD* 32.6 (1986): 39, 41; and Colonel (Ret.) E. H. Grayson, "Infantry and Army Aviation," *AD* 34.6 (1988): 4.

194. Major Keith H. Bardot, "AVNEC Part III—How to Fight: Defense," *AD* 24.8 (1978): 7. Most articles proposing HATAC had a distinct, defensive tone as they constantly reassured readers that Army Aviation was not becoming another air force.

195. Major Frank E. Babiasz and Captain Charles E. Daschke, "Anti-Helicopter Operations," *AD* 26.5 (1980): 24–27; and Edward J. Bavaro, "Soviet Helicopter Air to Air," *AD* 34.3 (1988): 2–7.

196. Snyder and Hancock, "Aviation Employment Conference," 8; and Barry, "Planning Aviation Cross-FLOT Operations," 38.

197. Major General Carl H. McNair, Jr., "Helicopter Air-to-Air Combat Operations: The Big Picture," *AD* 27.10 (1981): 5.

198. Brigadier General Parker in House of Representatives, *Hearings on Military Posture for Fiscal Year 1983*, Part 2, 259.

199. Major Michael L. Brittingham, "Army Air Combat—The Challenge Continues," *AD* 33.2–3 (1987): 1.

200. McNair and Reinsprecht, "Army Aviation Systems Program Review '82," 44.

201. Colonel E. H. Grayson, "Army Aviation 1984–2015," *AD* 30.11 (1984): 3; and FM 1–100, 3.

202. FM 100–5 *Operations* (1982), 7(5); and McNair and Reinsprecht, "Army Aviation Forces in the Airland Battle," 12.

203. FM 100–5 *Operations* (1986), 34; and Brigadier General D. F. Whalen in United States Congress, House of Representatives Committee on Armed Services, *Hearings on Defense Department Authorization and Oversight for Fiscal Year 1985*, Part 2, 98th Congress, 2nd Session (Washington, DC: USGPO, 1984) 562.

204. "Chief of Staff Verbatim Comments Reference Employment of Aviation Brigade," (unpublished: TRADOC, 1987) 1–2.

205. Ibid.

206. Brigadier General Rudy Ostovich III, "Airland Battle Part II: Views from Aviation Commanders," *AD* 32.12 (1986): 32. See also Saint and Yates, "Attack Helicopter Operations in the Airland Battle: Close Operations," 4–5, 7.

207. FM 17–47, ii, 3(2).

208. Snyder and Hancock, "Aviation Employment Conference," 6–7.

209. Bahnsen, "A New Army Air Corps or a Full Combat Arms Team Member?" 74; and FM 1–111, v.

210. See, for example, Brigadier General Ellis D. Parker, "Aviation Update," *AD* 28.1 (1982): 9, 11.

211. Captain Keith G. Scherrer, "Army Aviation—An Insider's View," *Infantry* 72.9–10 (1982): 7; and Major General John R. Galvin, "Army Aviation Systems Program Review '82—The Training Panel," *AD* 28.10 (1982): 6.

212. Mackmull and Reinsprecht, "Army Aviation Systems Program Review '82—Concepts, Doctrine and Tactics Panel," 13.

213. Galvin, "Army Aviation Systems Program Reivew '82—The Training Panel," 3–5; and Major General Bobby J. Maddox, "Army Aviation Branch Implementation," *AD* 29.8 (1983): 2–9.

214. Richardson, "Airmobility in the 1980s," 26; and Galvin, "Army Aviation Systems Program Review '82—The Training Panel," 5.

215. Stewart, "ACAB—Air Cavalry Attack Brigade," 6; and Major General James H. Merryman, "Air Cavalry Attack Brigade," *AD* 25.11 (1979): 4–5.

216. Merryman, editorial, *AD* 25.10 (1979): 1.

217. Cook, "Shaping the Airland-Echeloned Battlefield with Army Aviation," 11; and Major General Ellis D. Parker, "The Challenge of Winning," *AD* 31.9 (1985): 2.

218. Major General Bobby J. Maddox, "Landpower's Third Dimension," *Army* 34.3 (1984): 20; and Mackmull and Reinsprecht, "Army Aviation Systems Program Review '82—Concepts, Doctrine and Tactics Panel," 13.

219. Galvin, "Army Aviation Systems Program Review '82—The Training Panel," 4; and Bahnsen, "A New Army Air Corps or a Full Combat Arms Team Member?" passim.

220. Lt. Colonel John B. MacKenzie, Jr., "Challenges for the Aviation Branch," *AD* 30.12 (1984): 11.

221. Major General Bobby Maddox, editorial, *AD* 30.1 (1984): 1.

222. McKenzie, "Challenges for the Aviation Branch," 11–12; and Parker, "The Challenge of Winning," 4, 9.

223. Captain T. Cowart, "7 and Growing," *AD* 36.3–4 (1990): 14.

224. Various comments in Ostovich, "Airland Battle Part II: View from Aviation Commanders," 33–34.

225. Parker in ibid., 36.

226. Cowart, "7 and Growing," 14.

227. Romjue, *From Active Defense to Airland Battle*, 6–7, 17–21 summarized the debate from a viewpoint somewhat favoring Active Defense. For the firepower emphasis, see ibid., 83.

228. RisCassi, "Army Aviation in the 1980s," 6, for one example.

229. Galvin, "Army Aviation Systems Program Reivew '82—The Training Panel," 8.
230. "White Paper—Employment of Combat Aviation," 20.
231. FM 1–100, 3.
232. Saint and Yates, "Attack Helicopter Operations in the Airland Battle: Close Operations," 8; Lt. General Crosbie E. Saint and Colonel William H. Yates, "Attack Helicopter Operations in the Airland Battle: Deep Operations," *Military Review* 68.7 (1988): 9; and RisCassi, "Army Aviation in the 1980s," 6.
233. Ostovich, "Airland Battle Part II: Views from Aviation Commanders," 35.
234. "Employment of the Divisional Combat Aviation Brigade," 1–2.
235. The quote was from Parker, "The Challenge of Winning," 5. Compare it with the references to USAF operations in Saint and Yates, "Attack Helicopter Operations in the Airland Battle: Deep Operations," 3–5.
236. General Glen K. Otis, "Future Concepts and Capabilities in NATO's Central Region," *Journal of the Royal United Services Institute* [hereafter *JRUSI*] 133.4 (1988): 19.
237. Although much of the evidence for this political maneuvering is implicit, see for example the comments by Merryman, "Bringing Army Aviation through the 70s... Into the 80s," 4.

2

"Revolutions at Every Turn"— Developments in the Soviet Union

There is no sense to consider seriously plans for "flying," "jumping" or other tanks of unusual design.

Chief Marshal of the Armored Forces Rotmistrov, 1972.[1]

Helicopters have gained for themselves an unquestionable position ... [and] they are playing an increasingly greater role.

Colonel Scibiorek, 1987.[2]

Unlike the United States, the 1940s and 1950s saw only tentative progress with military helicopters in the USSR. In many ways their technical development was superior to that in America, but there was no matching innovation in doctrine. The first important doctrinal developments came in the early 1960s and continued until 1973. During this time, the "revolution in military affairs," a Soviet term denoting the massive upheavals in military thinking caused by the widespread deployment of nuclear weapons, strongly influenced helicopter doctrines. The "revolution" produced a need for greater mobility and the capacity to attack the enemy's rear areas as soon as possible; as a result the military helicopter became an important part of Soviet air-landing operations. It was to be employed mainly in tactical missions conducted by armies and fronts that enabled the paratroops to concentrate instead on strategic landings. Helicopters not only improved the mobility of the troops they carried but, by transferring combat power rapidly into the enemy's rear, allowed the army as a whole to advance more quickly.

Armed helicopters did not figure prominently in Soviet military planning at this stage except as a way of carrying out air landings more effectively. As the 1960s ended, however, it was proposed that they be employed in a direct attack role as a form of close air support; this idea was eventually accepted after the October 1973 Arab-Israeli War, thus initiating the next major phase in helicopter development. Following this conflict the Soviets realized that their conventional fire support capacity was inadequate, and attack helicopters began to be introduced as one of a number of measures to correct this deficiency. As before, the aim was to achieve a very fast rate of offensive. The Soviets believed that the more fire that could be placed on front-line targets, the more quickly the attacking units could move past or through them and reach the enemy's vulnerable rear areas.

However the October 1973 war had far more fundamental effects on helicopters because it caused a general reappraisal of Soviet military thinking about conventional conflict. This reevaluation was intensified by the USSR's interest in developing the capacity to wage war without nuclear weapons and, indeed, in a manner that would limit the likelihood that either side would resort to such measures. One result was that helicopters—both armed and transport—became more important than ever before. By the 1980s, they were a crucial complement to tanks and other armored vehicles in Soviet planning and particularly in relation to operational maneuver groups. While the change can be seen in technical developments of new aircraft, a better indicator is the reorganization of the Soviet Air Forces to improve the level of coordination between helicopters and ground units; special air assault units were also formed to increase the effectiveness of air landings. The aim was to develop a capability for tactical air-land warfare with helicopters, tanks and infantry—supported by artillery and other arms—fighting in close combination. These developments coincided with the fighting in Afghanistan, which was not a catalyst for change but did provide a testing ground for many new ideas.

Because of their speed, helicopters were always highly regarded by the Soviet army. Yet the exploitation of aerial mobility was, for many years, only indirectly linked to ground action. Only when the Soviets began to recognize that combat had become air-land at the operational and strategic levels of warfare did they seek to extend this combination to the tactical level. To do so, they utilized the characteristics of the helicopter—speed and flexibility—in close concert with other arms. By the end of the 1980s, helicopters occupied a position of great importance—so much so that the Soviets believed their general military concepts were not viable without them.

THE 1940s–1950s: TENTATIVE BEGINNINGS[3]

The USSR first showed an interest in rotary-wing military aircraft prior to the Great Patriotic War (World War II) and by 1941 about one hundred autogyros had been delivered to the army for observation and communications duties.[4] They were soon destroyed in the disastrous first months of the war—most without ever being used—and the development of this particular type of aircraft ended. One design bureau, OKB–3 (*opytno konstruktorskoe biuro–3*), continued experimenting with true helicopters but, after little success, its existing projects were curtailed in 1947. Instead, a competition was arranged between it and two other aircraft design bureaus—those led by Mil and Yakovlev—to produce a light helicopter for both military and civil operations. The winner of the competition was Mil OKB's Mi–1 Hare, which could carry three passengers, and this helicopter entered service with the *Voyenno-vozdushnyye sily* (VVS or Air Forces) in 1951. As with the autogyros, the Hare's missions were artillery spotting and light utility work.[5]

In October 1951 a new helicopter requirement and, implicitly, a new mission were given to the designers. In a typically blunt directive, Stalin personally ordered Mil and Yakovlev to design, build and fly twelve- and twenty-four-passenger helicopters respectively—all within twelve months. It is hard to judge Stalin's exact intentions or whether he was supported by senior military advisers. The Soviets were well aware of the use of helicopters in Korea and according to Lev Chaiko, an engineer with Mil OKB who has since emigrated to the West, this knowledge was the prime motivation.[6] However, since helicopters and airborne operations in general did not figure prominently in Stalin's military thinking and, as Chaiko admitted, the army was entirely unsure how to employ the new technology,[7] it would seem that a sense of national or ideological pride and the need to compete with America impelled Stalin as much as the deliberate pursuit of a military capability.

Impressed more by Stalin's demands for immediate action and occasional threatening visits by his henchman Beria than by the technical difficulties,[8] Mil and Yakovlev quickly produced the Mi–4 Hound and the Yak–24 Horse. The Horse proved less than satisfactory, although fortunately for Yakovlev its faults did not become fully apparent until after Stalin's death; although a few were ordered, neither it nor its creator played a prominent part in future helicopter developments. In contrast, Mil's Hound was much more successful, entering production by the end of 1952 and establishing him as the leading Soviet helicopter designer (although the Kamov OKB also produced helicopters, mainly for the Soviet navy, it had far less prestige). Mil further enhanced his reputation with the record-breaking, heavy-lift Mi–6 Hook. Its development showed, however, that the Soviets still did not fully appreciate the military value of the helicopter. Mil began designing the Hook on his own initiative and, although he soon received official approval, it was mainly to provide aerial assistance to mining and industrial activity in remote and otherwise inaccessible regions of the USSR.[9]

As demonstrated at the 1956 Tushino Air Show, when 36 Mi–4s airlifted a dismounted infantry battalian, the main mission for the Soviet helicopter fleet was logistic and troop transport, and they were therefore assigned to the *Voyenno-transportnaya aviatsiya* (VTA or air transport branch) of the VVS.[10] But Tushino was a demonstration of what was possible rather than of confirmed doctrine. It was not until the end of the 1950s that tactical thinking began to catch up with Mil's technical advances and definite conclusions were made on the application of the helicopter to military affairs. Many helicopters had been deployed by this time, including the large-capacity Hook, and this quantitative improvement began to allow a qualitative change in the use of the helicopters. More importantly, "the revolution in military affairs" was beginning to have a significant impact, and the new concepts that it brought to the army were to dominate helicopter development in the 1960s.[11]

THE 1960s: A REVOLUTION FOR HELICOPTERS

Helicopters and Tactical Air Landings

The phrase "revolution in military affairs" was used to describe the effect of the development of nuclear weapons, which, when combined with long-range delivery systems, provided the potential for devastating strikes on political, economic and military targets in any part of the enemy's territory. Although its most fundamental effects were felt at the strategic level of warfare, the Soviets assumed that all forces, from tactical to strategic, would be both potential victims and benefactors of nuclear attacks. There were two basic premises of this revolution and both encouraged further progress with air transport and airborne operations, including transport by helicopter.

First, a means had to be found to "eliminate the increasing disproportion between the [destructive] capacity of nuclear-missile weapons and the mobility of land forces."[12] Second, nuclear warfare both allowed and required combat within "the entire depth of disposition of the enemy simultaneously."[13] It was thought that future battles would be fought over a much broader and deeper area; that linear fronts would be rare and attacks made from front, flank and rear; that troops would operate in more dispersed and smaller groups; and that vast sections of the battlefield would be rendered impassable to ground vehicles by the aftermath of nuclear explosions.[14] Better mobility would be needed to take advantage of the effects of nuclear strikes, to cope with the increased depth of military operations and to operate under the threat of any enemy nuclear attack. In short, the answer was "speed, speed and more speed."[15] Although efforts were made to increase the speed of all ground forces, it was air movement that best offered the "flexibility, speed and surprise" necessary for fighting a nuclear war.[16]

While it was thought that all troops could be deployed strategically by air—given enough aircraft of sufficient capacity—it was the *vozdushno-desantnayye voyska* (VDV or airborne troops) who gained most from the new emphasis. The VDV "acquired the importance of a connecting link between nuclear strikes against the enemy and the influence upon him of *soyedineniya* (i.e., divisions or armies) attacking from the front."[17] Hence, in 1956 the VDV was removed from the control of the subordinate Air Forces and became a part of the Ground Forces, thus improving its status within the army as a whole. In the previous year the VTA, which provided the transport aircraft, had also been significantly upgraded in status; new aircraft were introduced and the number of parachute exercises increased dramatically.[18] The VDV's sphere of action moved from purely tactical support of ground forces to the operational and strategic levels of war.[19]

The revolution in military affairs also coincided with a renewed enthusiasm for many of the doctrinal ideas that Tukhachevskiy, Triandafillov and others had suggested in the 1930s concerning deep battle and the rapid advance of forces into the enemy's rear areas. Tukhachevskiy had favored the development of paratroops as a complement to tank forces, and the Soviet Union had excelled in this area prior to the Great Patriotic War. As these prewar theories reemerged at the expense of Stalin's military views, it was not surprising that the VDV and air landings became more prominent.[20]

Yet the tactical air-landing missions that the VDV had previously carried out remained important, and it was for these missions that helicopters became crucial. There were insufficient VDV units and aircraft to provide paratroop air landings at all levels and, as strategic and operational missions took precedence, the Soviets instead decided to use regular, dismounted motor rifle troops (i.e., infantry) for tactical air landings. This action was only possible because using helicopters did not require the special training or equipment of paratroopers, thus giving all normal infantry the potential to be airborne soldiers without rendering them unavailable for normal operations at other times.[21]

Helicopter landings offered a number of advantages. First and in particular:

> Whereas with the parachute method of landing the troops were greatly dispersed after their landing and their combat efficiency remained low for a certain period of time, the troops delivered to the landing region by helicopters are ready to go into combat immediately.[22]

Although delays in any type of air landing might cause difficulties, they were worst at the tactical level, where the enemy's response would be swiftest. Another benefit was that helicopters did not need to operate from prepared airfields, which in wartime would have been at a premium. The number of airfields would have been diminished by enemy interdiction and priority given to fighters and bombers; nor were they close enough to the leading ground troops who would have carried out the landings.[23]

There were other reasons why helicopters were more suitable for tactical air landings. Unlike the VDV units, which were controlled directly by the Soviet high command, the forward deployment of helicopters placed them directly under the front commanders who were responsible for the landings, thus helping to meet the Soviets' desire for strict control and coordination of all airborne operations.[24] Moreover, helicopters enabled a ground commander to conduct numerous landings—not only because of the availability of many more troops, but also because of the ease of helicopter operations in comparison to parachute drops. For all these reasons, helicopters became "the main means of dropping tactical airborne troops."[25] This development demonstrated the contradictions between technical innovations and new conditions, and existing capabilities; the Soviets responded to this situation using a dialectical approach in which a synthesis emerged from the conflict between the existing and the new. Without considering its failure as a political system, Marxist-Leninist philosophy, with its emphasis on the mixing of old and new, was well suited to coping with military innovation.[26]

The Soviets were well aware that enemy air defenses and interceptor aircraft posed a threat to airborne operations. While the need for fighter cover and close air support aircraft limited the extent of VDV paratroop missions to a few high-priority attacks, helicopters were not as restricted.[27] Helicopters were better able to evade air defenses and achieve surprise by "utilizing the relief of the terrain and the forests and by flying to objectives . . . at low altitudes."[28] Whether or not this optimism would have been realized is uncertain but in any case, since the Great Patriotic War, the Soviets had become accustomed to accepting high loss rates as long as the objective was attained. Air landings were of sufficient importance that the benefits outweighed the risks.

Like any attack in the depth of the enemy's defense, air landings had a number of aims of which the most important was the destruction or neutralization of enemy nuclear weapons, closely followed by command and control assets and air defense units. In particular, air-landing parties were thought capable of attacking targets that were too mobile or too small to be destroyed reliably by nuclear strikes. Air-landed troops would also seize objectives, especially river crossings, bridgeheads, mountain passes and defiles, or disrupt the movement of enemy reserves. Another task for helicopters was resupplying land units already operating in the enemy's rear, thus exploiting the helicopter's ability to operate without airfields and easily bypass areas of devastation and contamination.[29] These actions were meant to increase the tempo of the attack by stopping nuclear strikes, difficult terrain, reserve units or a lack of supplies from obstructing the rapid movement of Soviet armored forces into and in the enemy's rear areas. Thus, air movement was not only able to increase the tactical mobility of some troops directly but, indirectly, the speed of the whole offensive.[30]

The development of tactical air landings as part of a range of airborne operations was summed up in 1967 during exercise *Dnepr*. Elements of two VDV divisions were employed at operational level while one hundred Hound and several Hook helicopters were used to lift:

> a motorized rifle battalion [which] acted as a tactical airborne landing force.... Without preliminary practice, the battalion together with its organic weapons and combat equipment was loaded onto helicopters.... After suppression [of the enemy], when motorized rifle and tank battalions were approaching the water's edge, the landing force seized its designated area in the depth of the "westerners' " defense and thus facilitated a successful assault crossing.[31]

The Revolution Continues: The Conventional Phase

Dnepr was also significant because it began with a conventional phase of operations before later moving on to portray nuclear conflict. It thus symbolized a new trend in Soviet military thinking that gave renewed emphasis to conventional warfare.[32] A number of theories have been proposed to explain this change. Generally it has been asserted that Khrushchev, with the support of some members of the army and the Communist Party, used the "revolution in military affairs" as a way of decreasing expenditure on conventional forces, thus weakening the army's political power and helping to improve the USSR's economy. This meant the majority of the army, especially senior and more traditional officers in the Ground Forces, had political and bureaucratic motives to resist the change. They not only assisted in Khrushchev's removal, but afterwards promptly set about reasserting the theoretical and practical importance of nonnuclear forces.[33] One Soviet historian summed it up in typical official paraphrase, stating that after Khrushchev's departure, "Certain incorrect views within the military-scientific circles, concerned with the overevaluation of the potential of the atomic weapon" were rectified.[34]

However, this theory does not allow fully for the continuity between the Khrushchev and Brezhnev eras in terms of the importance of nuclear weapons.[35] Now, Soviet military writings indicated that, at first, the army genuinely believed the revolution in military affairs heralded drastic changes and, as in the United States, overstated these changes; but in the mid–1960s and before Khrushchev had actually been deposed, a balanced approach including both nuclear and nonnuclear weapons began to emerge. It is just not possible to make a logical division between a nuclear and nonnuclear period in the 1960s, especially if one considers that "the Soviets can live rather comfortably with contradictions.... It may be that in looking for the logic [about nuclear weapons] in Soviet military policy we are looking in the wrong dialectical haystack."[36]

It should also be remembered that Soviet writings concentrating on nuclear affairs did not necessarily represent a desire by the Soviet Union that, if war came, it should be nuclear, but rather an acceptance that the North Atlantic Treaty Organization (NATO) was likely to use nuclear weapons.[37] Therefore, when the balance of nuclear forces became more equal in the later 1960s and the adoption of the doctrine of flexible response by NATO increased the probability of at least a conventional phase in any major war, Soviet military doctrine was more than capable of adjusting:

> The new [conventional] possibilities of waging armed struggle have arisen not in spite of, but because of, the nuclear missile weapons.... [They flow out of] this revolution, continuing it, instead of contradicting it.[38]

Indeed, the Soviets found that air-landing and other helicopter operations, which had initially been developed for nuclear warfighting, were also well suited to conventional combat. In 1966 a detailed article on airborne warfare had concluded that "everything that has been discussed above concerns conditions of nuclear rocket warfare. However, airborne troops are also widely used to conduct combat operations with conventional means of destruction."[39] Even more revealing was an article in 1967 explaining that while the use of air transport, including helicopters, to deploy troops to the depths of the enemy's defense would be important in a nuclear war, it would be essential in nonnuclear conditions. Air landings had already been seen as a replacement for nuclear weapons in respect of certain targets: if there were no nuclear strikes at all, then air landings, along with other means of conventional destruction, would have to attack all targets. Moreover, without nuclear firepower to prevent enemy occupation and defense of "natural terrain obstacles," seizing these objectives rapidly became crucial, requiring the use of troops "simultaneously throughout the entire depth of their [the enemy's] position."[40]

Events abroad also played their part in the USSR's military helicopter developments. As Marshal Grechko stated, "The fighting in South-East Asia... provided much to think over and demanded fresh ideas" and, while the Vietnam War should not be thought of as the Soviets' sole inspiration, it certainly influenced their military thinkers.[41] The Soviets also learned from the successful Israeli air landings in 1967, but direct examples of Soviet or allied helicopter use were rarer, since the Arabs' large stock of helicopters was all but destroyed in 1967 by the Israeli air force's first strike and North Vietnam only occasionally used helicopters.

One idea that did emerge, mainly from U.S. experiences in Vietnam and its experiments with the TRICAP division, was the use of helicopters to provide troops with what might be termed secondary mobility—that is, mobility after they had made an initial landing. In operational and strategic airborne operations conducted by parachute drop, secondary mobility was essential since the units had to "conduct active independent and highly maneuverable operations in the enemy rear" and were unable to rely on a speedy link-up with advancing forces.[42] The VDV divisions were therefore equipped with light armored vehicles. Helicopters were unable to lift any meaningful quantity of these vehicles to support tactical air landings, but the lack of mobility was not an issue under existing doctrine because these missions were to be conducted at relatively short depths, with the helicopters delivering the troops directly to their objectives.[43]

However, if the helicopters themselves remained with the troops, tactical air landings might become much more powerful. According to Colonel (later Major General) M. Belov, the USSR's most prolific helicopter advocate in terms of published material, "Helicopters in this case do not return to their rear straight after landing troops but are used for maneuver on the battlefield and continued support of the land forces' subunits."[44] Belov published many articles on this concept, indicating at least some official interest, and one senior officer with responsibility for developing military theory, Lieutenant General Vasily Reznichenko, also supported it.[45] However, there were many difficulties: helicopters remaining behind enemy lines would have been much more vulnerable and harder to sustain; the ground and air components would have needed a very high level of understanding and coordination; and, above all, the gain from repeated airmobile operations was not clear enough to warrant the risks involved. Therefore the concept was not implemented. To some extent, the Soviets were simply trying out ideas copied slavishly from the West; certainly the Soviet military eagerly examined other countries' doctrine for possible solutions to their own problems, but the United States had hardly begun to examine the sort of operations Belov proposed when it too abandoned them. As important was the Soviet tank branch's opposition to any suggestion that helicopters might challenge tanks for the title of primary combat system. For these soldiers, imitation was certainly no reason for innovation.

The 1960s also saw technical developments to match the rapid advances in military theory. By 1967, the Mi–8 Hip was in production.

Developed directly from the Mi–4 Hound, which it replaced, Hip had turbine engines, enabling it to lift a whole platoon (twenty-four personnel)—double its predecessor's load. The Hip improved command and control by allowing an entire platoon to be carried by one helicopter; it could also lift a wider range of heavy equipment, including light vehicles. Mil also developed the Mi–12 Homer, with a vast twin rotor system and more than twice the payload of the Hook, and Kamov OKB experimented with a large compound helicopter, the Ka–22 Hoop, combining rotary-wing vertical lift with turboprops for longer range. However, the Mi–6 Hook continued to be the main heavy-lift helicopter. The Hoop never entered production, while the Homer's primary task was, in conjunction with the An–22 fixed-wing cargo aircraft, to move intercontinental ballistic missiles (ICBMs) from factories to launching silos in remote locations.[46]

By 1969 air-landing procedures had been confirmed and refined in numerous exercises; apart from a few significant exceptions discussed below, there was little later variation of these tactics. In most cases, either a motor rifle battalion or company would have been used for air landings, normally reinforced with anti-tank, air defense, artillery and engineer subunits. Motor rifle units that had prior experience and training in helicopter landings were used.[47] Great emphasis was placed on suppressing enemy air defenses, although the Hips and Hooks were also to fly low and at high speed to avoid air defenses and achieve surprise.[48] The likely loss rates for these helicopters would still have been high, especially since they were technically incapable of flying at very low, nap of the earth altitudes like Western helicopters. Fronts were more likely to have launched large air landings, with a fully reinforced battalion carried by over thirty Hip and Hook helicopters, usually in support of an exploiting armored unit from the second echelon. Smaller, company-sized landings with less heavy equipment were to be carried out at army or even divisional level, using no more than a few Hips. Differences between these and frontal air landings were also reflected in the depth to which they were delivered (from a few to about fifty kilometers) and the time before relieving ground forces arrived (varying between 15 minutes and 24 hours).[49]

The Development of Armed Transport Helicopters: 1967–1972

As in the West, the emergence of armed helicopters was initially sparked by the need for fire support during transport operations. However, while America, Britain and other countries soon proceeded to acquire dedicated attack and anti-tank helicopters with an entirely separate mission, Soviet developments remained closely linked to air landings.[50] The Soviets thought that transport helicopters needed fire support to assist in penetrating the enemy's air defenses, ensure a safe landing and help the troops after they had disembarked. While artillery or ground attack aircraft were able to give some of this support, there would have been problems of coordination, particularly in the depths of the enemy's defense, and it was thought that the best solution was to rely mainly on armed helicopters escorting the air-landing force.

While there had been experimental work with armed helicopters in the 1950s, it was not until the next decade that serious development began.[51] The first step was to equip existing transport helicopters with their own weapons: "In view of intensification of air defense, the new flying apparatuses used to transport personnel are better equipped with means of combating antiaircraft weapons on the ground, as well as the enemy in the air."[52] However, this article went on to foreshadow the creation of a new, specialized armed and armored transport helicopter, since "means and methods of landing troops should be improved in such a way as to ensure the least vulnerability of planes (or helicopters) in the air."

Belov gave further support to the idea of a heavily armed, air-landing helicopter in 1969 in an article of particular significance because, unlike almost all of his other writings, it was not based on foreign sources. This distinction signified that the article was not just an attempt to stimulate discussion but, more than likely, that its content was already accepted officially. Having established that troops air-landed behind enemy lines needed fire support both during and after their flight, Belov called for an "airborne APC [armored personnel carrier]," which would give these soldiers mobility, protection and firepower similar to that provided by the recently deployed BMP (*boyevaya mashina pekhoty* or infantry combat vehicle).[53]

Work on such a helicopter had begun in 1968. The final result was the Mi–24 Hind-A, which first appeared in 1972 and was operational by 1973. It was heavily armored by helicopter standards, carried a section of eight fully equipped soldiers and was armed with a variety of rockets, bombs, missiles and machine guns. The Hinds would have been used to lead and cover the rest of the air-landing group, mainly composed of Hips. They would have cleared the way to the landing zone, suppressing enemy air defenses, and then landed a small advance party to secure the area, thus reducing the dangers to the following, more vulnerable helicopters. The Hinds would have remained in the vicinity giving fire support.

There has been considerable confusion in the West over the real mission of the Hind-A. Observers often mistakenly applied their own military concepts to it and thus were hard pressed to explain the Hind-A's combined transport and armed capability. The misunderstanding was

Table 2.1
Soviet Helicopter Numbers and Organizations

Early 1970s

Unit	No.	Each With	Airforce Total
Transport Regiment	20	20 Mi-1 Hare/Mi-2 Hoplite	400
Administered by VTA		40 Mi-4 Hound/Mi-8 Hip	800
		20 Mi-6 Hook	400
Small utility units		Mi-1 Hare/Mi-2 Hoplite	?200
Total Helicopters			**1800**

Late 1970s

Unit	No.	Each With	Airforce Total
Transport Regiment	20	20 Mi-2 Hoplite	400
Probably still with VTA	40	Mi-4 Hound/Mi-8 Hip	800
		20 Mi-6 Hook	400
Assault Regiment	20	40 Mi-24 Hind-D	800
Frontovaya aviatsiya		20 Mi-8 Hip	400
Low level units		Mi-2 Hoplite/Mi-4 Hound	?400
Total Helicopters			**3200**

Mid-1980s

Unit	No.	Each With	Airforce Total
Transport Regiment	20	20 Mi-2 Hoplite	400
Frontovaya aviatsiya		20 Mi-8/Mi-17 Hip	400
		20 Mi-6 Hook/Mi-26 Halo	400
Transport Regiment	?20	20 Mi-2 Hoplite	400
Armyeyskaya aviatsiya	20	Mi-8/Mi-17 Hip	400
		?6 Mi-6 Hooks	120
Assault Regiment	?20	40 Mi-24 Hind-D/E/F	800
Armyeyskaya aviatsiya	20	Mi-8/Mi-17 Hip	400
Divisional Squadron	?50	6 Mi-24 Hind-D (or armed Mi-8/-4)	300
		6 Mi-2 Hoplite	300
		6 Mi-8/Mi-17 Hip	300
Low level units		Mi-2 Hoplite/Mi-4 Hound	?400
Total Helicopters			**3820**

Note: "Assault" does not refer to air-landing but attack units.
Source: *The Military Balance* (London: IISS, 1971–1990).

heightened because, just as the West was beginning to learn of the Hind in the mid–1970s (well after its design and development), its mission changed and a new version, the Hind-D, was acquired to act as a specialized attack helicopter. These developments, which stemmed from Soviet analysis of the October 1973 Arab-Israeli War, are described below. The most common assumption in the West—often made without any apparent evidence—was that Hind had always been intended for fire support and was given a transport capability because "a pure attack helicopter could not be justified" or because cost savings were required.[54] These reasons rang true in terms of Western weapons acquisition but were less credible in the Soviet Union. More recently Chaiko, who had worked for the Mil OKB, suggested that the Hind's hybrid design came from its hurried development from the large Mi–8 Hip following a demand by the general staff for immediate action. However, Chaiko's view is unconvincing. His descriptions of Soviet military planning were generally inadequate—perhaps reflecting the relatively small part that he played in them; he was also intent on showing that the Soviet weapons acquisition process was inefficient. Most problematic, however, was his admiration for Mil and his obvious desire to answer any criticism of the Hind's design by blaming the military, thus protecting Mil's reputation.[55]

But most critically, as will be seen below, the Soviets' interest in increasing their conventional firepower did not begin until after the October 1973 war—and after the Hind was first begun. At this time they started to acquire better artillery and ground attack aircraft, as well as converting the Hind to an attack helicopter. Moreover, the Hind-A bore an uncommon resemblance to a prototype Sikorsky helicopter, the S–67 Blackhawk, which was also intended to perform both assault and fire support missions. One Czech journal even stated that the Mi–24 was based on the S–67 and was "armed to clean up an area and unload a dozen troops there."[56] When Hind-A was being deployed in late 1973, Belov wrote an article purporting to examine Western attack helicopters, but which devoted itself mainly to the S–67—even though it had already been rejected by the U.S. Army and its further development was in serious doubt.[57] Finally, one of the earliest references to the Hind-A noted its use in an exercise for a small helicopter-borne "raid" in which both a fire strike and then an air landing were conducted by the same helicopter.[58]

Conclusion

By the early 1970s airborne operations of all types had become "a widespread phenomenon" in the Soviet army's planning and training, and many more helicopters and aircraft were being acquired. Air landings of all types were seen as a way of speeding up the main offensive and attacking the enemy's entire defensive depth simultaneously.[59] Undoubtedly, these operations would have entailed risks and some, or even many, would not have succeeded. However, the Soviets were so convinced of their necessity that they planned to employ them in spite of these problems. Moreover, as indicated by the decline in the number of general references to airborne troops in favor of more specific comments about helicopters and, as shown by the examples of the *Dnepr* and *Dvina* exercises in 1967 and 1970 and the development of the specialized Hind-A transport, the importance of helicopter operations increased far more than that of parachute landings; for the introduction of the helicopter meant that, instead of specialized troops, "motorized rifle subunits of the land forces, in acting as airborne troops, can carry out different missions in any form of combat."[60] Clearly the distinctions between the air and the ground components of land warfare were starting to break down. This change began the process by which close air-land cooperation became important at more than just the operational and strategic levels of warfare—a process that was to reach its peak in the 1980s.

SOVIET ATTACK HELICOPTERS

Initial Arguments about Armed Helicopters

While armed helicopters, principally the Hind-A, were being developed for transport operations, there was considerable discussion as to whether they also ought to perform specialized attack missions independent of air landings. Unlike the West, the group of helicopter advocates proposing this mission in the Soviet Union was not particularly visible; but it did exist, centered around the Mil OKB and occasionally seen in public in articles by Belov. These articles, as well as those by other commentators, often relied heavily on "foreign press materials," were expressed in terms of what was supposedly being thought in the West (though often was not) and were written with a less than assured tone—all of which suggests that there was continuing uncertainty and debate in Soviet military circles on this issue.[61] The careful phrasing also reflected the lack of organizational power of those arguing in favor of helicopters.

One strand of the debate concerned the use of armed helicopters for close air support, an activity for which the VVS had been ill-equipped since the 1950s when the wartime Ilyushin IL–2 *Shturmovik* and its successor, the Il–10, had become obsolete. As a more balanced nuclear-conventional approach emerged in the mid–1960s, it seemed that, as in the Great Patriotic War, tank forces attacking deep in the enemy's rear would require considerable close air support in addition to nuclear fire support.[62] In 1967 Marshal of Aviation Krasovskiy published an article stating that such support could best be provided by "special aircraft which operate directly with the ground troops," thus ensuring that there was close cooperation between the two. These special aircraft could, by exploiting short/vertical takeoff and landing (S/VTOL) technology, operate dispersed and without airfields.[63] The helicopter was not the only special aircraft available with these characteristics because the USSR has unveiled its first fixed-wing VTOL jet in 1967. However, it took the Soviet Union many years to achieve a useful VTOL capacity, and helicopters were more readily available. Like transports, armed helicopters did not have to use fixed bases, were easily dispersed and could operate close to the front line under the control of the ground force commander. The lower speed of a helicopter, when compared to high-performance jet aircraft, increased the precision of its strike, making cooperation with ground forces even easier. The Soviets keenly observed the success of Western attack helicopters, both in combat and in trials, and the actual deployment of the Hind-A, even though its main task was armed transport, allowed the USSR to begin its own experiments with fire support helicopters. While most Hind-As were deployed according to normal doctrine, those of the Lenin Regiment (consistently associated with innovative developments) tested out fire support missions with their "new equipment."[64]

The possible use of attack helicopters had first been aired in two important Ground Forces textbooks, Colonel A. A. Sidorenko's *The Offensive* (1970) and Colonel V. Ye. Savkin's *The Basic Principles of Operational Art and Tactics* (1972), as well as in articles by Belov and others.[65] Sidorenko listed a number of missions for armed helicopters,[66] but his equivocal tone and use of Western theories indicated that the development of attack helicopters was still uncertain in the Soviet Union—mainly because of doubts over vulnerability. Moreover there was no general trend as yet to improve the amount and quality of fire support available to the Ground Forces, and it was not until after the mid–1970s that large increases occurred in self-propelled artillery and ground attack aircraft. As will be seen, there was a very close link between the development of better artillery and attack helicopters. A Soviet army textbook written before the October 1973 Arab-Israeli War stated that helicopters "can carry out a broad range of missions such as neutralizing the enemy on the battlefield and in his rear" but, significantly, this comment appeared in the section on transport aviation and not in relation to combat aircraft.[67]

However, during the late 1960s a more extreme proposal arose that, instead of being "accompanying fire means," the helicopter might possibly be used as a "flying armoured weapon" in direct combat, not simply for fire support, Savkin suggested:

> The armed helicopter may turn out to be a means of fundamental change in the nature of ground combat. Although the helicopter is not ideally adapted for

> this role, it nevertheless possesses those characteristics which most ensure superiority in mobility. Helicopter *chast* are beginning to be employed abroad *for the accomplishment of tasks performed by tanks in World War II*, i.e., for the development of success and as a mobile reserve ... *sometimes even independently* [emphasis added].[68]

These ideas led to much discussion among senior members of the general staff, especially in relation to the future roles of tanks and helicopters. A recent article in the Soviet press—the first to deal openly with this issue—commented that, when Mil suggested military helicopters were more than just a support for armored forces, "many military leaders and aviation specialists received this proposal, to put it mildly, with bewilderment, some even very negatively."[69] The helicopter offered much more mobility than tanks and, given the Soviets' commitment to speed in military operations, it was not surprising that helicopters held some appeal. However, the suggestion that the helicopter might one day compete with, or even supplant, the tank as the main combat weapon was rejected. Even though the defector "Suvorov" stated that the Soviet military regarded the Hind as a "flying tank," his comment is most unconvincing in comparison to the pronouncement in 1972 by Chief Marshal of the Armored Forces Rotmistrov that "there is no sense to consider seriously plans for 'flying,' 'jumping' or other tanks of unusual design."[70]

The reasons for the rejection of combat helicopters in the 1960s demonstrate the failure of the Soviet senior planners to distinguish the important characteristics of helicopters that made them different from tanks and fixed-wing jet aircraft. Opponents to Mil's idea pointed out that helicopters did not fly as fast as aircraft and were not as well armored as tanks; nor could helicopters carry as powerful weapons. Instead of assessing the helicopter's unique combination of air-land capabilities, they made a simple and misleading comparison to weapons systems operating only on the ground or in the air.

Other difficulties were, however, more valid. The first difficulty was that, although helicopters were mobile, they were very vulnerable to the effects of nuclear blast and radiation. They also required considerable maintenance and much logistic support. Finally, as Chaiko noted, the Soviet helicopter industry was unable to build them in sufficient numbers to meet the military's demands for mass employment.[71] Thus, they were unsuitable to replace tanks entirely and still meet the needs of Soviet doctrine.

However, this explanation is not of itself sufficient: for all that Western political and organizational theories are not necessarily applicable to Soviet society, the impact of interest groups and especially the pressures innate to any large organization such as the Soviet army cannot be ignored entirely. The tank had become a potent and enduring symbol of the Soviet military's prowess during the Great Patriotic War. Moreover, the tank branch was paramount within the Ground Forces, and so its political power was challenged by suggestions that the helicopter might replace the tank. Given that the helicopters were part of the Air Forces, the balance of power within the army as a whole might change unfavorably away from the Ground Forces—a danger to which its senior commanders were especially sensitive after the reduction in their forces' status following the creation of the Strategic Rocket Forces by Khrushchev.

At least as important as political considerations was the fact that the traditions and routines of Ground Forces organization were all based on the tank, and this situation tended to impede innovation in other areas.[72] Moreover, since helicopters were well established as a means of conducting air landings, it was not surprising that new advances with helicopters were first of all applied in that sphere of operation. It was a small innovation to introduce armed helicopters to air-landing procedures; it was a very different matter to promote helicopters in a completely separate field, especially when dedicated ground attack aircraft had not been in service for some years. Unlike the West, there was no organization closely associated with helicopters. There was no natural constituency of helicopter advocates within the Ground Forces (since helicopters were operated by the VVS), and in that service, helicopters belonged to the least important branch, Transport Aviation. In any case, the VVS was dominated by the Ground Forces on issues of air support for land operations. This situation meant that the organizational imperatives for expansion and improved status that affected Western developments were, by and large, not present.

Therefore, in the early 1970s, both unarmed and armed helicopters were still seen predominantly as vehicles for carrying out and supporting air landings. While experiments were starting with the Hind-A, attack helicopters were not yet accepted by the Soviet general staff. It was not until the October 1973 war between Israel and its Arab neighbors (many of whom had Soviet equipment and doctrines) that the Ground Forces' assumption about the relative ease of offensive operations without nuclear weapons was rudely shattered and the need for conventional fire support was recognized. Armed helicopters were suddenly to gain newfound importance.

The October 1973 Arab-Israeli War

While the October 1973 war gave the Soviets considerable evidence that tactical air landings carried out by helicopters were both practical and necessary—if somewhat risky[73]—the war was more important for its general lessons on the nature of nonnuclear warfare. These lessons in turn drove helicopter developments, notably the acquisition of a new model of the Hind for use as an attack helicopter. The Arab-Israeli war demonstrated that conventional conflict was much more difficult for attacking armored forces than had previously been thought. Just as the power of the offensive had worried Western analysts, it was the strength of the defense that surprised Soviet observers—a paradox that reveals the importance of an existing mind-set in viewing wartime experiences. There were a number of reasons: the tank's own defensive abilities had not been fully recognized; the impact of ATGMs, which gave infantry a much better chance of resisting armored attacks, had been underestimated; there had been little understanding of the problems for attackers where defending forces were able to concentrate without being threatened by nuclear attack; and the Soviets had miscalculated the extent of conventional fire support needed to replace nuclear weapons.

At the time of the October 1973 war, Soviet military operations were essentially based on the offensive use of the tank, with limited support by other arms. This approach had first been suggested by Soviet theorists in the late 1920s and early 1930s, and had eventually been adopted with success in the Great Patriotic War. It had become even more important during the revolution in military affairs because of the ability of tanks to move rapidly and to withstand most of the effects of nuclear explosions. It was assumed that nuclear weapons would destroy an enemy's defense and render unnecessary the somewhat ponderous approach used during the Great Patriotic War, in which the main tank forces were held back in the second echelon while infantry and artillery made a methodical breakthrough. Following the revolution in military affairs, tanks were to lead the first echelon, entering the enemy's rear areas as soon as possible. And, despite the evidence of the war, even in nonnuclear combat it was planned that the tank forces would make an immediate breakthrough so as to reach and destroy nuclear weapons in the enemy's rear areas.[74] It was this latter assumption that the October 1973 war undermined.

Marshal Grechko revealed in 1975 that, after "the fighting in the Middle East," there was "a new approach to the correlation of offense and defense" because "modern defense has become more powerful and tenacious . . . tanks, the main striking power of the attackers, have become more vulnerable and their employment on the battlefield more difficult."[75] The need to overcome these new defenses had a profound impact on all areas of Soviet military thinking.[76] No longer could the tank rely on limited assistance from artillery and infantry; more and better fire support was necessary, especially if the fighting was to be nonnuclear. The Soviets immediately began to build self-propelled artillery—more powerful, mobile and survivable than existing towed equipment. They also introduced their first ground attack aircraft since the 1950s, the

Su–25 Frogfoot.[77] As well, the Soviets accorded the armed helicopter new importance as a source of fire support with which to neutralize or destroy all types of enemy anti-tank weapons (including enemy tanks): "All these factors [which were demonstrated in 1973] make the success of the offensive operations of the land forces doubtful unless mass use is made of helicopters, promoting neutralization of new obstacles [i.e., ATGMs]."[78]

No doubt still aware of criticisms about helicopter vulnerability, which had stymied earlier development, Belov later wrote that "there is no reason to suppose that this loss [predicted for helicopters] will be especially pronounced when compared to that of planes. Yet in spite of the strength of modern air defense forces neither of the sides will agree to reject the use of aircraft."[79] But in fact the October 1973 war also reduced the validity of these criticisms. It provided evidence that "modern air defense weapons are less efficient against helicopters," while the success of the Arabs' anti-aircraft weapons convinced the Soviets that fixed-wing aircraft were very vulnerable when used for close air support because they had to fly within the minimum range of many anti-aircraft weapons. The conclusion was not that helicopters were invulnerable but rather that they were much more survivable than had first been thought.

The importance of the October 1973 war as a catalyst for the USSR's development of attack helicopters can be judged from Soviet military authors who repeatedly mentioned them in the context of the war. For example, one wrote that "helicopters armed with antitank guided missiles, particularly the American TOW missile, were used *extensively* in the Middle East War in the battle against armored targets [emphasis added]."[80] Yet no anti-tank helicopters were ever used in the October 1973 war! What had in fact been important was the demonstration of the ATGM and of the potential importance of attack helicopters—rather than the helicopters themselves. The supposed use of anti-tank helicopters in the October 1973 war was a cover for the Soviets' initial tardiness in adopting the attack helicopter. The Soviets were already well aware of the evidence in favor of attack helicopters provided by U.S. combat trials in Vietnam and the experiments at Ansbach, but did not act upon it until after the October 1973 war.[81]

Applying the Lessons of the October 1973 Arab-Israeli War

The October 1973 war sparked a major reevaluation by the Soviets of the way in which they planned to employ helicopters. In reference to air landings, it had been thought in 1970 that "[they] were assuming increasingly definite features . . . and that in the future they will be used on an increasingly broader scale."[82] However, this gradual process was vastly accelerated by the war, as can be judged from the increase in articles on helicopters and air-landing operations in various military journals.[83] Nevertheless, the aim of such operations remained the same: to ensure "an uninterrupted advance of the main forces" and to increase the tempo of the offensive.[84]

The lessons of the October 1973 war had a much greater impact on Soviet armed helicopters and brought about a radical change in both their missions and their design. The transformation of the Hind from an assault transport to fire support helicopter was rapid. By 1976 the new Hind-D, which was not used to carry troops, had been put into production, and the Hind-A was soon replaced. Further technical improvements were made in the following years, with the Hind-E and -F being deployed in the late 1970s and early 1980s, but the fundamental change occurred with the introduction of the D model.[85] Although the Hip had already been armed for fire support during air landings, there is also evidence that the incidence and strength of Hip armament increased in about 1976 while references to Hips performing independent fire strikes became more frequent.[86]

Able to deal with "the most diversified fire problems with its own weapons," the Hind filled the gap between artillery and fighter-bombers, combining the former's precision and responsiveness to ground forces with the latter's radius of action, flexibility and maneuverability. Moreover, the Soviets thought that the attack helicopter could act as its own observation platform, thus eliminating some of the problems of coordination that limited the effectiveness of indirect artillery fire.[87]

Soviet planners returned to the Great Patriotic War to help explain the changing military situation. Thus, the Hind was imagined to be a direct replacement for the wartime Il–2 *Shturmovik*. Yet Hind was in fact superior: it was less vulnerable;[88] was not restricted by poor weather to the same extent as ground attack aircraft; and, most of all, since it did not need an airfield, could operate very close to the troops it was supporting. Its mobility enabled it to support troops in the enemy's rear who were out of friendly artillery range, whether they were air-landing parties or armored units that had broken through. Moreover, the Soviets' emphasis on speed and surprise attacks meant that encounter battles would be very prevalent in the next war, and the helicopter ideally fulfilled the need for rapid fire support, taking the place of artillery when the latter was still moving into position.[89]

Hinds were used extensively on exercises to aid ground attacks, striking anti-tank strongpoints or armored vehicles in defensive positions, and to assist air landings by neutralizing air defense units and providing covering fire while troops were landed. Attack helicopters also made independent strikes on targets in the enemy's rear, such as command posts or air defense assets.[90] Because the helicopter was such an effective platform from which to use ATGMs, the Soviets began including the Hind in their ad hoc "anti-tank reserve" groups, thereby releasing for the offensive operations the tanks normally assigned to this task.[91] As part of the anti-tank action, Hips or Hounds were to lay minefields that would then be covered by the fire support helicopters, creating a double barrier to armored attack.[92] In the 1980s fire support helicopters were also given the similar task of acting as an "anti-landing reserve," using their mobility and responsiveness to counterattack enemy air assaults.

Fire support missions were normally to be carried out at long range from behind friendly lines.[93] However, this technique did not suit all situations and was obviously incompatible with missions in the enemy's rear. In this case, like helicopters penetrating the front line to carry out an air landing, fire support helicopters were to evade or surprise enemy air defenses through high speed, terrain concealment and attempting to fly only in areas where there were few anti-aircraft weapons.[94] Although some Western commentators described the Soviet helicopters as using a stationary pop-up tactic similar to that of the West's anti-tank helicopters, the Hind's poor hovering performance (due to its size and its relatively large fixed wings) and its likely use in offensive situations meant that the normal method of attack was a low-level approach and then a "hump" or sharp rise in altitude from which a shallow, diving attack would be made.[95] Similarly, although a few references to combat flights of three helicopters in Soviet articles in the mid–1980s caused some premature Western assumptions that this had become the norm, throughout the 1970s and 1980s fire support helicopters in fact operated in variable numbers (from one to more than eight) according to the tactical situation and available helicopters.

While the October 1973 war immediately confirmed the Soviets' existing views on air landings and led to a reassessment of the need for attack helicopters and the speedy conversion of the Hind, the ramifications of the conflict were to continue throughout the 1970s. Soviet military thinking was to undergo dramatic changes during these years and, as the following comments indicate, helicopters were to fill an increasingly important role in the Soviet army. In 1976, Reznichenko stated, "The transformation of the helicopter into an armored combat vehicle . . . is having a substantial impact on the conduct of the combined arms battle"; while three years later another Soviet theoretician, General Ivan Vorobyev, wrote: "In modern combat there will be much more extensive maneuvering that can, in our opinion, be called 'envelopment by air' . . . [which] makes it possible to increase the depth of simultaneous influence on the enemy."[96]

THE RISE OF AIR-LAND WARFARE

Changes to Soviet Doctrine and Operational Art

As well as leading to an expansion in the size and quality of the Soviet army, the October 1973 war sparked off a major doctrinal debate. The period of debate essentially ran from 1973 into the mid–1980s; it was marked by the significant *Zapad–81* exercise in 1981, during which many new doctrinal ideas were tested, and by the release in 1984 and 1987 of new editions of a key Soviet military textbook, *Tactics.* The Ground Forces sought to revise their operational art and organization to take the best advantage of new weapons such as self-propelled artillery and attack helicopters, as well as established equipment like BMPs and tanks so as to achieve the sort of rapid victory required by Soviet military doctrine. One crucial change was that helicopters became so important that, during the 1980s, they became equal with tanks and other armored vehicles in the developing concept of air-land warfare. Moreover, because helicopter doctrines were less developed than those for tanks and BMPs, it was the air component that became a focus for Soviet military thinking during the 1980s.

Although the Soviets had been paying more attention to nonnuclear warfare since the late 1960s and the October 1973 war confirmed this process, in the late 1970s the debate over military planning was intensified by a further change in Soviet attitudes on the utility of nuclear weapons. Although one author on Soviet affairs has written that "it is generally agreed by Western experts that the USSR does not want nuclear war, if she can avoid it,"[97] there has been no unanimity on the extent to which the Soviet Union would, or would not, try to prevent a war from becoming nuclear. On the one hand, some argued that the Soviets were aware conventional fighting was likely but that, in a major war, "it would be difficult to prevent the use of nuclear weapons as well as conventional ones." Regularly published analysts William and Harriet Scott believed that "this view changed little in the 1980s."[98]

A different and more widely accepted thesis was that, in the event of a major war with the West, the Soviets would be more likely to fight a "theater offensive" solely with conventional arms (see below for discussion of the Soviet concept of theaters).[99] It was assumed that the USSR's ultimate goal was to gain its objectives at minimum cost, requiring the Soviet ground forces to advance as quickly as possible and to destroy or neutralize the enemy's battlefield nuclear arsenal at the same time. These two aims were interactive since achievement of the former goal assisted with the latter and, if the enemy had few or no nuclear resources, neither a threatened nor actual nuclear attack could seriously hinder further advances. Carrying out the offensive by conventional means would inhibit the enemy's political decision to use nuclear weapons; the Soviets also believed that using their own nuclear weapons could reduce the speed of the offensive because of the resulting devastation.

To some extent both theories suffered from the same problem. They emphasized either conventional or nuclear weapons but not both, thus paying insufficient attention to the Soviets' dual capability.[100] Both sides in this particular debate were seeking to warn the West about a danger posed by the USSR.[101] On the one hand, it was the old danger of the Soviets' apparent willingness to use nuclear weapons, prepare for their use and pay little heed to Western notions of limited war and nuclear escalation. On the other hand, the conventional proponents were arguing that the West should not become too fixed upon responding to the nuclear threat because a Soviet conventional attack might not give it enough time, nor enough justification, to use the nuclear weapons on which it was depending. On balance, however, the conventional offensive theory had more merit and better fitted the evidence of a very large buildup of Soviet conventional forces and numerous statements by Soviet military and political leaders.[102]

The arena in which the Soviets planned to fight a conventional offensive of this nature was the *teatr voyennykh deystivy* (TVD—theater of strategic military action). The TVD had been a part of Soviet military thinking at least since the 1960s.[103] However, because a TVD was, essentially, the largest area in which combined land, sea and air operations could be conducted to achieve a strategic victory by conventional means alone, it became much more important in the late 1970s. As a result, some TVDs were given a permanent existence, rather than being established only in wartime, and new techniques were developed so that the TVD's land operations could be conducted as effectively as possible without the use of nuclear weapons.[104]

The Operational Maneuver Group

As far as Western observers were concerned, a crucial indication of Soviet planning for a conventional offensive was the rise of what they termed the operational maneuver group (OMG).[105] These groups were first mentioned in a modern context by the USSR's military press in 1974. OMGs were referred to directly in 1978 and the discussion about them reached its peak in the next year.[106] After being tested in the *Zapad–81* exercise, OMGs became an accepted and essential part of Soviet military planning: they were the basic change in method and organization that the Soviets had been seeking to allow them to use existing weapons more effectively.

OMGs were ad hoc groupings whose most likely size would have been a reinforced division, controlled at army level and assisting the accomplishment of army goals.[107] The Soviets planned to push OMGs beyond the front line as soon as possible "to switch the foci of the fighting into the depth and rear of the enemy formation," thereby limiting "the freedom of maneuver and the effectiveness of enemy action."[108] In the absence of nuclear weapons this task was even more critical to maintain or increase the tempo of the main offensive. An OMG could act as a raiding force, attacking enemy nuclear weapons, air defense assets and command and control facilities, or as an advance force to capture important objectives and hold bridgeheads over or passages through terrain obstacles, and as an encircling force to cut off the enemy's forward positions.

The emergence of the OMG seemed to be the latest threat—the new challenge to be faced by the West in any future war. But the analysts who highlighted it would have been the first to point out that, in fact, the OMG was more a new emphasis of an old idea than any radical advance in military strategy. The deep battle that the OMGs would help to initiate and lead had first emerged in theory in the 1920s. It had eventually been implemented in the latter years of the Great Patriotic War, and much Soviet writing on the OMG was based on the experiences of their wartime predecessors, the mobile groups. Deep battle had also been a key part of Soviet military thinking throughout the revolution in military affairs when nuclear weapons had become the main means of conducting it; even in the conventional phase period in the late 1960s, nuclear weapons had retained this preeminence.

In the 1980s, however, conventional weapons became more effective while at the same time the Soviets also realized that it was no longer possible to rely so heavily on nuclear weapons. As noted by Hines and Petersen, by 1982, the term *ognevoy udar*, meaning a fire strike by conventional weapons, was being used by the Soviets, when previously the word *udar* had been reserved "almost exclusively to describe nuclear fires."[109] Even more significant was the emergence of another phrase, *udary voysk* or troop strike, with the clear implication that ground actions in the enemy's rear were an important substitute for nuclear weapons. Although airborne operations, which had always been an adjunct to nuclear strikes, retained their value, the OMG was adopted as the main means of conducting troop strikes in the "operational depth of the defense," that is, up to 150 kilometers into the enemy's rear.[110]

Helicopters and the OMG

In the Great Patriotic War air support had been central to deep operations, especially when the ground forces had penetrated far beyond the front line. The relationship between air power and deep ground attack had been weakened by the deployment of nuclear weapons but again became prominent with the advent of the OMG. It was concluded that OMGs would need "more in the way of [air] support and protection than do forces in the first echelon." Relying upon wartime experiences to express his views, Marshal Yefimov, commander of the VVS, was emphatic:

> To a significant extent [success] depended upon the effective cover and support from the air for the mobile groups and upon the clarity of their co-operation with the air formations. *Being the most mobile, powerful and longest-range means of combat, aviation played an exceptionally important role....* Under conditions when the tank formations drew 30–80 km away from the main forces of a front and were forced to operate in a dynamic situation, *aviation was actually the only means of their cover and support* [emphasis added].[111]

Although the use of OMGs was not conditional on achieving total air supremacy, fighter cover was needed because the Soviets expected enemy aircraft to be the main threat to an OMG, which would not benefit as much from the protective cover provided by the Soviets' ground air defenses.[112] However, aviation would also have been the major source of fire support to an OMG. Once deployed, the OMG would soon be beyond the range of most Soviet artillery. Moreover, in a "dynamic" battle characterized by numerous meeting engagements, artillery within the OMG itself would be unable to provide the required intensity of fire support.[113] Helicopters were also able to conduct "free hunt" missions in which they would discover and then immediately attack enemy targets, thus combining reconnaissance and fire support in a way that artillery could not. This facility was well suited to the confused and rapidly changing combat conditions in which an OMG would have to fight.[114] Thus, the advantages that had led the Soviets to favor helicopters for fire support in the early 1970s—their ability to operate in close physical proximity to ground troops—also recommended them to OMG operations.[115] Moreover, "qualitative changes in the system of defense [and] its growing resistance to classical fire-power [i.e., artillery] through the constantly increasing extent to which tanks, transport vehicles and artillery are being armored" meant that helicopters, with ATGMs, were becoming a more appropriate source of fire support.[116]

Fire support missions in the enemy's rear in support of OMGs would not have been different from those conducted as part of first echelon combat (i.e., at the front line) and as described above. Long lists of potential targets were routinely discussed in the Soviet military press, suggesting that almost any target was suitable for helicopter attack. However, the inference can be drawn that some tasks had a higher priority than others, and the helicopters were mainly seen as a means to attack "enemy weapons posing a threat to the group," especially counterattacking enemy armored forces but also missile artillery with remote mine-laying capability, reconnaissance-strike fire complexes (the Soviet term for long-range, terminally guided munitions and their delivery system) and enemy attack helicopters.[117]

The vital role of armed helicopters in the OMG led the Soviets to begin developing a new fire support helicopter to supplement and eventually replace the Hind. Known as the Mi–28 Havoc, it was remarkably similar to the U.S. AH–64 in appearance, and it had similar performance and weapons load. It improved on the Hind in its armament, avionics and numerous other details, but particularly in its performance, thus enabling it to operate easily in nap of the earth conditions where required.[118] The Havoc's apparent design for NOE flight caused some analysts to speculate that the Soviets were abandoning their usual practice of carrying out diving attacks from low altitude and adopting Western practices of much more stealthy ambushes using maximum concealment at all times.[119] While the diving attack was not the only method used in the 1970s, it was the most common; for helicopters operating in the enemy's rear area while attached to an OMG, however, NOE hovering attacks were much more suitable.[120] Concealed NOE flight was also more suitable for the initial OMG attack where it sought to get into the enemy's rear as quickly as possible by avoiding combat.

Although the Soviets had always been prepared, as with all their forces, to use their helicopters in combat at night, the 1980s saw an intensification of interest in operating twenty-four hours a day, first to ensure that the pace of the offensive was maintained and second to use darkness to cover the movement of forces such as OMGs. While existing helicopters were fitted with night-vision equipment retrospectively, the Havoc was the first to be designed with a full range of equipment to enable it to fly and fight at night and in bad weather.[121] Although technical problems delayed the actual installation of night-vision equipment, nevertheless the Soviets clearly recognized the need for a twenty-four hour combat capability.[122]

An OMG would have been supported by helicopters carrying out air landings, either from within the group or from the controlling army or front. Despite the different circumstances in which an OMG would fight, the Soviets believed that the Great Patriotic War—and the Manchurian campaign especially—had provided convincing evidence of the need for air landings in deep operations.[123] As with the attack helicopter mission, air landings in support of OMGs would be very similar to those conducted by the first echelon. They aimed to "maintain a high momentum of advance" by neutralizing enemy targets, forestalling the enemy's attempts to establish new lines of anti-tank defense ahead of the OMG and capturing bridges and mountain passes.[124]

Helicopters were also seen as answering the OMGs' logistic problems. During the Great Patriotic War "the capability of the front rear services to support the mobile groups was significantly reduced" despite the use of aircraft.[125] However, now that they had large helicopters, which, of course, did not require landing fields, the Soviets were much more confident that at least some supplies would reach units in the enemy's rear.[126] Helicopters would also have been used by OMGs for communications relay and radio-electronic combat, and as command posts and for liaison.

While armored vehicles constituted the core of an OMG, the Soviets believed that ground forces alone were not sufficient. Helicopters, both armed and transport, were equally necessary. They provided additional mobility, increasing the speed with which an OMG could have achieved its goals. Not only did helicopter units themselves operate more rapidly but, by forestalling counterattacks and carrying troops to capture enemy and natural obstacles, they could raise the speed of an OMG as a whole. At the same time the use of helicopters became more important in other units because of the need to win the main battle quickly, thus allowing other forces to exploit the successes of the OMGs. Yet the use of helicopters in closer coordination with ground troops required organizational changes to ensure that the two components—air and land—could integrate effectively.

The Reformation of Army Aviation

In the mid–1970s most Soviet helicopters were moved from the VTA to *frontovaya aviatsiya* (frontal aviation), with the exception of the Mi–6 Hook heavy cargo lifters and some light utility helicopters such as the Mi–2 Hoplite.[127] Although there was some uncertainty about exact organizations, most Western observers agreed that two distinct units were formed, the attack and transport helicopter regiments, and both were controlled by fronts.[128] These groupings to some extent reflected the desire to retain scarce assets, particularly Hinds, at the highest level of operational command. Yet, even as more aircraft became available, the Soviets still thought that the "tremendous scope" of modern combat made it vital that "control of aviation was kept centralized" so as "to fully utilize the maneuvering possibilities of aviation . . . and to mass them on the axes of the main strikes of the fronts."[129]

However, in the latter part of the 1970s, the Soviets began looking for ways in which the Ground Forces and the VVS could work together more closely. General Merimskiy stated in 1976 that "the problem of a close co-operation among tanks, helicopters and BMPs deserves particularly careful attention. Once this problem has been solved one can count on a high pace of the attack."[130] By 1978 a potential solution had emerged, although it was yet to be fully developed. While published in the Polish military newspaper *Zolnierz wolnosci*, the two-part article by a member of the Polish General Staff Academy was clearly applicable to the Soviet Union. It made direct comment on Soviet military art and cited from the restricted Soviet journal *Voyennaya mysl*; it also echoed Merimskiy, referring to "the synchronisation of combat actions and the co-ordination of the capabilities of combat tanks, [and] BMPs with the tactical-combat virtues of attack helicopters."[131]

Hinting that the current system was inadequate, the article described a new type of grouping, which placed helicopters and ground units under the control of a single combined-arms commander with the assistance of an air force liaison officer.

> On the surface, it might seem that commanding is greatly complicated by the organization of an air-ground fire and attack group. Actually, things are quite different, because by creating this sort of group the commander can closely co-ordinate the operations of all the forces and resources working on the ground and in the air, as he so decides. *The focusing and co-ordination of the effort of all forces . . . is decisive in achieving success* [emphasis added].[132]

The air-ground group was to consist of "the forces and equipment of the ground echelon, tank and mechanized subunits, and assault [i.e., meaning attack] helicopters." It was designed to improve the responsiveness of aviation so that it was better suited to the speed and sudden changes of situation that the Soviets assumed would characterize a future war. Although conceived of "mainly at the tactical level," the implication was that the same theories could also be applied to larger units. The publication of a favorable critique in a later edition of the same paper indicated that the concept was well regarded by the military authorities.[133] It did not matter that the new concept of closer air-ground co-operation was expressed in terms of a specific grouping: the article and those who supported it were plainly advocating decentralized control of aviation assets.

The actual implementation of this idea was signaled in 1980 in an article by a deputy commander of the VVS. He described in detail the early stages of the Great Patriotic War, during which VVS assets for support of the Ground Forces had been split between *frontovaya aviatsiya* and what was termed *armyeyskaya aviatsiya* (army aviation). The latter was mainly composed of ground attack aircraft squadrons, which were subordinate to individual armies. "However, although this [structure] may have been true in theory, the idea did not correspond with our actual capabilities at the time"; the result was a dispersion of air effort and "a reduction in fighting strength." Thus the air assets were taken away from armies and the front became the lowest level at which aviation was controlled. Yet, by the end of the war, many ground attack and some fighter units were routinely being assigned from *frontovaya aviatsiya* to ground formations, particularly mobile groups conducting deep operations, serving in effect as *armyeyskaya aviatsiya*. The article concluded that this close association of ground and air forces played an important part in the success of these deep operations and reminded readers that such experience had "lost none of its importance, even in contemporary conditions."[134]

The renewed interest in the OMG and its need for effective air power led to the reformation of *armyeyskaya aviatsiya*; moreover, the helicopter provided the capability that had been missing in 1941. The new structure would appear to have been introduced in 1981, coinciding with exercise *Zapad–81*, which was also the first major test of the OMG. In the same year another Polish article confirmed the organization, treating it as an accepted part of Warsaw Pact (and therefore Soviet) military theory.[135] The next year saw the first direct reference to *armyeyskaya aviatsiya* in modern conditions:

> Air men gained many useful things in exercise "Zapad–81." The experience of this exercise is still being carefully studied today in our units and subunits.
>
> Army aviation demonstrated a high level of combat training . . . [and] different tactical groups successfully completed their missions, operating in extremely complex [i.e., new] conditions. This became possible . . . [in part because of] *their ability to co-ordinate well with gunners, motorized riflemen and tankmen* [emphasis added].[136]

The Soviets' decision to rely on the helicopter for much of their close air support requirement considerably assisted the reformation of *armyeyskaya aviatsiya*, and the vast majority of its assets were helicopters, principally Hinds and Hips. A problem with this organization in the Great Patriotic War had been that army commanders who, unlike front commanders were not responsible for overall air operations, were too eager to commit their fighter bombers to ground attack missions before they had assisted the fighters of *frontovaya aviatsiya* to achieve at least some degree of air superiority.[137] However, helicopters were not capable of being used to win air superiority (except, perhaps, against other helicopters) and so this problem now did not arise.

By 1985 each Category A (i.e., full readiness) tank or motor rifle division had its own helicopter squadron with six each Hinds, Hips and Hoplites—less important divisions were probably allotted a smaller unit without Hinds; armies had an assault (by which the Soviets meant "attack") helicopter regiment, similar to the one previously deployed at frontal level and another regiment for transport, electronic warfare and command and control. Fronts lost their assault regiments but retained some Hips and a majority of the Hooks, supplemented by the new Mi–26 Halo heavy-lift helicopters, for air-landing and logistics missions.[138]

The main distinction between *armyeyskaya* and *frontovaya aviatsiya* was that the former would provide "air accompaniment"—that is, direct assistance with its Hind or Hip helicopters (either fire support or tactical air landings)—while *frontovaya aviatsiya* would furnish "air support" consisting of fighter cover, longer range reconnaissance and strikes against targets at some distance from the OMG. It would also carry out operational-tactical air landings, command and control, and resupply missions.[139] *Frontovaya aviatsiya*'s action was more general, aiming "to create favorable conditions" for the ground troops while *armyeyskaya aviatsiya*'s air accompaniment "includes combat operations of *air force units attached to combined-arms formations*. Its purpose is to render *continuous support* to advancing troops in the depth of the enemy's defenses [emphasis added]." [140] Thus, while fixed-wing aircraft were to attack targets deep in the enemy's defense, helicopters were to operate in close proximity to the OMG they were supporting and, for this reason, were to accompany the group into the enemy's rear area. Even though this method entailed considerable logistic and maintenance difficulties, it was necessary because of the limited range of helicopters and the OMG's need for immediate air support.[141] The increased danger of destruction on the ground while based in the enemy's rear was offset by the reduced need to pass through the heavy air defenses while crossing back and forth over the front line.

The new VVS organization was also intended to address the difficult problem of ensuring effective cooperation between air and ground units. In the late 1950s and 1960s close air support was downplayed in favor of air defense, air superiority, and operational and strategic strike missions. When this emphasis changed again in the 1970s, the Soviets had to rediscover the techniques of coordinating tactical air and ground missions as they began deploying the Hind and fixed-wing close air support aircraft.[142] Indeed, although the popular press might have reported that Soviet forces had the "co-ordination of a symphony orchestra,"[143] keen-eared military listeners could tell that they were playing out of tune.

In 1977, the army newspaper *Krasnaya zvezda* reported frankly that "combined-arms commanders don't have an in-depth knowledge of the combat capabilities of aviation sub-units and aviators, in turn, can only judge the development of the ground battle in the most general terms."[144] The article suggested that ground troops needed "to get together with aviators more often," while aviators were criticized for not knowing enough about the ground troops they were meant to be supporting. Pilots were reminded that "missions in the interests of ground-based troops *in direct proximity* to attacking or defending subunits [emphasis added]" were very different from the independent strike missions or aerial combat, which, by implication, had been their main tasks up until that time.[145]

During the war the Soviets had employed "the principle of operational subordination of the air formation to the tank army commanders" to ensure "dependable interaction," and mobile groups were "usually assigned those air formations which had already co-operated with [the armies] in previous operations." In the reconstituted *armyeyskaya aviatsiya*, air units were even more integrated with their particular army, creating a "close relationship and mutual understanding" between armies and their aviation units.[146] This in turn made it much easier to have "joint technical conferences, tactical briefings, and discussion of results" as well as joint training and personal contact, all of which were required to actually achieve successful cooperation.[147]

However, setting up *armyeyskaya aviatsiya* did not solve all the problems of ground and air interaction. New emphasis was placed on another wartime measure—forward air controllers. Although they had been used before, the significance of forward air controllers grew, especially at battalion level, as the decentralization of aviation was implemented in the 1980s. In 1981, participants in an exercise reported that "forward air control is needed not only to ensure the most effective action by aircraft, but also was pursued throughout to provide a sense of contact between pilots and motor riflemen."[148] Yet the Soviet military press continued to reveal that air-ground coordination was not being achieved.

In fact *armyeyskaya aviatsiya* and all that it implied created more difficulties than it solved. It certainly brought air and ground troops into closer contact but, consequently, demanded more of them than the less coordinated missions of the 1970s.[149] It has been suggested by some analysts that this problem was political and stemmed from the fact that helicopters were not an integral part of the Ground Forces but belonged to the VVS. However, this view was based more on perceptions of Western armed forces and their internal disputes between air force and army.[150] In the Soviet army, the Ground Forces were paramount and the VVS existed only to assist them. Indeed, the same problems of coordination were also suffered in respect to artillery support, and in both cases the real problem was a lack of training. Not only was the training of forward air controllers haphazard, but if an exercise were "organized by combined-arms commanders, army aviation must be included. . . . On the other hand, if an exercise is set up by aviation commanders, no-one is obligated to provide [ground forces] support."[151]

But, although effective training first would have promoted coordination by allowing both air and ground forces to practice their activities, it was even more important as a way of developing a culture, or assumption, of integrated action. As will be seen below, combat experience in Afghanistan continued to remind the Soviets of the difficulties of achieving effective coordination, and the Soviets were not confident that the link between air and ground forces would be strong enough to stand the strains of actual combat. The critical problem was that helicopters were misused because of "land . . . psychology of combined arms commanders."[152] The answer required aviators to take a far more active part in the development and implementation of land warfare doctrine—to develop a genuine air-land approach.

Air Assault Units

Another major change in Soviet force structure related to the OMG was the creation of air assault units. Air assault brigades consisting of four battalions of airborne rifle troops, two equipped with the *boyevaya mashina, desantunaya* (BMD or airborne combat vehicle), were formed in the early 1980s for use by front commanders; separate air assault battalions, probably equipped with BMDs, were later deployed with some individual armies. They were not regular infantry units but belonged to the VDV, as indicated by the distinctive uniform of their troops and the presence of parachute equipment and BMDs. However, unlike airborne divisions, which were controlled by the Supreme Command, the air assault brigades and battalions were deployed at the operational level of command (Figure 2.1).[153]

Air assault units should not be confused with airmobile brigades, three of which were formed in the late 1960s for operations in mountainous regions on the USSR's southern borders: these units did not have BMDs and may well have had a defensive orientation.[154] The air assault brigades and battalions were patently a new phenomenon and associated with the formation of OMGs. An article in *Krasnaya zvezda* in 1980 described the landing of an "airborne battalion" by helicopters, and a reference to a previous "parachute drop" made it clear that it was not a motor rifle unit simply acting in an airborne role. The helicopters came from the Lenin Regiment, which had a history of being used to develop new techniques, and the crews had been practicing for some time on "exercises of heightened complexity."[155] Then, in *Zapad–81*, helicopter-borne troops secured an airfield and were then reinforced by large transport aircraft carrying armored vehicles (presumably BMDs).[156] By 1983 the air assault brigades were well established. An after-exercise report revealed that the commander of the "assault force" was "experienced," having "participated in more than one tactical exercise assault landing and [with] over 100 parachute jumps to his credit."[157]

As discussed above, the OMG was to work closely with troops air-landed in the enemy's rear area. Although it was possible that paratroops might have been used for this task, they were not controlled by the

Figure 2.1
Soviet Air Assault Brigade

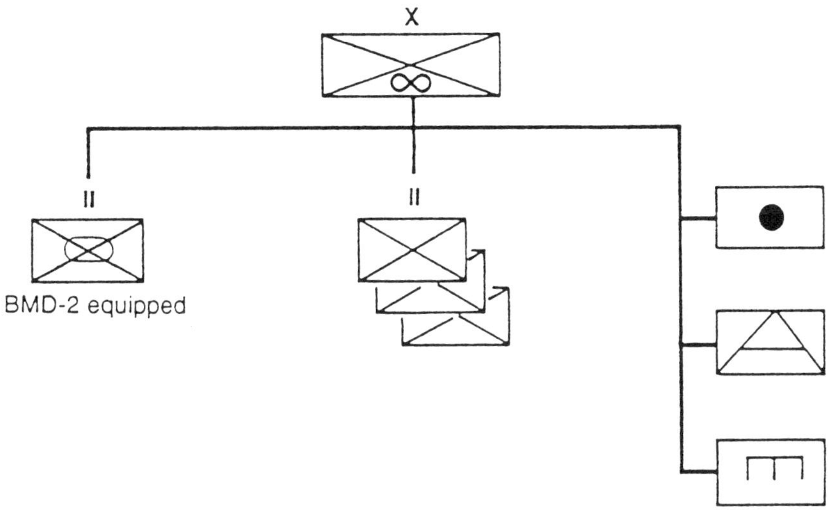

Source: Various articles including Bort, "Air Assault Brigades: New Element in the Soviet Desant Force Structure," and Isby, "Soviet Airmobile and Air Assault Brigades."

commanders who would launch an OMG; they were also to act well beyond the OMG, probably three hundred kilometers or more into the enemy's rear. Further, paratroop operations required considerably more advanced planning than those carried out by helicopter, a restriction that did not suit the OMGs' fast-moving style of combat.[158] Helicopter-borne air landings were therefore more appropriate. The crucial question was, why were air assault units needed when it was established Soviet practice to use motor rifle troops and, indeed, when the Soviets had turned to helicopters to reduce the need for special air-landing troops?

One organizational motive may well have been that the VDV was seeking to increase or maintain its size within the Ground Forces. Although paratroop operations remained important, there is no doubt that they had lost some significance since the 1960s. Even if there was no overt political pressure from the VDV to be more closely involved in helicopter-borne operations, as an organization it would have been keen to expand into new areas of airborne activity to increase its status and simply to improve its professional competency.

Such factors probably affected the specific manner of the creation of air assault units, ensuring that it was the VDV—and not some other organization—that was responsible for them. The actual decision to develop this new type of organization was, however, based primarily on the fact that OMGs were to operate much deeper in the enemy's rear than the distance to which dismounted motor rifle troops could be used effectively for air landings. In exercises the Soviets practiced landing these troops up to fifty kilometers into the enemy's rear area—the deepest point of what the Soviets defined as the tactical zone.[159] While motor rifle troops were adequate for this level of operations, air landings in support of OMGs needed to be conducted at depths of one hundred kilometers or more ahead of the main forces. Belov wrote in 1979 that "a sort of redistribution of zones of action is taking place," with what he termed the tactical zone extending out to one hundred kilometers.[160] The new edition of the *Soviet Military Encyclopaedic Dictionary* in 1983 clarified the issue by defining an operational-tactical level of warfare, at which troops in tactical units operated at a depth and with objectives in mind that were operational in scope.[161] It was this bordering zone at depths of between 50 and 150 kilometers in which air assault brigades and battalions were to be used (Figure 2.2).

Air assault units were more suitable than motor rifle troops because, like the regular parachute regiments, they were very highly trained—both physically and mentally—and were equipped with a higher percentage of heavy weapons. They were more able to undertake difficult missions in the enemy rear and to survive longer before the main forces arrived to relieve them. The value of the air assault units can be judged from the comment by a *mujahadin* commander, Abdul Haq, that in Afghanistan, "the loss of one [Soviet] heliborne soldier is worth the loss of ten conscript soldiers."[162] Indicating the importance of specialized training, the Soviet navy began to develop its own air assault troops for use in amphibious operations in addition to naval infantry units.[163] Indeed air assault units were a counterpart of *armyeyskaya aviatsiya*: in both cases the particular problems of air-land coordination were solved by special organizations.

Air assault brigades were significantly stronger because they were equipped with BMDs. A weakness of air landings conducted by dismounted infantry was that they lacked the firepower and, more importantly, the mobility afforded by armored vehicles. In tactical landings this limitation was not as important since the main aim was usually the capture of a specific location. However, when supporting an OMG, the airborne party needed to be more mobile after landing. Its targets would be more widely dispersed, or themselves mobile, and it had to match the mobility of the group it was assisting. The value of such secondary mobility had been clearly demonstrated in the Ogaden War in 1978 when, under Soviet command, the Ethiopians and Cubans twice used Hips and Hooks to transport troops and light armored vehicles into the Somalis' rear and in each case gained a crushing victory. The next year

Figure 2.2
Battlefield Depths and Soviet Airborne Operations

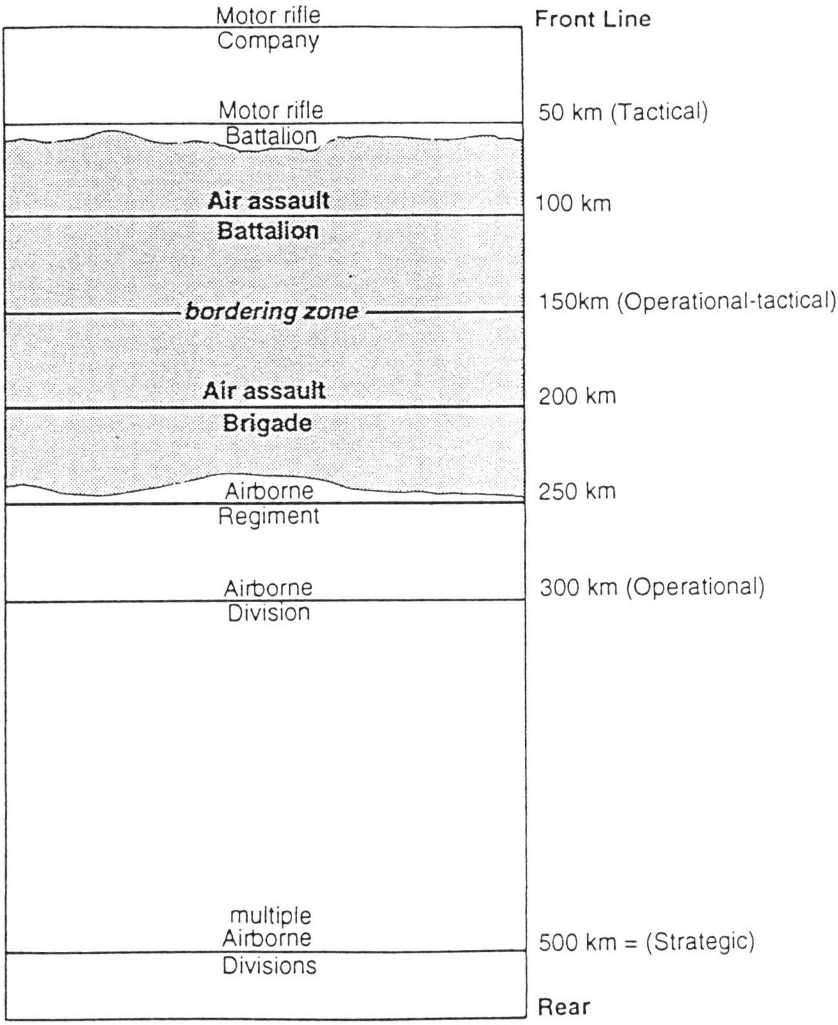

Source: Holcomb, *Recent Developments in Soviet Helicopter Tactics*, 30, and Wardak and Turbiville, *The Voroshilov Lectures*, 362–63, 381.

Belov concluded that "during [air-landing] operations in the enemy rear priority is given to the maneuverable forms of battle."[164]

However, the widespread use of BMDs was not matched by an equally lavish deployment of delivery platforms. The Soviets might have employed fixed-wing aircraft to parachute-drop the BMDs or land them directly at captured enemy airbases, but more likely would have been the use of large helicopters, particularly the Mi–26 Halo, which was introduced in the early 1980s and could carry three BMDs. However, these methods would have been restricted by enemy air defenses and by the lack of transport assets. Each front had just enough helicopters to lift only one battalion with BMDs; inevitable losses would have severely restricted further operations and, despite their interest in giving air-mobile troops secondary mobility, most Soviet air landings would have had to be conducted without BMDs. Nevertheless, the formation of the air assault brigades meant that, at least for critical air-landing operations, the Soviets had the capacity to deploy powerful, mobile forces one hundred kilometers ahead of friendly forces. It would have been a potent complement to the OMG.

Helicopter Air-to-Air Combat

The USSR recognized, as early as 1973, that its own helicopters might come into combat with those of the enemy,[165] but it was not until the 1980s, when the helicopter was more established within the armed forces as a whole, that air-to-air combat received more attention. The Soviets experimented with a three-helicopter formation in which two machines were to perform "search and attack" with a "single helicopter to provide cover." Although the "principal task" of the latter helicopter was to command the group, "If the attack element is threatened with attack by 'enemy' helicopters, the commander will cut them off and ensure that the principal mission is accomplished."[166] The edition of *Tactics* published in 1984 defined one of the tasks of fire support helicopters as "destroying enemy helicopters in the air," and two detailed articles on Western helicopter air-to-air combat practices appeared to inform Soviet pilots about the tactics of their potential enemies.[167] Although training for helicopter aerial combat had become routine in the VVS by the late 1980s, the Soviets were not developing the dedicated fighter helicopter capability that some Western analysts imagined. The Soviets preferred destroying enemy helicopters on the ground rather than in the air; they continued to accord great importance to ground air defenses, and the use of offensive helicopter aerial combat remained "a last resort."[168]

At about the time of this doctrinal interest in air-to-air combat with helicopters, a fire support helicopter was being developed by the Kamov OKB and reported in the West as the Ka–36 (or –136) Hokum. Of course, the Mi–28 Havoc had also been identified by this time, and the development by the Soviets of another armed helicopter caused considerable speculation about the former's purpose. The most extreme argument was that Hokum was a fighter helicopter, whose contrarotating blades gave it very high speed to chase and destroy enemy helicopters, thus giving the Soviets a "rotary-wing counter-air" capability.[169] The latter idea was mainly proposed by U.S. intelligence analysts and was as much a product of that country's own interest in helicopter aerial combat. Certainly, it was enthusiastically accepted and given wide prominence by U.S. Army Aviation proponents as they sought to influence political and military decision makers on the continued funding and design requirement for their own new helicopter, the LHX.[170]

However, most analysis of Hokum has been, at best, informed conjecture with assertions about its mission based on supposed technical characteristics. For example, the rotor-blade design does in fact give it high speed. A more reasonable assumption is that the Ka–36 is like all Kamov helicopters in having contrarotating blades. This design makes helicopters more compact and was thus favored by the OKB because its aircraft were almost always used by the Soviet navy, and small size reduced the deck space needed for landing pads. Certainly the most recent news report on Hokum assumed that it was simply the naval counterpart to the Havoc.[171] Perhaps the most likely result is that, in the new Commonwealth of Independent States (CIS), the continued production of the Ka–36 Hokum will have little to do with air-to-air combat doctrine and a lot to do with the new political and economic situation that is emerging. Despite all the intelligence effort expended on it by the West, maybe Hokum will end up being the most appropriate name for this elusive helicopter.

Afghanistan

A general—but mistaken—reaction of Western observers to the Soviet Union's war in Afghanistan was that it "has been one of the prime stimuli for change in the Soviet rotary-wing force."[172] While, as with the Americans in Vietnam, helicopters were essential to the Soviets' military counterinsurgency campaign, there was less direct impact of Afghanistan experience on the major developments outlined above than might be imagined. Many changes were made to Soviet practice and procedure so that helicopters could operate more effectively, but these changes were almost always specific to the particular geographic and military circumstances of the war. Only in a few cases did the Afghanistan experience extend to general tactics and operating concepts; even then it tended to confirm or reinforce ideas that were already developing.[173] Part of the reason for this was that helicopters were already well established in the Soviet army and, as noted above, major doctrinal innovations had already started before 1979. Although their implementation occurred at the same time as the Afghanistan conflict, the timing was as much coincidental as causal. Equally, although the Soviets were aware of the value of helicopters from their study of other counterinsurgency wars, the use of helicopters in Afghanistan was dictated far more by their existing employment concepts than by imitation of Western experiences. As the United States found when rethinking its doctrine in the early 1970s, counterinsurgency operations provided a poor base of experience from which to develop doctrine for more intense forms of warfare.

At first the Soviets deployed one helicopter regiment, with some sixty Hinds and Hips, in Afghanistan. The regiment, which in 1979 was still part of *frontovaya aviatsiya*, was all that was available within the military district (an administrative equivalent of a wartime front) from which the Soviet contingent was drawn. It was not until 1981 that three more regiments, one a transport unit with a mix of Hips and Hooks, were dispatched to Afghanistan, bringing the total number of helicopters to about three hundred (approximately 140 Hinds, 105 Hips, 40 Hooks and 40 Hoplites).[174] This number of helicopters represented the normal allocation to a major front, and it was at this level of command, based on the Turkestan Military District, that the campaign was controlled.[175] Undoubtedly the helicopter's usefulness in Afghanistan was obvious to the Soviet army but the increase in numbers in 1981 owed as much—if not more—to the general reorganization of helicopters that began in that year coinciding with the development of *armyeyskaya aviatsiya*. To claim that the Afghan experience drove Soviet helicopter developments ignores the fact that, even as the fighting against the *mujahadin* intensified, helicopter numbers remained constant throughout the 1980s at the level set by Soviet doctrine for all types of combat operations. The level of deployment reflected—paradoxically—both the inflexibility of Soviet doctrine in preventing added deployment of helicopters but also its ready adaptability to nonintensive warfare.

Although new missions such as providing a flying escort to road convoys[176] emerged, armed helicopters were used mainly as had been planned for a major war. With their low speed and hovering ability they provided accurate close air support. Helicopters transported airborne and motor rifle troops in operations to capture mountain passes ahead of advancing ground troops and to cut off enemy units. They were also used for logistic support, evacuating casualties, reconnaissance, and command and control. Although the exact nature of these missions was obviously specific to Afghanistan, the types of missions and their relationship to the prosecution of the war by ground-based troops was just the same as it would have been in a conventional conflict. For example, they were used defensively in reaction to enemy attacks just as they would have been employed in anti-tank or anti-landing reserves in a more intense conflict. Equally, no matter who the enemy, the Soviets would have used air assault troops against their command and air defense systems because these targets were beyond the range of ground-based attacks: the Soviets treated *mujahadin* headquarters and anti-aircraft units in exactly the same way.[177]

Moreover, Afghanistan did little to alter the course of helicopter development. Certainly, special tactics and procedures were developed, the design of new and modernized helicopters such as Havoc and the Mi–17 (an Mi–8 with more powerful engines) was shaped by the war and many VVS pilots gained considerable experience, but these tactics, lessons and experiences were usually specific to Afghanistan.[178] For example, because the guerrillas' only effective air defense weapons up until 1987 were machine guns, Soviet helicopters were most vulnerable at low altitudes and safe at high altitudes—exactly opposite to the environment of a major war. As was demonstrated when the first Stinger SAMs (surface-to-air missiles) began to arrive, this experience was not only of little relevance to the conventional high-intensity conflict but was in many ways counterproductive. Its actual involvement in local wars notwithstanding, the Soviet army almost exclusively prepared itself for major conflicts, thus negating much of what applied in Afghanistan.

Of course, the confirmation that missiles such as the U.S. Stinger and British Blowpipe were deadly unless the correct tactics were employed was important. Extravagant press reports, often based on official U.S. sources that sought to emphasize the importance of the American contribution to the *mujahadin*, claimed these missiles virtually changed the course of the war.[179] However, they did not of themselves drive Soviet helicopters and aircraft from Afghanistan's skies, nor did they win the war. One of the most knowledgeable Western reporters on the war, Mark Urban, concluded that "based on accounts of Westerners who did see them fired, it is unlikely that more than 10 percent of the Stingers fired destroyed their target."[180] At most, they made the Soviets much more aware of the need to fly as low as possible, to use a wide variety of passive and active infrared countermeasures and to employ artillery to suppress air defenses.

The conflict with the *mujahadin* was more important for providing evidence that supported two other developments in Soviet thinking on helicopters. It showed that air landings were not as simple as had first been thought, thus confirming the development of specialized air assault units (although it must be stressed that the eventual deployment of such units was not caused by the war). In particular, air landings placed platoons or companies of Soviet troops further from centralized control than was normal, demanding much more initiative from their junior officers. Indeed, it was the mental preparation of the air assault brigades for their mission as much as any other factor that differentiated them from motor rifle troops. A more important lesson was that air and ground operations had to be far more closely coordinated, as Afghanistan provided a constant reminder and testing ground for measures such as the use of forward air controllers.[181] Although operational experience was not necessarily critical to successful innovation,[182] the knowledge of air-ground coordination procedures gained by those who actually served in Afghanistan was the type of experience most easily transferred to a major war.[183]

Perhaps the most crucial effect of Afghanistan is the least quantifiable. There can be no doubt that the development of the helicopter as an equal partner in air-land warfare during the 1980s was affected by the campaign against the *mujahadin*. The conflict meant that the most influential officers within the army were those who had served in Afghanistan; at the same time it had shown these people—and 80,000 officers went to Afghanistan at one time or another—that tanks, BMPs and helicopters were able to combine and that, in doing so, compensated for one another's weaknesses.[184] While not directly relevant to a major war, it seems that at least some of those with experience in Afghanistan were able to make the conceptual leap from helicopter operations in counterinsurgency campaigns to their application in full-scale war. Moreover, because of their combat experience, these soldiers were well placed to overcome any doubts or opposition to a broader role for rotary-wing aviation in planning for major warfare. As in America after Vietnam, it was exposure to the possibility of helicopter operations in conventional warfare, not the actual details of low-intensity operations, that was crucial.

CONCLUSION

In the 1960s helicopters were developed by the Soviets to extend their existing capabilities for air-landing operations, and they became the basic means of dropping troops up to fifty kilometers behind the enemy's front line. Armed helicopters began to be deployed to assist these air landings. This interest in helicopters stemmed mainly from the need to speed up the tempo of offensive operations: since nuclear weapons were the main source of firepower, it was thought that the ground forces needed great mobility to exploit the effects of atomic strikes on the enemy. However, although air landings were important, tanks remained the key element in the combat force. The October 1973 Arab-Israeli War not only reaffirmed the importance of helicopter transport missions but also led to Soviet acceptance of the use of attack helicopters for fire support—mainly to help neutralize enemy anti-tank positions or to stop armored counterattacks. As the perceived utility of nuclear weapons decreased, so the importance of conventional fire support increased. Both attack and air-landing tasks were tactical in nature and, because helicopters were independent from fixed bases, the Soviets could bring them much closer to the ground forces and improve cooperation between the two to some extent. Thus, by the mid–1970s, armed and transport helicopters had become an important part of the Soviet army.

Helicopters were still regarded as part of the aerial component of warfare, which although intersecting with the ground battle, was nevertheless separate from it. However, as the lessons from the October 1973 war were analyzed, especially in the context of the USSR's growing reluctance to rely on nuclear weapons, the ground and air aspects began to be combined much more than before. Indeed, in the 1980s the Soviets had begun to think of warfare as having "a clearly pronounced air-land character." As a result there had to be "firm and continuous command and control of diverse forces, clear organization of co-ordination of the troops [i.e. ground units] with aviation, tactical airborne forces and helicopter subunits . . . [for] attaining success in battle."[185] Although it was recognized that air forces had played a "great part" in winning the Great Patriotic War, "its interaction with the land forces was effected chiefly within individual *operations* [emphasis added]."[186] Now it was thought that air and ground were connected at the tactical level as well. Furthermore, this change was essentially caused by the introduction of the helicopter.

> Air-land operations in the tactical dimension appeared principally only from the moment of the generalization in the forces of helicopters. . . . There is no doubt that under the influence of the rapid development of helicopters, air-land tactical operations will become the dominating procedure at the threshold of the 21st century.[187]

Consequently the helicopter, which previously had been regarded as having a supporting role rather similar to artillery, now became, with tanks and BMPs, "the third basic means of combined arms combat."[188] The Soviets reestablished a tactical organization for their helicopters. *armyeyskaya aviatsiya*, which gave control to army and division commanders and thus helped to ensure combined air-land action rather than simply joint air and ground operations arranged at frontal level.

Helicopters brought about this fundamental change because they could "carry out combat missions *under conditions and in places which are disadvantageous for traditional aviation*. They are in a position *to directly support* fighting troops [emphasis added]."[189] In both attack and transport missions, the helicopter was able to operate close to the front line; it was also able to use terrain for cover and security. In attack it combined observation and direct fire, and was able to engage the enemy at less risk from air defenses. In air assaults, the helicopter enabled less well-trained troops to be used and to land them undispersed and close to the objective. Yet, while having some of the characteristics more usually associated with ground vehicles, helicopters retained the independence from terrain from which all aircraft benefited; they were thus much faster than tanks and trucks and, for the Soviets, speed was vitally important. Capable of combining tactically with tanks, infantry and artillery in a way that jet aircraft were not, attack helicopters were still able to be switched rapidly from one sector to another or concentrated in a decisive area, thereby fulfilling the "important principle" of "fast maneuver by fire."[190] Transport helicopters provided the same speed advantage for the movement of infantry or other ground forces.

Although such movement was appropriate to lateral shifts, reinforcing successful attacks or meeting enemy counterstrokes, the key application was to the rapid attack throughout the depth of the enemy's defense required by Soviet military theory. First, helicopters reduced reliance on nuclear weapons for deep strikes into the tactical and tactical-operational depth of the enemy, and were better than fixed-wing aircraft since they were more responsive and were able to search for and attack targets with greater freedom. Second, helicopters gave the Soviets the ability to conduct troop strikes with air assault or motor rifle troops as part of the tactical battle. As Reznichenko concluded: "Tactical airborne forces . . . have become an important component of modern combined-arms combat and they have introduced significant changes into its content, the methods of its conduct and its scope."[191] Air-land combat and the close association of helicopters and ground vehicles was developed in tandem with the evolving OMG concept that developed at the same time and, while the OMG was itself a most important advance in Soviet operational art, the helicopter was crucial to its tactical success. Without the helicopter, the OMG concept would have developed very differently indeed and perhaps not at all. The future, it would seem, will be governed equally by the tank and the helicopter. It is thus not surprising that one article suggested the army of the future will need officers "capable of perceiving and evaluating the battlefield not only from the height of a turret of a tank or BMP, but also, and perhaps most of all, from the height of a helicopter flight."[192]

But it is unlikely that this view from the helicopter will be of Germany or its NATO allies. The CIS, which nominally and for the moment controls the remains of the Soviet army, will undoubtedly be wracked by internal strife and unrest, even civil war. Already Georgia, Azerbaijan and Armenia have experienced the sort of violence that has marked the breakdown of Yugoslavia into its constituent republics. In this environment, OMGs and air-land warfare may be as far away as the old certainties of the cold war and Soviet military doctrine. Indeed, it may be that Afghanistan will be the sort of experience, and will provide the doctrine, that is far more suitable for military operations in the CIS. The development of Soviet military helicopters is as finished as the Soviet Union itself. The massive military-industrial complex that supported the

USSR's quantitative and qualitative advantage over all but the United States is in decline, as is the need for doctrine to guide a rapid offensive into northwest Europe. The changed circumstances are no better displayed than in the eager marketing campaign being run in Western military magazines to promote the Mi–24 Hind and Mi–28 Havoc to potential buyers and the appearance of the latter at the premier international aerospace show in Paris, where it was "the calling card of our combat helicopter construction industry."[193]

NOTES

1. Chief Marshal of the Armored Forces P. A. Rotmistrov, "The Postwar Development of Tank Forces," *Time and Tanks*, in William F. Scott (ed.) *Selected Soviet Military Writings 1970–1975* (Washington, DC: USAF, 1977) 289.

2. Lt. Colonel Z. Scibiorek, "Helicopters," *Zolnierz wolnosci*, 9 September 1987: 3 (trans.) (Ft. Leavenworth, KS: Soviet Army Studies Office [SASO], 1988) 2.

3. John Everett-Heath, *Soviet Helicopters: Design, Development and Tactics* (London: Jane's, 1988), is a good general source for early Soviet helicopter developments.

4. For consistency and in line with Soviet terminology, in this chapter "army" means armed forces, with the phrase "Ground Forces" describing what Western countries would call the army; "army" is also the formation between a division and a front and equivalent to a Western corps. Similarly, "Great Patriotic War" is preferred to Second World War.

5. Major General M. I. Cherednichenko, "On Features in the Development of Military Art in the Postwar Period," *Voyenno istoricheskiy zhurnal* [hereafter *V-IZ*], June 1970, in Scott, *Selected Soviet Military Writings 1970–1975*, 109. For ease of identification, the NATO reporting name for Soviet helicopters is used throughout.

6. Lev Chaiko, *Helicopter Construction in the USSR* (Falls Church, VA: Delphic Associates, 1986) 3–4.

7. Lt. Colonel David M. Glantz, *The Soviet Airborne Experience* (Ft. Leavenworth, KS: CSI, 1984) 136; and Chaiko, *Helicopter Construction in the USSR*, 20–21.

8. Chaiko, *Helicopter Construction in the USSR*, 7.

9. Ibid., 71–72.

10. S. A. Tyushkevich, *The Soviet Armed Forces: A History of their Organizational Development* (1970; Washington, DC: USAF, 1973) 415, 421.

11. Colonel V. Ye. Savkin, *The Basic Principles of Operational Art and Tactics* (trans.) (1972; Washington, DC: USAF, 1974) 197; and Colonel M. Belov, "New Factors in the Development of Modern Armies," *Soviet Military Review* [hereafter *SMR*], February 1974: 10.

12. Colonel M. Belov, "Air Mobilization of Modern Armies," *SMR*, October 1975: 13. See also Colonel N. F. Miroshnichenko, "Changes in the Content and Nature of Modern Combat," *Voyennaya mysl* [hereafter *VM*], October 1966, in

William R. Kintner and Harriet Fast Scott (eds.) *The Nuclear Revolution in Soviet Military Affairs* (Norman, OK: University of Oklahoma Press, 1968) 371.

13. Colonel K. Andrukhov and Colonel V. Bulatnikov, "The Growing Role of Airborne Troops in Modern Military Operations," *VM*, July 1966, in Joseph D. Douglass, Jr., and Amoretta M. Hoeber (eds.) *Selected Readings from Military Thought 1963–73* (Washington, DC: USAF, 1982) 119.

14. General V. D. Sokolovskiy (ed. and trans. Harriet Fast Scott), *Soviet Military Strategy* (London: Macdonald and Jane's, 1975) 249–50, 291–93.

15. Colonel I. A. Grudinin, "The Time Factor in Modern War," *Kommunist vooruzhennykh sil*, February 1966, in Kintner and Scott, *The Nuclear Revolution in Soviet Military Affairs*, 253.

16. Major General S. F. Begunov, "The Maneuver of Forces and Material in an Offensive," *VM*, May 1968, in Douglass and Hoeber, *Selected Readings from Military Thought 1963–73*, 223.

17. Savkin, *The Basic Principles of Operational Art and Tactics*, 180, 194. See also Lt. General P. Pavlenko, "The Development of the Tactics of Airborne Troops in the Postwar Period," *V-IZ*, January 1980: 27, in Joint Publications Research Service, *Report on USSR Military Affairs* [hereafter *JPRS-UMA*] 1511 (1980): 31–38.

18. Kenneth R. Whiting, *Soviet Air Power* (Boulder, CO: Westview, 1986) 113; and Grudinin, "The Time Factor in Modern War," 253. While the members of the VDV were airborne troops, the word "airborne" was also applied to all units moving by air, whether they were naval infantry, paratroopers or motor rifle troops.

19. Sokolovskiy, *Soviet Military Strategy*, 250.

20. Chris Bellamy, "Red Star in the West," *JRUSI* 132.4 (1987): 66; and Richard E. Simpkin, *Deep Battle: The Brainchild of Marshal Tukhachevskii* (London: Brassey's, 1987) 53–73.

21. Colonel A. A. Sidorenko, *The Offensive* (trans.) (1970; Washington, DC: USAF, 1973) 102.

22. Andrukhov and Bulatnikov, "The Growing Role of Airborne Troops in Modern Military Operations," 117.

23. Marshal S. Krasovskiy, "Trends in the Use of Aircraft in a Nuclear War," *VM*, March 1967, in Douglass and Hoeber, *Selected Readings from Military Thought 1963–73*, 159.

24. Savkin, *The Basic Principles of Operational Art and Tactics*, 104. "Front" is the Soviet military organization in which a number of air and ground armies are combined under one commander. It is roughly equivalent to a Western army group.

25. Sokolovskiy, *Soviet Military Strategy*, 294.

26. Colonel S. I. Krupnov, "According to the Laws of Dialectics," *Krasnaya zvezda* [hereafter *KZ*], 7 January 1966, in Kintner and Scott, *The Nuclear Revolution in Soviet Military Affairs*, 24.

27. General G. Kublanov, "Airborne Assaults and Combat with Them in Modern War," *VM*, August 1965, quoted in Major James F. Holcomb, Jr., and Graham H. Turbiville, Jr., *Exploiting the Vertical Dimension: Continuing Development of the Soviet Desant Force Structure* (Ft. Leavenworth, KS: SASO, 1987) 11.

28. Colonel D. M. Samorukov, "Combat Operations Involving Conventional

Means of Destruction," *VM*, August 1967, in Douglass and Hoeber, *Selected Readings from Military Thought 1963–73*, 184.

29. Krasovskiy, "Trends in the Use of Aircraft in a Nuclear War," 159.

30. Major General Kh. Dzelaukhov, "The Infliction of Deep Strikes," *VM*, February 1966, in Douglass and Hoeber, *Selected Readings from Military Thought 1963–73*, 113.

31. Sidorenko, *The Offensive*, 193.

32. Throughout this work, the term "conventional" is understood to include the potential use of chemical weapons and is used simply to distinguish between nuclear and nonnuclear warfare.

33. Edward L. Warner III, *The Military in Contemporary Soviet Politics* (New York: Praeger, 1977) 137–47.

34. Tyushkevich, *The Soviet Armed Forces*, quoted in J. G. Hines, A. Petersen and N. Trulock III, "Soviet Military Theory from 1945–2000: Implications for NATO," *The Washington Quarterly* 8.3 (1986): 121.

35. Harriet Fast Scott and William F. Scott, *Soviet Military Doctrine* (Boulder, CO: Westview, 1988) 42–51.

36. Matthew P. Gallagher and Karl F. Spielmann, Jr., *Soviet Decision-Making for Defense* (New York: Praeger, 1972) 14.

37. P. A. Peterson and J. G. Hines, "The Conventional Offensive in Soviet Theater Strategy," *Orbis* 27.3 (1983): 695–99.

38. Lt. Colonel V. M. Bondarenko, "The Modern Revolution in Military Affairs and the Combat Readiness of the Armed Forces," *Kommunist vooruzhennykh sil*, December 1968, quoted in Harriet Fast Scott and William F. Scott, *The Armed Forces of the USSR* (Boulder, CO: Westview, 1984) 53.

39. Andrukhov and Bulatnikov, "The Growing Role of Airborne Troops in Modern Military Operations," 121.

40. Samorukov, "Combat Operations Involving Conventional Means of Destruction," 174–77, 181 and 184.

41. Marshal A. A. Grechko, *The Armed Forces of the Soviet Union*, (Moscow: Progress Publishers, 1977) 159.

42. Holcomb and Turbiville, *Exploiting the Vertical Dimension*, 7–8.

43. Savkin, *The Basic Principles of Operational Art and Tactics*, 192. Although the Hook was able to lift the BMD, this vehicle was only available to the VDV. Moreover, no reference to the use of Hooks to lift armored infantry vehicles can be found in this period, in contrast to many articles on their use for other purposes.

44. Colonel M. Belov, "Helicopters and Land Force Tactics," *SMR*, December 1976: 23.

45. Lt. General Vasily G. Reznichenko, "Tactics: Development Trends," *KZ*, 5 October 1975, in Harriet Fast Scott and William F. Scott, *The Soviet Art of War* (Boulder, CO: Westview, 1982) 281.

46. Chaiko, *Helicopter Construction in the USSR*, 8–9; and Everett-Heath, *Soviet Helicopters*, 40–41, 112–117.

47. Lt. Colonel L. Chuyko and Sr. Lieutenant D. Getmanenko, "An Assault Landing Aboard Helicopters," *KZ*, 8 June 1974: 1 (trans.) in *JPRS-UMA* 1050 (1974): 23.

48. Lt. Colonels A. Ladin and V. Usoltsev, "Mutual support," *KZ*, 17 October

1986: 1 (trans.) in *JPRS-UMA* 87–012: 31–33; and Colonel A. Khorobrykh, "Above the Iori Plateau," *Aviatsiya i kosmonavtika* [hereafter *AIK*], October 1985: 4ff (trans.) in *JPRS-UMA* 86–011: 3.

49. Compare the many smaller exercises that Soviet sources described with the bigger operations, such as *Berezina* in 1978, dealt with by Colonel I. Dynin, "Tactical Heliborne Desant," *Voyenny vestnik*, April 1978: 35ff (trans.) in UKTRANS No. 79 (RMA Sandhurst: Soviet Studies Research Centre 1979) [hereafter UKTRANS].

50. Colonel V. Lebedev. "In the Lenin Regiment," *SMR*, August 1987: 13.

51. Savkin, *The Basic Principles of Operational Art and Tactics*, 183.

52. Andrukhov and Bulatnikov, "The Growing Role of Airborne Troops in Modern Military Operations," 118.

53. Colonel M. Belov, "Helicopters and Combined Arms Combat," *KZ*, 7 August 1969: 3 (trans.) in *JPRS-UMA* 549 (1969): 25–27.

54. For example, John W. R. Taylor, "Combat Helicopters of the Warsaw Pact," *Jane's Defence Review* February 1980: 117.

55. Chaiko, *Helicopter Construction in the USSR*, 9–10, 18–19.

56. John Everett-Heath, *Design Development and Tactics of Soviet Helicopters* (London: Jane's, 1983) 87. This comment is in the first edition of Everett-Heath's work and was omitted from the second version, thus reflecting some of the confusion among Western observers over the purpose of the Hind-A.

57. Colonel M. Belov, "Combat Helicopters—A New Weapon," *Krylya rodiny*, December 1973: 30–33 (trans.) in *JPRS-UMA* 1002 (1974): 8–12. The S–67 ought not to be confused with Sikorsky's other and completely successful Black Hawk, the UH–60.

58. Major A. Bychkov, "Helicopters Move Over the Hills," *KZ*, 15 August 1973: 1 (trans.) in *JPRS-UMA* 954 (1973): 6–8. Not naming the helicopters was unusual (suggesting some censorship); moreover the article appeared about the same time as the Hind's introduction and the "raid" was unlike the air-landing missions normally described by Soviet authors.

59. Savkin, *The Basic Principles of Operational Art and Tactics*, 183.

60. Colonel General N. A. Lomov (ed.), *Scientific-Technical Progress and the Revolution in Military Affairs* (trans.) (Washington, DC: USAF, 1977) 14; and Colonel V. Izgarchev, "With a Landing Force on Board," *KZ*, 8 February 1978 1 (trans.) in *JPRS-UMA* 1350 (1978): 48.

61. For issues about which there had been no final decision, a discussion in "Western terms" was more ideologically sound for the Soviet army.

62. Army General P. A. Kurochkin, "Operations of Tank Armies in Operational Depth," *VM*, November 1964, in Douglass and Hoeber, *Selected Readings from Military Thought 1963–73*, 79–80, 83–84.

63. Krasovskiy, "Trends in the Use of Aircraft in a Nuclear War," 157.

64. Lt. Colonel V. Izmylov, "Continuity," *AIK*, July 1973: 11 (trans.) in *JPRS-UMA* 950 (1973): 14–17; and Y. Veraska, "An Aircraft over the Battlefield," *SMR*, September 1970: 43.

65. These two books were published as part of a series that each officer was meant to read. For articles see Colonel M. Belov, "Problems in Making Modern Armies Airmobile," *KZ*, 7 March 1972: (trans.) in *JPRS-UMA* 802 (1972): 1–5;

and Colonels N. Byskov and N. Grishin, "Helicopters against Tanks," *KZ*, 30 May 1973: 3 (trans.) in *JPRS-UMA* 935 (1973): 4–7.

66. Sidorenko, *The Offensive*, 130–32; and Savkin, *The Basic Principles of Operational Art and Tactics*, 183.

67. Lomov, *Scientific-Technical Progress and the Revolution in Military Affairs*, 110; and G. D. Wardak and Graham H. Turbiville, Jr., *The Voroshilov Lectures—Materials from the Soviet General Staff Academy* Vol. 1 (Washington, DC: National Defense University Press, 1989) 386. It should be noted that Sidorenko's doubts concerned the use of helicopters for fire support and not for air landings.

68. Savkin, *The Basic Principles of Operational Art and Tactics*, 184.

69. A Radin, "Implementing the Concept of the Mi–24," *Krylya rodiny*, May 1989: (trans.) in *JPRS-UMA* 89–024 (1989): 27. Until this article appeared there were tantalizing hints about this reaction in Chaiko, *Helicopter Construction in the USSR*, 11, and Colonel M. Belov, (no title), *KZ*, January 1974: 3 (trans.) in *JPRS-UMA* 1007 (1974): 3.

70. "V. Suvorov," *Inside the Soviet Army* (London: Book Club Associates, 1982) 191–92; and Rotmistrov, "The Postwar Development of Tank Forces," 289.

71. Chaiko, *Helicopter Construction in the USSR*, 19–20.

72. See A. J. Alexander, *Decision-Making in Soviet Weapons Procurement*, Adelphi Paper 147–48 (London: International Institute for Strategic Studies [hereafter IISS], 1978) 25–26, for a summary.

73. Lt. Colonel (Ret.) E. Eyal, "Combat Helicopters in the Yom Kippur War 1973," *Defense Update International* 21: 32–43.

74. Major General V. I. Zemskov, "Characteristic Features of Modern War and Possible Methods of Conducting Them," *VM*, July 1969, in Scott and Scott, *The Soviet Art of War*, 215.

75. Grechko, *The Armed Forces of the Soviet Union*, 160.

76. Lt. General Vasily G. Reznichenko, "Tactics—A Component Part of Military Art," *VM*, December 1973, in Scott and Scott, *The Soviet Art of War*, 238.

77. David C. Isby, *Weapons and Tactics of the Soviet Army* (London: Jane's, 1988) 223–24.

78. Belov, "Helicopters and Land Force Tactics," 22. It is notable that, unlike many other articles by Belov, this piece was based on Soviet as well as foreign sources. While Belov might be thought to have voiced an extreme view of the helicopter's importance, it must be remembered that it could not have been published (especially in the widely distributed *SMR*) without having official sanction. See also Major General V. Larionov, "Certain Questions of Military Art from the Experiences of Local Wars," *V-IZ*, April 1984: 46ff (trans.) in *JPRS-UMA* 84–048 (1984): 43.

79. Colonel M. Belov, "Helicopters Used by Ground Troops," *SMR*, April 1976: 34. See also Colonel M. Belov, "How to Fight Helicopters," *SMR*, September 1979: 19.

80. Colonel N. Nikitin, "Some Operational and Tactical Lessons from the Local Wars of Imperialism," *V-IZ*, December 1978: 60ff (trans.) in *JPRS-UMA* 1425 (1979): 9. This comment was not supported by a citation to Western press materials, even though elsewhere the article made extensive use of such references. See also Belov, "Helicopters and Land Force Tactics," 23. The unusual emphasis in Soviet literature on helicopters in the October 1973 war was raised

more generally by William F. Scott, "Changes in Tactical Concepts within the Soviet Forces," in Lawrence L. Whetten (ed.), *The Future of Soviet Military Power* (London: Macdonald and Jane's, 1976) 91.

81. The exchange ratios produced in the Ansbach trials (see chapter 1) were repeatedly quoted in the Soviet press. See, for one example, Belov, (no title), 2.

82. Lt. General I. G. Zavyalov, "The New Weapon and Military Art," *KZ*, October 1970: 2–3, in Scott, *Selected Soviet Military Writings 1970–1975*, 211.

83. In *Voyenny vestnik* (a practical journal for junior officers), before 1973 the average had been between two and four articles a year on helicopters; after 1973 it jumped to twenty-four or twenty-six per annum: Scott and Scott, *The Armed Forces of the USSR*, 170 fn.51. The upsurge in interest can also be seen from an analysis of the SSRC Data Base of Soviet Military articles.

84. Colonel P. Simchenkov, "Manoeuvre is the Key to Victory," *Voyenny vestnik*, April 1977: 68ff (trans.) in UKTRANS 50.

85. Everett-Heath, *Soviet Helicopters*, 120. The first Soviet reference to the Hind as a fire support helicopter was Colonel A. Khorobrykh, "The Helicopter Gunship Attacks," *KZ*, 6 June 1976: 4 (trans.) in *JPRS-UMA* 1245 (1976): 51, which, with its comment on "tear-drop" canopies, could only have meant the Hind-D. An extensive search of Soviet literature has not revealed the Hind-D or later models being used for air-landing operations, although there were a few references to its use for casualty evacuation in Afghanistan.

86. Lynn Hansen, "Soviet Helicopters: History and Employment Doctrines," in P. J. Murphy (ed.), *The Soviet Air Forces* (Jefferson, NC: McFarland, 1984) 247.

87. Lt. Colonel A. Zakharenko, "Lessons in Coordination," *KZ*, 5 August 1977: 1 (trans.) in *JPRS-UMA* 1304 (1977): 62; and Lt. Colonel G. Kashuba, Lt. Colonel A. Zakharenko and Major G. Miranovich, "Tanks Move into the Breach," *KZ*, 8 February 1978: 1 (trans.) in *JPRS-UMA* 1336 (1978): 54.

88. Lt. General P. V. Bazanov (Deputy Chief, VVS) quoted in Colonel V. Izgarchev, "The Helicopters Are Attacking," *Pravda*, 20 September 1979: 6 (trans.) in *JPRS-UMA* 1480 (1979): 50.

89. Major V. Konek, "They Provided Air Support," *KZ*, 18 July 1984: 1 (trans.) in *JPRS-UMA* 84–063 (1984): 36; and Major Yu. Tomarev, "Experience from the Carpathian Exercise," *KZ*, August 1977: 2 (trans.) in *JPRS-UMA* 1304 (1977): 61.

90. Major A. Yurkin, "Helicopters on Combat Course—The Main Thing Is Effectiveness and Quality," *KZ*, 28 May 1977: 7 (trans.) in *JPRS-UMA* 1291 (1977): 83–84; Captain A. Popov, "Tactical Air Exercise: Pluses and Minuses," *AIK*, November 1986: 8ff (trans.) in *JPRS-UMA* 87–019 (1987): 13; and Lt. Colonel I. Onishchenko, "Sniper Attack," *KZ*, 28 February 1979: 1 (trans.) in *JPRS-UMA* 1428 (1979): 34.

91. Lt. Colonel L. Chuyko, "Tanks in the Sight," 14 July 1977: 1 (trans.) in *JPRS-UMA* 1297 (1977): 34–36; and Lt. Colonel Ye. Polyakov, "Commended by the Commander-in-Chief," *KZ*, 25 January 1981: 1 (trans.) in *JPRS-UMA* 1592 (1981): 21.

92. Lt. Colonel Yu. Chumakov, "Swiftness and Continuity of Attack," *Voyenny vestnik*, January 1983: 22ff in USAF, *Soviet Press Translations* 200–1, 11 December 1983: 208.

93. L. Vladimirov, "The 'Westerners' Triumph in the Encounter Battle,"

Kommunist, 4 February 1976: 3 (trans.) in *JPRS-UMA* 1226 (1976): 40; and Lt. Colonel V. Shevtsov, "Surprise Attack, Tactical Exercise Report," *KZ*, 18 February 1984: 1 (trans.) in *JPRS-UMA* 84–028 (1984): 62–63.

94. Lt. Colonel N. Bogatkin, "Bearing Lenin's Name," *KZ*, 3 June 1977: 2 (trans.) in *JPRS-UMA* 1297 (1977): 11; and Sr. Lieutenant V. Lavrenyuk, "Helicopters Attack at Night," *KZ* 6 December 1984: 1 (trans.) in *JPRS-UMA* 85–014 (1985): 46–47.

95. Yurkin, "Helicopters on Combat Course—The Main Thing Is Effectiveness and Quality," 83; and Izgarchev, "The Helicopters Are Attacking," 50.

96. Reznichenko, "Tactics: Development Trends," 281–282; and Major General Ivan N. Vorobyev, "The Tactics of Fire and Maneuver," *KZ*, 5 June 1979: 2 (trans.) in *JPRS-UMA* 1458 (1979): 58. Throughout the 1960s, 1970s and into the 1980s both Vorobyev and Reznichenko were highly influential theorists: note the comment in John Erickson, Lynn Hansen and William Schneider, *Soviet Ground Forces* (Boulder, CO: Westview, 1986) xviii, and Scott and Scott, *Soviet Military Doctrine*, 241.

97. Peter H. Vigor, *Soviet Blitzkrieg Theory* (London: Macmillan, 1984) 8–9.

98. Scott and Scott, *Soviet Military Doctrine*, 110–111, 113, and Scott and Scott, *The Armed Forces of the USSR*, 65, 68.

99. This idea was advanced in particular by two U.S. analysts: see Petersen and Hines, "The Conventional Offensive in Soviet Theater Strategy"; Hines, Petersen and Trulock, "Soviet Military Theory from 1945–2000: Implications for NATO"; and Hines, *Encirclement*, Pamphlet AA11, (RMA Sandhurst: SSRC, 1985).

100. See particularly the comments of John Erickson, "Soviet Theater-Warfare Capability," in Whetten, *The Future of Soviet Military Power*, 163–65.

101. For example, see Scott and Scott, *The Armed Forces of the USSR*, preface and 68, and Hines, Petersen and Trulock, "Soviet Military Theory from 1945–2000: Implications for NATO," 117.

102. For a good summary, see Edward L. Warner III, "New Thinking and Old Realities in Soviet Defence Policy," *Survival* 31.1–2 (1989): 15–18. A good Soviet source is Marshal Nikolai Ogarkov, "Deathless Glory of Soviet Arms," *Kommunist vooruzhennykh sil*, November 1984: 16 (trans.) in ADVAB 229 (RMA Sandhurst: Soviet Studies Research Centre, no date) [hereafter referred to as ADVAB]. This is not to say that the Soviets were more eager for war, rather, that they were changing the emphasis of their preparation.

103. This was revealed in the previously secret Soviet General Staff Academy materials recently brought to the West by an Afghan army colonel and published in Wardak and Turbiville, *The Voroshilov Lectures*, 93ff.

104. Marshal Nikolai Ogarkov, "Military Strategy," *Soviet Military Encyclopaedia* (1979) in Scott and Scott, *The Soviet Art of War*, 249; and Wardak and Turbiville, *The Voroshilov Lectures*, 19.

105. Chris N. Donnelly, "The Operational Manoeuvre Group," *International Defense Review*, 15.9 (1982): 1177–86 was the first detailed Western examination of the OMG. Originally derived from Polish, the name "operational maneuver group" has been recently superseded by "operational mobile group" to which the acronym OMG can, fortunately, still be applied. See Chris N. Donnelly,

"Future Soviet Military Policy: Part II Where and How," *International Defense Review* 22.2 (1989): 143.

106. John H. Hansen, *Correlation of Forces* (New York: Praeger, 1987) 134, 142, 149, 153 and 171.

107. Charles Dick, "Soviet Operational Manoeuvre Groups," *International Defense Review* 16.6 (1983): 774–76; and Chris N. Donnelly, *The Place of the Helicopter on the Modern Battlefield*, Pamphlet A51 (RMA Sandhurst: SSRC, 1983) 24–31.

108. Colonel J. Lachiewicz and Lt. Colonel R. Rajmanski, "Helicopter Reconnaissance and Fire Missions in Support of Land Operations at Night," *Przeglad wojsk lotnicznych i wojsk obrony powietrznej kraju* [hereafter *PWLIWOPK*], June 1981 (trans.) in UKTRANS 136: 4. See also J. G. Hines and P. A. Petersen, "The Warsaw Pact Strategic Offensive—the OMG in Context," *International Defense Review* 16.10 (1983): 1393.

109. Petersen and Hines, "The Conventional Offensive in Soviet Theater Strategy," 710–14, citing Vorobyev, "Weapons and Tactics: The Commander and Modern Combat," *KZ*, 12 January 1982: 3. It should be noted that Soviet military theory placed considerable emphasis on the exactitude of its terminology.

110. Major General I. Krupchenko, "Methods of Developing Success in Operational Depth with the Forces of Tank Armies and Tank and Mechanised Corps," *V-IZ*, July 1981, 13ff (trans.) in UKTRANS 140: 7.

111. Marshal A. N. Yefimov, "Soviet Military Art," *V-IZ*, August 1986: 14ff (trans.) in *JPRS-UMA* 87–004 (1987): 1. See also Major W. Michalak, "Aviation in the Raid-Manoeuvre Operations of Ground Forces," *PWLIWOPK*, February 1981 (trans.) in UKTRANS 139: 2.

112. Colonel A. Musial, "The Character and Importance of Air Operations in Modern Warfare," *PWLIWOPK*, March 1982 (trans.) in UKTRANS 138: 1–3. Concerning operations without air supremacy, see Konek, "They Provided Air Support," 36–37; and Colonel V. Moroz and Lt. Colonel A. Polyakov, "The Battalions Force the Laba," *KZ*, 13 September 1984: 1 (trans.) in *JPRS-UMA* 85–008 (1985): 93.

113. Lachiewicz and Rajmanski, "Helicopter Reconnaissance and Fire Missions in Support of Land Operations At Night," 4, 5 (although concerning the particular effectiveness of helicopters at night, the implication was that they would also be very effective at other times); and Major James F. Holcomb, Jr., *Recent Developments in Soviet Helicopter Tactics*, Pamphlet B34 (RMA Sandhurst: SSRC, 1988), 7.

114. Lt. General Vasily G. Reznichenko (ed.), *Tactics* (np: Moscow, 1987) (trans.) in *JPRS-UMA* 88–008-L–1 (1988): 103. Lachiewicz and Rajmanski, "Helicopter Reconnaissance and Fire Missions in Support of Land Operations at Night," 4; and Scibiorek, "Helicopters," 2.

115. Reznichenko, *Tactics* (1987), 98.

116. Colonel T. Wojcik, "Combat Vehicles and Helicopters on the Modern Battlefield, Problems in Military Thought," *Zolnierz wolnosci*, 4 January 1978: 3 (unpub. SSRC trans.).

117. Yefimov, "Soviet Military Art," 3. See also Colonel L. Konopka, "Prob-

lems of Air Force Support for Ground-Air Strike Teams," *PWLIWOPK*, April 1981 (trans.) in UKTRANS 146: 2; and Reznichenko, *Tactics* (1987), 76, 145.

118. The Mi–28 Havoc was revealed publicly in 1989: "Mil Mi–28 Attack Helicopter in Final Tests Prior to Full-Scale Production," *AWST*, 5 June 1989: 78–79.

119. For example, Everett-Heath, *Soviet Helicopters*, 151.

120. For example, compare the different techniques reported in Captain M. Stepanov, "A Maneuver Yields Success," *KZ*, 1 March 1984: 1 (trans.) in *JPRS-UMA* 84-037 (1984): 47–48; and Major V. Sekhin, "Attack from Ambush," *AIK*, April 1984: 5 (trans.) in *JPRS-UMA* 84–061 (1984): 6–8.

121. Popov, "Tactical Air Exercise: Pluses and Minuses," 16. Compare with the description of the techniques needed for night flying in Lavrenyuk, "Helicopters Attack at Night," 47.

122. " 'Havoc' Learns the Afghan Lesson," *Jane's Defence Weekly*, 17 June 1989: 1235.

123. General D. S. Sukhorukov, (no title), *V-IZ*, December 1985: 14ff (trans.) in *JPRS-UMA* 86–032 (1986): 15–16. See also Colonel O. K. Frantsev, "Our Interviews," an interview with Marshal P. S. Kirsanov, *V-IZ*, August 1985: 20ff (trans.) in *JPRS-UMA* 86–003 (1986): 24–25. For the importance of Manchuria see Reznichenko, *Tactics* (1987), 75.

124. Lt. General Vasily G. Reznichenko (ed.), *Tactics* (trans.) (Washington, DC: USAF, 1987) 85.

125. Yefimov, "Soviet Military Art," 3.

126. Michalak, "Aviation in the Raid-Manoeuvre Operations of Ground Forces," 6.

127. Tyushkevich, *The Soviet Armed Forces*, 455; and Colonel V. Denisov, "Combat Helicopters: 1977," *KZ*, 16 March 1977: 3 (trans.) in *JPRS-UMA* 1271 (1977): 22.

128. See, for example, Taylor, "Combat Helicopters of the Warsaw Pact," 113–18.

129. Colonel G. Bryukhovskiy, "Features of Employing Aviation in the Manchurian Operation," *V-IZ*, August 1979: 177ff (trans.) in *JPRS-UMA* 1472 (1979): 29. Although it was an article on the Great Patriotic War, it stated, "It is important to study this experience . . . [it] has theoretical and practical significance to air force strategy and tactics."

130. Colonel General V. Merimskiy, "BMPs in Combat," *Voyenny vestnik*, March 1976: 19ff (trans.) in UKTRANS 68: 4.

131. Wojcik, "Combat Vehicles and Helicopters on the Modern Battlefield, Problems in Military Thought," and Wojcik, "In Two Dimensions: Air Ground Co-operation," *Zolnierz wolnosci*, 5 January 1978: 8 (unpub. SSRC trans.).

132. Ibid.

133. Lt. Colonel Z. Szvaczak, "Air-Ground Synchronisation of Operations," *Zolnierz wolnosci*, 25 January 1978: 3 (unpub. SSRC trans.). See also Lt. Colonel R. Mankowski, "Rear Support of Ground Forces Aviation Acting as Part of and for the Benefit of an Operational Maneuver Group," *PWLIWOPK*, April 1985: 35ff in Harry S. Orenstein, *Selected Translations from the Polish Military Press* Vol. 2 (Ft. Leavenworth, KS: SASO, 1988) 127.

134. Colonel General N. Dagayev and Colonel V. Chernetskiy, "Development

Trends in the Organization Forms of Front and Long-Range Air Forces in the Second World War," *V-IZ*, October 1980: 18ff (trans.) in UKTRANS 220: passim.

135. Konopka, "Problems of Air Force Support for Ground-Air Strike Teams," 1–4.

136. Major General D. Folomeyev, "Predicting the Combat Situation," *AIK*, August 1982: 4ff (trans.) in *JPRS-UMA* 1714 (1982): 10. See Lt. Colonel A. Lapshin, "Group, Take Off!" *AIK*, July 1982: 20ff (trans.) in *JPRS-UMA* 1727 (1986): 30, for the importance of *Zapad–81*.

137. Lt. General V. Gorbachev, "The Employment of the Soviet Air Forces in the Initial Period of the Great Patriotic War," *V-IZ*, November 1983: 25ff (trans.) in *JPRS-UMA* 84–009 (1984): 26.

138. A number of Western sources were consulted, the most useful of which were Holcomb and Turbiville, *Exploiting the Vertical Dimension*, 37B, and Lt. Colonel Manfred Neisters, "Soviet Helicopter Doctrine," *International Defense Review* 20.10 (1987): 1324. About fifty divisions were Category One, that is, with high levels of equipment and staffing.

139. Reznichenko, *Tactics* (1987), 11; and Donnelly, *The Place of the Helicopter on the Modern Battlefield*, 58.

140. Colonel M. Igumentsev, "Support from the Air," *SMR*, December 1982: 23–24.

141. Mankowski, "Rear Support of Ground Forces Aviation Acting as Part of and for the Benefit of an Operational Maneuver Group," 128–31; and Colonel Yeliseyev, "Experience of Army Offensive Operations in Mountainous Conditions," *V-IZ*, January 1986: 16ff (trans.) in ADVAB 245: 5.

142. Belov, "Helicopters and Land Force Tactics," 24.

143. V. Vasilets and P. Studenikin, "Developing the Success," *Pravda*, 15 July 1977: 6 (trans.) in *JPRS-UMA* 1296 (1977): 63.

144. Zakharenko, "Lessons in Co-ordination," 63.

145. Lt. Colonel Yu. Gudymenko, "Supporting the Motorized Infantrymen," *AIK*, May 1978: 6ff (trans.) in *JPRS-UMA* 1364 (1978): 45.

146. Yefimov, "Soviet Military Art," 5–7; and Dagayev and Chernetskiy, "Development Trends in the Organization Forms of Front and Long-Range Air Forces in the Second World War," 4.

147. Major G. Losev, "Air Attack," *SMR*, August 1981: 19; and Lt. Colonels G. Ivanov and V. Karpov, "Firing against Abandoned Trenches: Why Helicopter Pilots Failed to Fulfill Their Mission," *KZ*, 16 April 1982: 1 (trans.) in *JPRS-UMA* 1698 (1982): 24.

148. "Forward Air Controller in Action," *KZ*, 28 March 1981 (unpub. SSRC trans.). The exercise concerned support for a forward detachment, and the comments were by the squadron and motor rifle division commanders respectively.

149. As suggested by Colonel V. I. Shurov in "From the Reader Conferences of the GSFG," *V-IZ*, October 1983: 83ff (trans.) in *JPRS-UMA* 84–003 (1984): 61.

150. For example, Neisters, "Soviet Helicopter Doctrine," and Hussain et al., *The Future of the Military Helicopter*, 60. Compare these to the more informed Erickson et al., *Soviet Ground Forces*, 181; and Scott and Scott, *Soviet Military Doctrine*, 187.

151. Colonel V. Pismennyy, "Discrepancy in Guidance Documents Causes

Lack of Coordination between Army Aviation and Motorized Rifle Troops," *KZ*, 2 September 1989: 2 (trans.) in *JPRS-UMA* 89–023 (1989): 8–9; and Major A. Zhilin, "Contracts in Co-ordination," *AIK*, October 1985: 6ff (trans.) in *JPRS-UMA* 86–011 (1986): 11–12. For the artillery, see Chris Bellamy, *Red God of War* (London: Brassey's, 1986) 207–209.

152. Captain V. Lavrenyuk "Officer Training. A Look at the Problem. If there was a desire . . . ," *KZ*, 10 January 1989: 1 (trans.) in *JPRS-UMA* 89–006, (1989): 25–26.

153. Major Robert E. Bort, "Air Assault Brigades: New Element in the Soviet Desant Force Structure," *Military Review* 63.10 (1983): 21–38, was the best analysis and used by a number of other commentators. Much of the information on air assault brigades appears to have come from intelligence sources that are yet to be declassified.

154. David C. Isby, "Soviet Airmobile and Air Assault Brigades," *Jane's Defence Weekly*, 14 September 1985: 561–65. "Suvorov," *Inside the Soviet Army*, 74–75; and, following him, M. Galeotti, "*DShB*—Assault Landing Brigades from Afghanistan to 'Defensive Defense,' " *Jane's Soviet Intelligence Review* 2.9 (1990): 386, are good examples of superficial analyses that confuse the two units.

155. Major A. Drozdov, "Helicopters Conduct an Airborne Assault Landing," *KZ*, 18 December 1980: 1 (trans.) in *JPRS-UMA* 1575 (1981): 43–44. Colonel M. Belov, "Air Landing Forces," *SMR*, January 1979: 23, made an unusually specific reference to the "mass introduction" of helicopters to transport troops and heavy equipment in 1980.

156. Bort, "Air Assault Brigades: New Element in the Soviet Desant Force Structure," 31.

157. Lt. Colonel V. Shevchenko and Major N. Panyukov, "Battle for the Bridgehead, from the Dniester Exercise Area," *KZ*, 10 September 1983: 1 (trans.) in *JPRS-UMA* 84–002 (1984): 63.

158. For a discussion of some of these problems see Belov, "Air Landing Forces," 23. While this article was supposedly based on foreign sources, it was so very similar in most respects to detailed Soviet reports of air landings and so unlike Western concepts that its "foreign press" disclaimer can be safely disregarded.

159. Izgarchev, "With a Landing Force on Board," 48; and V. Zhitarenko, "Combat Study of the Army and Navy," *KZ*, 9 August 1979: 1 (trans.) in *JPRS-UMA* 1495 (1980): 14–15.

160. Belov, "Air Landing Forces," 23.

161. Noted by Holcomb and Turbiville, *Exploiting the Vertical Dimension*, 19–20.

162. Quoted in Ian Kemp, "Abdul Haq: Soviet Mistakes in Afghanistan," *Jane's Defence Weekly*, 5 March 1988: 381.

163. Lebedev, "In the Lenin Regiment," 12–13; and Graham H. Turbiville, Jr., and C. G. Pritchard, "Soviet Airborne Assault," *Marine Corps Gazette* 71.10 (1987): 52–53.

164. Belov, "Air Landing Forces," 23. For the Ogaden War, see Everett-Heath, *Soviet Helicopters*, 86.

165. Belov, (no title), 3.

166. Belov, "How to Fight Helicopters," 18; and Colonel B. Nesterov, "The

Quest for New Techniques," *AIK*, June 1983: 11 (trans.) in *JPRS-UMA* 1798 (1983): 29–30.

167. Lieutenant N. Zelenskiy, "In the Screening Group," *KZ*, 24 January 1984: 1 (trans.) in *JPRS-UMA* 84–029 (1984): 61; and Major General M. Fesenko, "Helicopter against Helicopter," *AIK*, March 1984: 46ff (trans.) in *JPRS-UMA* 84–046 (1984): 71–76.

168. Major Yu. Grechukhin, "In Aerial Combat—Helicopters," *KZ*, 7 July 1987: 1 (trans.) in ADVAB 310. See also Holcomb, *Recent Developments in Soviet Helicopter Tactics*, 18–19, citing articles in *Voyenny vestnik* and *KZ*.

169. *Soviet Military Power* (Washington, DC: US Department of Defense, 1985), 65.

170. William Dickinson, "Fighter Helicopter Needs," *Defence Helicopter World*, 5.4–5 (1986): 16 (the author was a member of Congress from Alabama, the state in which U.S. Army Aviation has its main base, and thus a fervent supporter of more funding for Army Aviation); and Edward J. Bavaro, "Soviet Helicopter Air-to-Air," *Aviation Digest* 34.3 (1988): 2–7.

171. H. Dodds, " 'Hokum' Takes to the Air," *Jane's Defence Weekly*, 28 September 1991: 551. See also Mark Lambert, "First Havoc, now Hokum—what Role for the New Soviet Helicopter?" *Interavia*, July 1985: 799.

172. For one example, see Isby, *Weapons and Tactics of the Soviet Army*, 431.

173. This was generally the case: M. J. Orr, *The Soviet Army's Image in Afghanistan*, Pamphlet P4 (RMA Sandhurst: SSRC, 1985) 3–4.

174. Mark Urban, *War in Afghanistan* (London: Macmillan, 1988) 230. See also A. Karp, "Blowpipes and Stingers in Afghanistan: One Year Later On," *Armed Forces Journal International*, September 1987: 40.

175. Urban, *War in Afghanistan*, 66.

176. Colonel Ye. Besschetnov, "Our Marks Will Remain . . . ," *AIK*, October 1986: 12ff (trans.) in *JPRS-UMA* 87–017 (1987): 21.

177. Various articles in Soviet journals were used, particularly Lt. Colonel Yu. Kryukov, "Final Exam," *AIK*, June 1986: 15ff (trans.) in *JPRS-UMA* 86–053 (1986): 20–24; and Colonel Ye. Besschetnov, "A Friend Will Lend a Helping Hand," *AIK*, August 1986: 8ff (trans.) in *JPRS-UMA* 87–007 (1987): 15–21.

178. " 'Havoc' Learns the Afghan Lesson," 1235.

179. Compare Gilles Bertin, "Stingers Change the Face of War in Afghanistan," *Jane's Defence Weekly*, 10 October 1987: 785, with Karp, "Blowpipes and Stingers in Afghanistan: One Year Later On," 36–40.

180. Mark Urban, "Soviet Operations in Afghanistan—Some Conclusions," *Jane's Soviet Intelligence Review* 2.9 (1990): 367.

181. For example, General Y. Maximov, "Mountain Training," *SMR*, December 1984: 5–7.

182. See, for example, the rather extreme conclusion to an otherwise excellent analysis of this issue by Y. Bodansky, "Afghanistan: The Soviet Air War," *Defense and Foreign Affairs*, September 1985: 14.

183. Orr, *The Soviet Army's Image in Afghanistan*, 4, made a similar point in regard to lessons about small-unit leadership and command.

184. Bodansky, "Afghanistan: The Soviet Air War," 13; and Urban, "Soviet Operations in Afghanistan—Some Conclusions," 369–70 both point to the importance of the Afghanistan experience within the officer corps.

185. Reznichenko, *Tactics* (1987), 75.

186. Ibid.

187. Major General Ivan N. Vorobyev, "Tactics Today," *SMR*, March 1985: 11.

188. Colonel S. Koziej, "Anticipated Directions for Change in Tactics of Ground Troops," *Przeglad wojsk ladowych*, September 1986: 5ff (unpub. SASO trans.).

189. Scibiorek, "Helicopters," 1.

190. Reznichenko, *Tactics* (1987), 98.

191. Ibid., 11, 103.

192. Koziej, "Anticipated Directions for Change in Tactics of Ground Troops," 6.

193. Colonel V. Morov, "Mi–28 in the Sky," *KZ*, 1 June 1989: 4 (trans.) in *JPRS-UMA* 84–015 (1984): 31. See also J. R. Wilson, "The JDW Interview," an Interview with General V. Pavlov (Commander, Soviet Army Aviation), *Jane's Defence Weekly*, 22 June 1991: 1156.

3

"Double Trouble"—Developments in the United Kingdom

> Ignorance and prejudice have combined, notoriously in the British services, to delay the extraordinary value [of the helicopter].
>
> *Anthony Farrar-Hockley, 1988.*[1]

> Though manoeuvre is the more difficult art to acquire . . . it provides for greater flexibility. . . . Manoeuvre is not confined to ground based forces. It has an air dimension in which helicopters play an increasing part.
>
> *British army doctrine, 1989.*[2]

The development of military helicopters by the British Army and the Royal Air Force (RAF) can be divided into two periods: the first began in the last years of the Second World War and concluded with the withdrawal from East of Suez in 1966–67, thereby initiating the second period, which continues to the present day. Prior to the mid–1960s, and while the United States and USSR were concerned with nuclear war-fighting, Britain's use of helicopters was focused almost entirely on counterinsurgency operations. But, following the end of most of Britain's colonial conflicts, the army set about introducing and then improving its Army Air Corps (AAC) anti-tank helicopter force. The corps, which had been created to help in counterinsurgency campaigns, meant a division between it and the RAF of the responsibility for providing helicopter support to the land battle. There was considerable opposition at first to anti-tank helicopters from those who believed them to be too vulnerable and too expensive. Some members of the air force also criticized army helicopters because they feared that funds would be diverted from procurement of RAF aircraft. As a result there was a long running political and organizational battle that not only hindered the first deployment of anti-tank helicopter units but also threatened the future expansion of this force. By the mid–1980s, however, such doubts had passed and many promising advances in both doctrine and equipment were being proposed. In particular, the role played by helicopters was changing so that instead of supporting the armor and infantry arms, the AAC would have equal status with them. And, instead of being a reserve force, anti-tank helicopters were starting to be seen as playing a part in all phases of battle. Although by 1991, the AAC was still to finalize these important developments, the coming reduction in the size of the British army will leave the AAC all but intact, thereby increasing significantly the percentage of personnel and resources associated with helicopters. The new strategic situation that confronts Britain after the cold war and the reduction in forces throughout NATO also seems to promise a bright future for helicopters.

Airmobile infantry will also benefit from this new scenario. Developments in this area had lagged well behind anti-tank doctrine. Helicopters were used for the aerial transportation of small numbers of troops in low-intensity campaigns in the 1960s, but attempts to produce a more innovative theory of airmobility for use in conventional warfare had foundered. In the 1970s the army had been concerned mainly with mechanized warfare and had shown little inclination for light, airmobile infantry; financial constraints and military conservatism defeated airmobility. Some of this skepticism remained, even after a successful four-year trial by an experimental airmobile brigade in the mid–1980s and the formation of a permanent unit in 1989, but it seems to be dissipating as the airmobile brigade overcomes the difficulties that had attended its birth.

Despite the intellectual advances in helicopter doctrine, acceptance and implementation of these ideas were for many years greatly restricted in the British services. The primary cause was financial, with a series of economic crises leading to reduced budgets for the armed forces. However, inextricably linked with the financial factor was the difficulty of overcoming the established hierarchy of spending priorities within the services. In both the army and RAF, helicopters were usually well down the list of requirements. This low priority resulted from the institutional and political strength of established groups and their general disinterest in helicopters. Only when helicopters were supported by these powerful organizations, first for armed action and then for airmobility, were helicopter proponents successful in gaining the resources required to implement their theories. Yet, in the future the support of outsiders may no longer be necessary. It is a sign of the success of the AAC and RAF helicopter units in embedding themselves completely within the accepted and most highly valued practices of the armed forces that the recent cutbacks did not greatly affect them.

Because of the distinct paths along which airmobility and the use of helicopters for anti-tank combat developed, this chapter, after beginning with a brief introduction of British helicopters between 1945 and 1967, discusses each topic in turn, tracing its history from the late 1960s to the present. This approach reflects the division of helicopter ownership between the two services, with the RAF being concerned with airmobility and the AAC with anti-tank combat. It also helps to emphasize that, unlike some other countries, no all-encompassing theory for helicopter employment was established in Britain.

FROM 1945 TO 1967

Remnants of Empire

The employment of military helicopters by the British began with the delivery and evaluation of seven Sikorsky R–4s (known to the RAF as the Hoverfly Mk. 1) in 1943. Britain also observed the use of an American R–4 to provide casualty evacuation in Burma in 1944–45 and, immediately after the Second World War, began conducting trials on how helicopters might be used in a future European war. But Britain was soon engaged in the first of many counterinsurgency campaigns throughout the remnants of its empire and, although these wars gave many examples of the value of military helicopters and provided much of the impetus for their development, they limited the resources available for the trials and drew attention away from conventional warfare preparations.

Helicopters first saw active service in 1950 when three Westland WS–51 Dragonflies were dispatched to Malaya to perform casualty evacuation. Despite the limited numbers, they proved remarkably efficient. The ability of the helicopter to hover made it indispensable in the jungle environment and, seeking to expand the number of helicopters under his command, the Director of Operations, General Sir Gerald Templer, told his superiors that helicopters were "essential to success" in the Malayan Emergency.[3] However, delays in the development and production of helicopters restricted this expansion for some years, and no more than thirty-two helicopters were ever deployed in Malaya. The British were also hampered by the poor performance of these early models. For example, the first helicopter bought to fulfill the troop lift mission could, in the Malayan climate, only carry a single soldier and caused the RAF theater commander to inform London that "your wretched Whirlwind is a complete washout."[4] Nevertheless, helicopters played a vital role in the fighting, being used for diverse missions—from crowd control to herbicide spraying—that took advantage of their ability to fly vertically. Troop and supply lifts were the most important task, with over 110,000 troops and 2,500,000 pounds of freight being transported between 1952 and 1960. One participant summed up their value by saying, "We could never have cleared the guerrillas from the deep jungle without helicopters."[5]

The Suez Canal crisis in 1956 also provided the British armed forces with an opportunity to employ military helicopters; twenty-two of them were used to fly troops and supplies ashore during the attack on Port Said. This first amphibious helicopter-borne assault was, essentially, experimental and did not materially affect the outcome of the operation. Indeed, by demonstrating the value of military helicopters it had more impact in Whitehall than in Egypt. The Royal Navy and the Royal Marines jointly established a permanent force of troops and helicopters for amphibious warfare, based aboard "Commando Carriers," which could then deploy rapidly to the world's trouble spots[6]—such as the next conflict, the Confrontation with Indonesia from 1962 to 1966.

Helicopters played an even greater part in this counterinsurgency campaign partly because of the rapid pace at which the war was conducted,[7] but mainly because of the greater number and better quality of helicopters available to the British, including the Whirlwind Mk. 10, with its more powerful turbine engine, and the very large Bristol Type 192 Belvedere. The recently formed AAC was also involved using its smaller Westland (ex-Saunders-Roe) P.531 Scout and Bell 47G–3B–1 Sioux. Eventually, over seventy helicopters were deployed in Borneo.[8] Centralized helicopter deployment, which had worked effectively in Malaya because of the extremely limited numbers of available helicopters, was the RAF's preferred method of operation in Borneo. However, the theater commander, General Walter Walker, overruled the RAF. He required its helicopters to operate in proximity to the soldiers in the field like those of the AAC, which were decentralized to battalion and brigade levels. This change provoked some tension between the services. But something that was rather forgotten in the heated, interservice debates of the early 1970s was that RAF officers were not personally to blame for the inadequacies of centralized control, but their system was inappropriate. With the new method, not only were valuable flying hours saved by closer deployment to the fighting, but the vital element of rapid reaction was also preserved: the helicopter gave Walker the "speed, mobility and flexibility" with which he conducted a highly effective campaign, despite the small numbers of British troops under his command and the very large area of operations.[9]

Although casualty evacuation, liaison and command flying, observation, and logistic support tasks were also carried out, especially by the army's light helicopters, troop and equipment lift remained the most important mission. For example, in 1963 Indonesian raiding parties were defeated as they retreated towards the safety of the border by using helicopters to land troops behind them to carry out an ambush. Equally, helicopters were essential for transporting surveillance radars and 105mm guns to otherwise inaccessible mountaintop sites deep in the jungle. General Walker concluded that the helicopter "proved itself over and over again to be the real battle winner in Borneo. . . . A clever company commander with a few choppers to hand could so plague guerrillas at every turn that they would think that the whole Army was on their heels."[10]

There were considerable technical problems to be overcome by the fledgling British helicopter fleet in its overseas commitments, not least because low production meant that numbers were always limited. Moreover, the harsh conditions East of Suez meant low serviceability, which was exacerbated by inappropriate or immature designs and a shortage of spare parts. Yet, in spite of these problems, helicopters had a fundamental impact on Britain's numerous decolonization conflicts. Above all, they provided mobility for troops and equipment, which allowed the British, with only limited forces, to overcome the difficult terrain in which they operated. In Asia helicopters were used to "defeat the jungle" as much as the enemy, and similarly they were used to negate the soaring mountains in other areas to which they were deployed—Kenya, the Middle East and Cyprus.[11] These overseas activities hampered practical progress with helicopters in Britain and the British Army of the Rhine (BAOR),[12] but did serve as an example of the importance of military helicopters and helped nonaviators to grow familiar with the use of helicopters.[13]

Helicopters and the Formation of Army Aviation

Prior to Britain's involvement overseas, trials had been conducted with Sikorsky R–6s (called the Hoverfly Mk. 2) to test the ability of helicopters to perform liaison flying and air observation duties, then performed by the RAF's army cooperation squadrons. By 1947 it had been demonstrated that helicopters had the potential to replace the fixed-wing aircraft currently used. Attempts to purchase helicopters for these missions were hampered by technical problems with the Cierva (later Saunders-Roe) W–14 Skeeter, but the army persuaded the RAF to continue. In 1952 an operational requirement was issued for the Skeeter and, by 1957, this helicopter had begun to enter service. However, its low engine power restricted it to the colder climate of Europe and, East of Suez, liaison and observation continued to be performed by fixed-wing aircraft.[14]

The introduction of the helicopter to meet the army's aviation needs

Figure 3.1
Army Perceptions of RAF Intransigence

Source: Brig. R. A. Norman-Walker, "The Formation of the AAC–1955 to 1957," *Army Air Corps Journal* (1982): 13.

coincided with, and was partly responsible for, the establishment of the AAC. Formed in 1957 "after prolonged haggling in Whitehall," the AAC was set up by the army because it believed the RAF was reluctant to procure helicopters in sufficient numbers for the army's needs.[15] With hindsight it is hard to see what more the RAF might have done, since funds were limited and, in any case, the manufacturing capacity did not exist to build the extra helicopters. However, at the time, the RAF was perceived as uncooperative (see the Army cartoon reproduced in Figure 3.1) and it did little to convince the army otherwise. Since the RAF owned all aircraft, its attitude was seen as the major obstacle to fulfilling the army's aviation requirement and one that could only be overcome by forming a separate aviation force controlled by the army.[16] Of course, the army did not endear itself to the RAF by this approach; certainly, it distracted and hindered those air force officers who were themselves struggling to overcome the resistance of their more conservative senior officers.

Although the RAF was not exactly eager to surrender ownership of light aircraft to the army, it was forced by Minister of Defence Duncan Sandys' defense review to look for cost savings while still protecting what it thought to be its key responsibilities.[17] Light fixed-wing aircraft were not particularly expensive, but there was the prospect of having to bear the burden of very costly helicopter acquisitions such as the Skeeter. The RAF preferred to concentrate on strategic air warfare and thus was amenable to the army taking over ownership of light aviation assets for unarmed liaison and observation missions. In an attempt to restrict any army encroachment into what the RAF thought was its exclusive realm of air power, an upper limit on aircraft weight of four thousand pounds was imposed and the army's aircraft were not allowed to be armed. The army and RAF soon agreed to waive the weight limitation and the armament restriction was removed with the procurement of anti-tank helicopters. Yet the basic division of the British helicopter force remained, with light aircraft owned by the army performing utility and observation tasks and heavier aircraft owned by the RAF providing major troop-lift and logistic support.[18] This division of responsibility was to have an important impact on the character of future developments.

The AAC made considerable progress in the decade or so following its creation, fulfilling observation, reconnaissance, liaison and general utility missions, especially when the underpowered Skeeters were replaced by the more modern and efficient Scout and Sioux helicopters. A plan known as the "Integration Scheme" began in 1963 to expand the number of helicopters in the army by forming smaller, more decentralized aviation units at battalion level. It was designed to thwart a takeover bid by the RAF, which was seeking to reverse its earlier decision on helicopter ownership, by making the dispersed army aircraft units too difficult to transfer back to the air force.[19] The personnel for these units came from other corps, notably the Royal Armoured Corps (RAC), and resulted in the AAC becoming just one part of what was termed Army Aviation. Perhaps the army had learned its tactics of dispersion from the elusive guerrillas of Southeast Asia.

With a reduction in central management of training and flying techniques, "Integration" caused an alarming increase in accidents and posed considerable maintenance difficulties because of the dispersion of technical ground staff and the difficulties in supplying spare parts to widely separated small units. Thus, the process was reversed in 1968, with all assets reverting to the AAC by 1973 except a small number of RAC helicopters, which returned in 1975. However, "Integration" was successful in that it had helped the army to better recognize the value of its own aviation force; it had protected the army's ownership of aircraft and had allowed Army Aviation to expand in size.[20] By 1973 the AAC had also gained the right to recruit and train ground staff directly into the Corps and to expand the number of officers permanently assigned to it. While the AAC remained a small part of the army, these measures undoubtedly improved its institutional strength. Its increased size also prepared it for the most important phase of helicopter development: the acquisition of an anti-tank capability.

THE DEVELOPMENT OF THE ANTI-TANK HELICOPTER

The Armed Helicopter Emerges

Armed helicopters first developed in response to the needs of Britain's counterinsurgency campaigns. Trials had been conducted by the Joint Experimental Helicopter Unit (JEHU—see below) in preparation for its deployment to Cyprus in 1958 but had been hampered by the lack of official approval and suitable equipment. By 1963–64, however, helicopters were regularly—if rarely successfully—being fired on in Borneo and the Near East, and so all three services began to fit their helicopters with machine guns for defensive fire. But the RAF officially continued to oppose helicopters with offensive armament, despite some experiments it carried out with SS–11 missiles on Whirlwind helicopters.[21] The RAF's decision was based partly on the shortage of helicopters for existing, unarmed tasks, partly on its view that helicopters were unsuitable for armed action, and partly because the RAF was worried that the army might take on the mission with its own helicopters and thus usurp what was seen as a purely air force responsibility.

The army was frustrated because RAF ground attack aircraft were unable to respond sufficiently quickly to deal with fleeting guerrilla targets. It was suggested that AAC helicopters ought to be used for offensive armed action, and a proposal was made that some of the new Westland-Aérospatiale WG–13 Lynx being developed for the 1970s be equipped with a turret-mounted cannon. The army's position was indirectly strengthened by what the official RAF history revealingly called "a highly embarrassing" incident in Borneo, in which successful—but unauthorized—fire support was given to an infantry attack by Royal Navy Westland WS–58 Wessex helicopters, armed with machine guns and SS–11 missiles.[22] However, although the army through Major General Napier Crookenden, the Director of Land/Air Warfare (DLAW), kept up the pressure for offensively armed helicopters and the RAF agreed to technical trials and experiments of helicopter weaponry, the air force fended off any immediate policy decisions by proposing a broad review of close air support aircraft including helicopters and the Hawker P.1127, as the Harrier jump-jet was then known.[23]

The Anti-tank Helicopter Evolves

This review coincided with the end of most counterinsurgency operations in 1968, and thus the army was no longer able to use these campaigns to justify an armed action mission for its helicopters. Yet it remained interested in armed helicopters and instead argued that they be used for anti-tank combat, an idea that had first been proposed in the mid–1950s but now took on new significance as Britain concentrated on its NATO commitment.[24] While the RAF remained skeptical about anti-tank helicopters, it was more willing to accede to the army's demands. The specific purpose of such helicopters meant they posed less of a threat to the RAF's monopoly on general close air support. The requirement for anti-tank helicopters sprang from the combination of a vast preponderance of enemy armored formations, the shortage of friendly forces and the need to disperse units because of the threat of nuclear attack. In this situation, as Major General Frank King (Crookenden's successor as DLAW) concluded, "penetration [of the front line] will be the rule rather than the exception."[25] It was believed that one of the main established methods of countering such penetrations—close air support by the RAF—would no longer be readily available; many in the army—and occasionally in the air force as well—believed that the RAF did not recognize the importance of this mission and did not provide sufficient aircraft for it. These fears were exacerbated by the impending withdrawal from service of the Hawker Hunter ground attack aircraft, even though a replacement for the Hunter (the Harrier) was being arranged.[26] Thus, because the new helicopter and the Harrier were in competition, army claims about the lack of air support seemed to be driven by ulterior motives concerned solely with weapons procurement.

But misperception occurred on both sides and was exacerbated by the tight financial situation. As far as the army was concerned, the reasons it had wanted armed helicopters in Borneo were equally compelling in Europe. Even if RAF close air support was available it could not respond quickly enough, nor with sufficient accuracy, nor in effective support of small army units. Senior RAF officers protested that close air support ought to be provided by RAF fixed-wing aircraft since these assets could also be used in the counterair battle when necessary. This attitude simply reinforced the perception that the RAF was jealously conserving its traditional roles and was prepared to ignore the army's needs on the battlefield.[27] Therefore, facing a considerable Soviet threat, the army decided that it ought to have its own close air support in the form of a highly mobile and responsive reserve of anti-tank helicopters, armed with accurate anti-tank guided weapons (ATGW).[28]

Institutional politics also played their part within the Army. The AAC was a new and small organization in the late 1960s, and helicopters were still a very minor part of the army's activities. Although the integration scheme had helped to increase the number of aircraft in the army, it had weakened the AAC, since it was no longer solely responsible for aviation support. Now, amid rumors that it would be disbanded, the AAC was struggling to establish an identity.[29] The best means for the AAC to ensure its continued existence was to propose for itself an important mission, such as anti-tank action, and argue that it could not be done as well by any other branch of the army or by the RAF.

The most serious threat to the AAC's existence was posed by the RAC. Kenneth Macksey noted that "senior RAC officers, anxious to save regiments from abolition due to the rundown of the Army, were proposing that the RAC should absorb the AAC."[30] As part of this plan and because it controlled the army's ATGW at that time, the RAC took a great interest in the development of anti-tank helicopters, helping the AAC with its trials and taking a keen interest in the results. However, because anti-tank helicopters conflicted with armored warfare—its main interest—the RAC eventually decided that it ought not to assume responsibility for them. Thus, although its very existence had been threatened by the RAC's plans, the AAC owed a good deal of its success to the other corps.

The RAC's lack of interest in helicopters caused some resentment among enthusiasts who were more concerned for the helicopter than for the AAC and wished for the political support of one of the more powerful army branches.[31] Crookenden believed that "the Royal Armoured Corps is not paying sufficient attention to the development of the armed helicopter," to which a senior RAC officer replied, "I assume that they are not doing much about it because there is a well-established branch of the Army [i.e., the AAC] which is supposed to be looking after this for them."[32] Similarly, the biting criticism of the RAC by other helicopter advocates, such as Colonels John Moss and John Waddy, was founded on the view that the helicopter would have developed more quickly but for the RAC's decision in the late 1960s to relinquish interest in anti-tank helicopters. Moss and Waddy were willing for other corps, such as the RAC, to be more involved—perhaps at the AAC's expense—because only they could have provided the additional personnel and money required to expand the helicopter force.[33] This exemplified the complex mix of financial and organizational factors that affected such innovative developments.

However, the RAC was not necessarily better placed to promote the anti-tank helicopter. Although it was more established and influential than the AAC, the latter corps' small size and youth meant that its members tended to pursue innovation with greater enthusiasm. Usually members of the AAC had come from other corps because of an interest in aviation and most were helicopter advocates, whereas only a few in the RAC fit that category. Anti-tank helicopters were the AAC's main responsibility, while the RAC naturally regarded them as supplementary to its armored vehicles. Thus, much as the vigor of the early armored theorists had helped to establish the place of the tank, the organizational dynamism of the AAC aided the early development of anti-tank helicopters. The only disadvantage was that, along with their enthusiasm, some advocates displayed insufficient common sense and tact by suggesting that the helicopter would soon replace the tank, thereby increasing the opposition that they wished to overcome.[34]

The deployment in the late 1950s of the first ATGWs was also very important. As far as the British were concerned, a guided missile was the only weapon of the time capable of being fired accurately and effectively from a helicopter. The long range of missiles such as the SS–11 was vital if helicopters were to stand off from the target and thus protect themselves. Finally, the helicopter seemed to be a very good vehicle on which to mount these new weapons because of its speed and flexibility. Thus, not only did ATGW provide an opportunity for armed helicopters, but they also made helicopters appear necessary if the new missiles were to be fully exploited.[35]

Undoubtedly pressures within Britain were the main factors behind the development of the armed helicopter, but observation of other countries' developments also played a part. Britain's initial interest in armed helicopters had been fueled by the experiments conducted in France and the United States;[36] events in Vietnam were also closely watched. Although comments were made in some articles on the war such as "the value of the armed helicopter in Vietnam has been quite startling in its impact,"[37] the conflict did not provide real proof with which to support or attack the value of helicopters. First, the war was a Southeast Asian counterinsurgency conflict and thus unrepresentative of a potential war in Europe. Second, the evidence—if it can be called that—from the war was so unclear that it provided arguments both for and against helicopters. For example, while figures on helicopter losses could be quoted to support armed helicopters, they could also be used to damn them.[38]

Trials, proposals and arguments continued throughout the 1960s. There was some thought at first that helicopters could simply carry ATGW infantry teams around the battlefield. Yet it was soon realized that the helicopters themselves were a suitable vehicle from which to launch the missiles and, by 1968, successful firing trials of the SS–11 had been conducted from the Scout helicopter. This method was better because it did not require the coordination of helicopters and infantry, reduced the number of personnel needed per missile and rendered the missiles more mobile. More importantly, the direct fire mission made the AAC more likely to survive and prosper in the army. If anti-tank helicopters were simply to be transports, then logically they belonged with all the other transport helicopters in the RAF; even if the helicopters remained in the army, they would be subordinate, not to the AAC, but to the corps providing the anti-tank missile troops.[39] There was also the excitement and prestige of being a direct fire arm. Although the Director of Aviation, General Dare Wilson, wrote in 1971 that "observation and reconnaissance remain the backbone of army aviation's contribution on the battlefield," many members of the AAC preferred to shoot directly at the enemy rather than just help the artillery do it.[40]

Because the early trials were either concerned with proving the technical ability of helicopters to fire ATGW or were rudimentary and unscientific, they gave little support to the army's efforts to establish the tactical validity of its aerial reserve anti-tank force. Thus, a series of extensive and apparently rigorous trials were conducted in 1966–67, going under the name of Helltank. From this title it might be presumed they were meant to demonstrate that things were going to get very hot for tanks on the future battlefield; in fact the trial did not produce convincing evidence but only the opportunity for more argument: "*Whatever the conclusions*, it is felt that Helltank has had a big impact on all involved. Much new thinking has already begun—*possibly* the armed helicopter has come to stay [emphasis added]."[41] Perhaps, however, this view reflected the potential for continued opposition to anti-tank helicopters despite Helltank rather than misgivings about the trial itself. These trials were in fact a turning point for anti-tank helicopters—even if they only altered people's perceptions and left an impression that "the anti-tank guided weapon helicopter . . . was a 'good thing.' "[42] Therefore, in 1969 it was decided to equip twenty-four Scouts with SS–11 missiles, prompting the editors of the *Journal of Army Aviation* to write that "the long awaited ATGW system is no longer a dream of forward looking Army Aviators, but a reality."[43]

Consolidation of the Anti-tank Helicopter Force

Although the lead article for the same journal in 1973 stated that "aviation's counter armour role has been fully endorsed and a greater number of ATGW helicopters will be seen in the future,"[44] considerable opposition remained and "there [were] still doubters" outside the AAC.[45] First, there were practical concerns because of the very small number of anti-tank helicopters and the unsuitability of helicopter and missile. At least some opposition stemmed from the fact that the AAC was seeking a more active role within the army, which made other, established arms wary about endorsing a potential drain on their own resources.[46] To counter these critics the corps pursued a deliberate policy of "spreading its influence" in an attempt to ensure that the helicopter became well recognized and accepted.[47] As in America, technical and doctrinal innovation had to be supported by a "selling campaign."

The dominant theme in opposition to helicopters was that they were too vulnerable and, given the somewhat unlikely appearance of helicopters in the early days, there was some justification for this view. Yet the

Photo 3.1
Scout AH Mk. 1

The AAC's first anti-tank helicopter. The need for vision and courage to support its introduction can be seen if it is compared to the far more effective-looking Lynx. (Westland Group plc. Used by permission.)

debate on vulnerability was usually conducted in rather unrealistic isolation. For example, many critics put forward the view that helicopters would become easy prey for fixed-wing aircraft when the enemy had air superiority. For example, "The stock reply of the skeptics is invariably: 'What happens if we don't have air superiority?' " However, "The short answer to that question is 'What happens to the whole of NATO's ground forces?' "[48] Moreover, outlandish assertions such as, "[Helicopters] are vulnerable to anything," indicated that, for the moment, emotion was the basis for this opposition—a view confirmed by the fact that the same RAF officer wrote an article some years later that strongly supported helicopters.[49] Helicopter proponents answered these "pundits, [who were] full of their own wisdom and a noticeable lack of practical experience" by stressing that helicopters could survive in the same manner as any other fighting system, be it tank, aircraft or infantryman—by using the correct tactics.[50]

The opposition went beyond survivability. Some argued that transport rather than armed attack was a more suitable mission for helicopters; for helicopter pilots within the RAF this was a particular concern, for spending on armed helicopters may have led to a diminution in their own role. Moreover, no matter what the AAC might claim, armed helicopters were more properly regarded as a part of air power for which the RAF was solely responsible. Helicopter advocates replied that the RAF saw helicopters as part of the air power spectrum because they were looking at them "through light blue tinted spectacles" when they were in fact part of the land battle.[51] Although there was some historical justification for the comparison of rotary- and fixed-wing aircraft, because of the initial army interest in having its own close air support, the two were performing very different missions.

Without unanimous approval in the armed forces for helicopters, the biggest threat came from the increasingly desperate financial situation of the armed forces. Although the introduction of anti-tank helicopters would appear to have been settled in 1969, their future remained open in the mid–1970s because of the reductions in defense spending initiated by the defense review of 1974.[52] In the ensuing cost cutting, the number of Lynx being ordered to replace the Scout was reduced from 162 to 100 and, although it was said that "the Army continues to place a very high priority on the helicopter-borne ATGW system," the TOW missile, chosen in 1976 to arm the Lynx, was not procured until after 1981.[53]

Cost effectiveness was a key issue. One opponent argued that "defence-cuts are leaving us with precious little money to buy the weapons we need. Let us make sure we are putting it on the right 'horse.' "[54] Some in the RAF argued even more forcefully along these lines, claiming that Britain's defensive posture in NATO meant reacting quickly to widely spaced Warsaw Pact attacks. Although conceding that helicopters were not completely worthless, they thought in these circumstances that offensive armed action by fixed-wing aircraft would be able to perform missions over a large area and therefore would be more cost effective.[55] The RAF was beginning to suggest that close air support was less effective than air interdiction missions deeper into the enemy's rear and that there were insufficient resources for both helicopters and ground attack aircraft. However, these arguments, which were essentially concerned with securing the procurement of the Harrier for the RAF at the expense of armed helicopters, failed because of the growing recognition that the two fulfilled different functions; nor was the army prepared to give up its form of close air support (helicopters) when the RAF appeared reluctant to perform the mission with the Harrier.

The AAC had not even entered this time of hardship with secure financial resources. The Lynx had been chosen to replace the Scout more for cost-saving reasons than for any inherent suitability. It was a part of the politically and financially motivated Anglo-French helicopter package and, to save money by increasing the production run, the army's Lynx had been derived from the naval version. Small enough to fit onto the stern of a frigate, the Lynx was therefore limited in the number of men it could carry; yet it was also too big for the light utility role for which it was first intended.[56] It was only good fortune that the Lynx's power and performance were eventually put to good use in the anti-tank mission. Spending cuts also affected the AAC's other new helicopter, the Aérospatiale SA.341 Gazelle, which was not given "a sight that matches that of the Lynx/TOW system,"[57] thus hampering effective cooperation between the two. It was not until 1984, after many years of argument, that funds were finally made available and a suitable sight procured.

During the debate on anti-tank helicopters, considerable attention was paid to the rival merits of purpose-built attack helicopters as opposed to utility helicopters with "button-on" armament. The RAF was opposed to the former, claiming that such helicopters would not be as flexible. However, an unspoken but equally important desire was to limit army involvement in what the RAF saw as its own task—specialized aerial fire support. Some army officers—even within the AAC—felt the same way, also wishing to maintain the utility of the small army helicopter fleet, especially since dedicated anti-tank helicopters were of little use in Northern Ireland and the other colonial commitments that remained.[58] Once again, cost was an issue: "[There] are powerful reasons for advocating specialized machines. However they are expensive . . . [and] few can afford them in large numbers. Meanwhile the 'button-on' type still offers real value as a substitute."[59]

The period also saw the first developments in anti-tank tactics for helicopters. Prior to Helltank, and despite early work that suggested that the best method was to fire while stationary and fly at very low altitudes,[60] an idea had developed that "the helicopter would need to keep moving to avoid presenting an easy target." Yet, as Macksey recorded, "Because of their movement and noise, they [the helicopters] were nearly always detected and often engaged."[61] These early tactics may have reflected some notion that helicopters were a replacement for RAF close air support: during Helltank, the anti-tank helicopters conducted "penetration" missions, which included flying over enemy positions, much as fixed-wing close air support aircraft might have done.[62] However, by 1975 it had been established that the most effective tactic was "engaging [the enemy] unseen while hovering in stationary fire positions behind our own forward defences."[63] In that year AAC anti-tank helicopter tactical doctrine was codified and published. The six Scout helicopters in each divisional squadron were to be formed into two teams of three. Their actions, based above all on self-survival, called for "concealment" and attacking "at maximum range"; movement was "reduced to the absolute minimum." Both reconnaissance of firing positions and surveillance of the enemy forces were stressed, and the use of an observation helicopter as an integral member of the fire team was proposed, depending on the introduction of the Gazelle; it was not possible to use the current helicopter, the Sioux, because of its low speed.[64] As a result of these tactics, the helicopters were usually able to remain undetected. At one exercise in 1978, "the fire positions selected by both sections were so good no-one on the [spectators'] stand, actually looking for them in a given area with binoculars, was able to spot them."[65] The care that no doubt went into staging this exercise reflected the importance of such demonstrations in establishing the helicopter's credibility with the army: good tactics worked in both in wartime and in peacetime.

The importance that the British attached to helicopter survival was evident in these defensive tactics. The proponents of helicopters had been forced to address vulnerability because of the opposition on the issue and were not about to endorse risky, offensive action. Concerns about survivability were reinforced by the small number of anti-tank helicopters available. Concealment was not, however, designed solely to minimize the possible losses their of anti-tank helicopters; it also allowed them to surprise the enemy and to increase the shock value of their attacks: "The best method of destruction is by concealed fire power over a short intense period."[66]

The anti-tank helicopters were held at divisional level because of their anti-tank reserve mission and the limited numbers available. It was thought that the impending arrival of the Lynx would then allow brigade anti-tank squadrons to be created, reflecting a continuation of the disenchantment with the RAF and the emphasis on decentralization that emerged from Borneo. However, the AAC's anti-tank helicopters remained a divisional force when Lynx was introduced. While this was at first because the brigade organization had temporarily been abandoned by the British, even when the brigades were reformed, the anti-tank helicopters stayed at division, indicating a changed view about the value of centralized control.[67] The larger, centralized units meant more control for the AAC—the opposite of the old integration approach—which helped fulfill the Corps' desire for a more prominent place in the army.

By 1982 the opposition was beginning to dissipate and, certainly, it did not threaten the future of the anti-tank helicopter as it had a decade previously. With the introduction of Lynx/TOW into full service, the director of the AAC confidently asserted: "We [have] . . . finally arrived in the business of killing tanks."[68] However, support for the anti-tank helicopter was based on the initial doctrine under which it had been developed: that the anti-tank helicopter force was a last reserve with which to counter enemy penetrations. This reasoning led to the view that the anti-tank helicopters would operate independently of ground forces, delaying the enemy until reinforcements arrived, at which time the helicopters would withdraw. Thus, in 1982, when the Chief of the General Staff (CGS), General Sir Edwin Bramall, stressed the growing importance of the armed helicopter, he described it only as "an ideal anti-tank reserve."[69] Thus, at the start of 1983, when the AAC's director, General Bill Withall, set the future paths of development for the anti-tank helicopter force, he declared that:

> We enter a new era of combat effectiveness and our position in the order of battle is assured. *The scope has now widened enormously for us to apply our thoughts towards new tactics and extensions of our present roles* and we must live up to these new capabilities [emphasis added].[70]

Withall's comment revealed the twin preoccupations of the AAC for the 1980s. Not only were aviation proponents seeking more effectiveness; they also tried to improve their standing among the rest of the army.

First Attempts to Expand the Armed Helicopter's Role

Initially developments in the 1980s were mainly concerned with specific helicopter tactics, an aspect for which the AAC was already directly responsible and could thus influence immediately, but towards the end of the decade more attention was paid to the way in which helicopters were used in relation to other arms—a more delicate area in which to pursue change because it affected other corps as well.

The first change was to try and increase the destructive power of helicopters, to make them the most effective tank-killing force on the battlefield. These changes involved no new weapons—which the British

Photo 3.2
Lynx AH Mk. 1

The impressive Lynx anti-tank helicopter, which is the main equipment of AAC aviation units at present. Note the roof-mounted sight to assist in concealment. (Westland Group plc. Used by permission.)

Photo 3.3
Gazelle AH Mk. 1

The Gazelle acts as the observation helicopter when the Lynx attacks enemy targets. The Gazelle is also used in France, where it is armed with cannon, anti-tank and air-to-air missiles. (Westland Group plc. Used by permission.)

army could not afford—but more aggressive tactics. In contrast to the earlier, deliberate approach, the "quick ambush" was now proposed, in which the helicopter force "flies forward, out of sight of the enemy and, without delay, from one or more locations behind cover rises up in unison . . . and immediately engages the enemy."[71] This approach meant that large-scale anti-tank helicopter action became more common. Previously, three to six helicopters were the maximum in any attack, but now anti-tank helicopters were regarded as being "at their most effective en-masse," increasing the chance of achieving a "shock . . . sufficient to induce a degree of paralysis" in the enemy; in one experiment fifteen Lynx/TOW fired fifty missiles in just four minutes.[72] Finally, it was envisaged that, instead of operating independently in reserve, anti-tank helicopters were capable of operating in all phases of the battle; thus in the major 1984 exercise Lionheart, anti-tank helicopters "were used in the . . . 'advance to contact' " and guarded against enemy counterattacks while a friendly offensive was launched.[73] Throughout the new approach ran a single theme: for anti-tank helicopters to operate with the same degree of involvement in battle, although with very different tactics, as armored units.

Favorable conditions for these changes were produced by the reorganization of the British Corps in Germany (1 (BR) Corps) into three larger divisions, which increased the size of the AAC's front-line units. There were to be three squadrons of anti-tank helicopters, two with eighteen Lynx/TOW and one with twenty-four (see Figure 3.2). Massed action was far more appropriate to the new organization.[74] The AAC was also able to develop its new ideas because of the introduction of the Gazelle and Lynx. Even before it received its TOW missiles, the vast improvement shown by the faster and more agile Lynx made a great impression, and the Corps was stimulated to find ways to exploit the increased capability offered by the new helicopter.[75] As important as the qualitative improvement was the quantitative increase.[76] Lynx was not just the medium of expression for these ideas but also a catalyst since it had first been ordered in the early 1970s. The new equipment clearly helped to create changes in doctrine and was not just procured in response to them. A number of the tasks that it was eventually assigned had been under discussion in the AAC for some time,[77] but with the Lynx the Corps was able to perform and demonstrate these tasks rather than just talk about them (Table 3.1).

During this period the AAC also began preparing for aerial combat. Given the increasing number of Soviet armed helicopters, it was thought that "helicopter vs helicopter warfare is inevitable in any future conflict."[78] Therefore, the AAC sought to acquire a self-defense capability to protect itself from "the chance encounter and . . . potential ambush" while carrying out its main task of anti-tank action.[79] At first there were

Table 3.1
Comparison of Scout/SS–11 and Lynx/TOW

	Helicopter/Missile Combination	
Criteria	Scout/SS-11	Lynx/TOW
Number and Organisation		
No. in service	24	100 (later 114)
Unit size	Squadron of 6	Regiment of 18-22
Helicopter		
Speed	211 km/h	259 km/h
Engine	single, 710 shp	twin, 900 shp ea.
Endurance	30 min	90 min
Range	505 km	630 km
Load	4 Missiles	8 Missiles/8 Reloads
Vertical Rate of Climb	183 m/min	472 m/min
Missile		
Guidance	Manual	Semi-Automatic
Range	500 - 3000m	65 - 3750m
Speed	115 m/s	200 m/s

Source: Various references cited in this chapter.

misgivings from the RAF, which saw the AAC as encroaching on its assigned tasks,[80] but by the late 1980s the RAF, especially those officers actually flying helicopters, also began to favor air-to-air armament.[81] More problematic was the technical difficulty of acquiring suitable weapons for aerial combat, with the Blowpipe SAM (surface-to-air missile) being tested unsuccessfully in the mid–1980s.[82] The solution will probably await the arrival of the replacement for the Lynx since it will be easier to acquire air-to-air capability as part of a new weapon system than to overcome the financial and technical constraints on modifying existing helicopters.

The changes just described began to make the AAC's anti-tank helicopters a more potent force, one that might be used in a wider range of combat situations; and by 1985, the AAC director could report to his command that "plans for the next generation of helicopters and the organization and concept of operations for them has also received general MoD [Ministry of Defence] endorsement."[83] The logical next step for the AAC was to examine and change the combat relationship between helicopters and the other arms. Now that the AAC's units were more powerful, so aviators reckoned, they should not be limited merely to a supporting role; indeed this change in status would have to occur if

Figure 3.2
British Army Aviation Regiment (late 1980s)

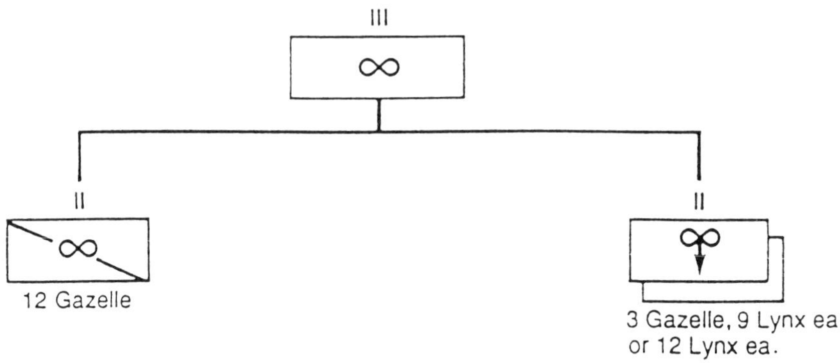

Source: Hussain et al., *The Future of the Military Helicopter*, 35.

there were to be any further development of armed helicopters. While the AAC wanted to retain its tactical independence from ground-based units,[84] it wished to become more integrated into the overall scheme of battle at the operational level, rather than simply being used in the last resort.

Perhaps following the American model, two levels of employment developed. While there was a "continuing need for the employment of anti-tank helicopters organic to the Division," a greater number could profitably be assigned to "reserve divisions [which] each need a brigade of about 50." The reserve airmobile infantry brigade was also thought to need a significant allocation of anti-tank helicopters.[85] This degree of mass employment would enable helicopters to fight tactically separate from other arms while being integrated at the operational level; the earlier units of three to six helicopters could only integrate tactically. The helicopters in the reserve division would be available to the corps commander for offensive missions up to fifty kilometers into the enemy's rear area in cooperation with long-range artillery units, and for this reason, so aviation proponents concluded, "The helicopter force should be seen as a *manoeuvre Arm fully trained for, and integrated into, the Corps Commander's design for battle* [emphasis added]."[86] Such operations would attempt to utilize helicopter mobility not just to react to enemy activity but, by striking at enemy second echelon forces before they reached the front line, to interrupt and upset the enemy's operational planning.

Maneuver status, as in America, meant that aviation would be more equal in stature to armor and infantry, especially as the British army was beginning to move away from firepower-based, attrition warfare.[87]

The interest in employing helicopters beyond the Forward Edge of Battle Area (FEBA) paralleled developments in other NATO countries, especially the United States (with which the AAC had considerable contact). The British were perhaps more reluctant about crossing the FEBA,[88] but there was a growing feeling that such operations were unavoidable: "Opportunities for deep attack will happen and when they do they must be seized."[89] In any case it was realized that the FEBA would be so ill-defined that it was foolish to impose a total restriction on operating across it. The impending decision on a more capable, dedicated attack helicopter to replace the Lynx provided the opportunity for further innovation; at the same time, its acquisition had to be justified in terms of new doctrine (see Figure 3.3).[90] Because the new helicopter was to be inherently more survivable, better armed and fully night capable, perhaps of the same standard as the U.S. Apache, the AAC's need to reduce vulnerability by cautious, defensive tactics diminished, thus allowing it to contemplate operations across the FEBA.[91]

Once again, part of the inspiration for these developments was a desire from within the AAC, "a young dynamic corps . . . involved in the pioneering of anti-tank philosophy and tactics,"[92] to make a more decisive contribution to the army's overall mission and thus to improve the status of the corps within the army.[93] The AAC officers were, as a group, more and more innovative—especially as their predecessors, who may have been innovators in the 1960s but had become conservative in the 1970s, retired.[94] Previously, the AAC had stressed its supporting role to retain the favor of the powerful armor and infantry branches whose members were most likely to be the army's senior commanders. The AAC of the 1980s had grown up with the existing view of anti-tank helicopters and now wished for more than just a limited reserve role. It believed that aviation was "part of the all-arms battle" and so had to be "fully integrated" with artillery, armor, infantry and so on.[95] AAC personnel were exhorted "to fit in and be able to advise the commander . . . not only as specialists flying sophisticated aircraft, but as competent soldiers with a deep understanding of the all arms battle."[96] Moreover, the general support that the anti-tank helicopter had gained, including the recognition that, like armor and infantry, army aviation was a direct-fire "teeth" arm, made a subservient role less politically necessary.[97] Indeed, a more decisive role was needed to justify the acquisition of new and expensive attack helicopters. The balance of restraint (to satisfy existing expectations about helicopters) and forward thinking (to bring innovation into practice), which had characterized the AAC's approach, was tipping even more in favor of the latter.

Yet internal pressure from the AAC was not of itself sufficient to bring about change, and the increasing importance of the Corps and its anti-tank helicopters was very much due to the views of key army officers

Figure 3.3
Proposed British Attack Helicopter Employment

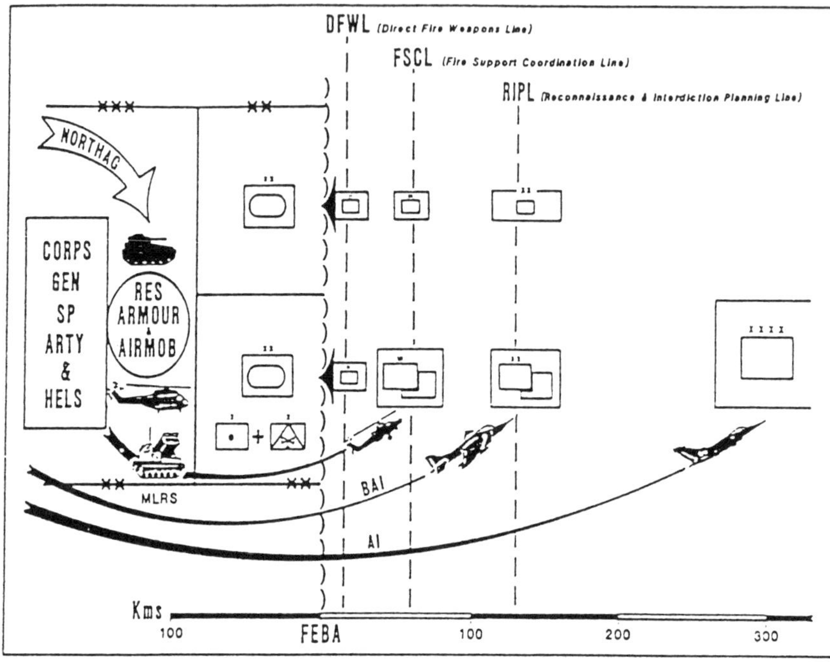

Source: Lt. General Sir John Learmont, "Military Theory and the Helicopter," *Army Air Corps Journal* (1989): 10.

such as Generals Sir John Learmont and Sir Martin Farndale (both Colonels Commandant of the AAC). Farndale, who had commanded both 1 (BR) Corps and then BAOR and NATO's Northern Army Group (NORTHAG), stated that "it is only by the creation of a powerful helicopter attack capability that the other combat arms, tanks, infantry and artillery can do their jobs."[98] He had commanded American forces—including the first operational Apache battalions—in exercise Reforger '87 and was clearly impressed by the United States' "decisive use of helicopters," afterwards calling for "more and better attack helicopters."[99] Other senior officers such as General Sir Brian Kenny, who succeeded Farndale in BAOR and NORTHAG, and Lt. General Sir David Ramsbotham, commander of UK Field Army, also lent their support. Learmont, who had been commandant of the Staff College, reported that his successor there was also convinced of the importance of helicopters.[100]

The actual organization of the AAC's anti-tank helicopters is yet to satisfy its current commander, Major General Sir Robin Grist, who recently argued that the development toward larger units must continue so the deep attack ideas outlined above can be implemented. In doing so, Grist employed the key weapon for helicopter advocates of the 1990s: the experience of the Gulf War against Iraq. Not only was Grist able to point to the work of the AAC in that conflict but was also able to highlight how much more effective were the U.S. Army's combat aviation brigades, equipped principally with the Apache. As he concluded, "There are many in Whitehall who have come to understand the value of attack helicopters in recent months."[101] While there may be a conservative approach to helicopters in the British army, Grist's readiness to criticize it so openly might be a ploy to help force the pace by producing a reaction in favor of helicopters as the army's leaders try to demonstrate that they are not as conservative as Grist depicted them.

Certainly, the army is about to decide on a key component in the AAC's future plans—the replacement for the Lynx—and helicopter advocates are keen to see the British acquire the Apache. The Lynx replacement was initially thought likely to be a joint development between Westland and the Italian company Agusta (the Tonal project). However, the cost and length of time involved in developing a new helicopter, even one less capable than the AH–64, has tipped the balance in favor of the Apache. It would seem that it is not a matter of which helicopter, but only when the Apache is procured: certainly the RAF has supported the acquisition of a replacement for the Lynx.[102]

Grist's apparent pessimism, and the eagerness for further development that it betrays, should also be contrasted against the sound picture of the AAC in terms of personnel. In 1986 there was an increase in the AAC's size at a time when other branches were suffering cutbacks, and in 1988 a new aviation regiment was formed.[103] Now, in 1991, as the British army begins a planned personnel reduction of over 25 per cent, the only major arm left intact will be the AAC—a point that the Minister of Defence stressed when announcing the plan.[104] Moreover, the reduction to just two divisions that Britain will maintain in Germany might well enable the creation of the larger helicopter units that the AAC has sought. Indeed, the AAC and its armed helicopters had come a long way from 1969 and the Scout/SS–11 combination; it had been seen then as a last resort and consequently regarded as not very much use at all by the other arms, who were intent on preventing the breakthroughs to which the helicopters would respond. While in the past "senior officers had failed to grasp the potential offered by the helicopter,"[105] in the late 1980s commanders both inside and, more importantly, outside the AAC were publicly supporting the notion that, because of its mobility and firepower, and resulting speed and shock action, the attack helicopter was "battling winning equipment."[106]

THE STRUGGLE FOR AIRMOBILITY[107]

The Precursor—Air Transport

While steady progress was being madey Britain in establishing its anti-tank helicopter force and associated doctrine, the same cannot be said for the theoretical development of airmobility. Before 1968 the overriding concern was the fighting in Britain's overseas colonies, followed by the technical development of future helicopter designs. In the late 1940s and 1950s there was academic interest in the possibilities of using helicopters for the transport of men and supplies in Europe, but there were simply no helicopters available to develop these ideas. The RAF was more concerned with the existing combat situation in Malaya. However, the Joint Experimental Helicopter Unit (JEHU) was set up in 1955 to carry out some basic tests.[108] Financed and commanded by the army but relying on the RAF for its technicians and many of its pilots, JEHU spent four years gathering information from trials and from practical experience at Port Said and later in Cyprus. Yet it was no more than the first stage of development. The RAF resisted attempts to make JEHU a permanent part of the army, which would have given the army responsibility for troop transport by helicopter. This mission was instead adopted by the RAF, and JEHU became a part of the RAF Short Range Transport Force (SRT Force—predecessor of the Support Helicopter (SH) Force), which was formed in the 1960s.[109] With the establishment of the force there was an increase in the numbers of helicopters available for exercises and trials in Europe, thus providing an opportunity for more developments. There was also recognition in principle of the value of air transport in Europe, despite the small number of helicopters available.[110]

But trials were greatly restricted by the fact that the helicopters were only capable of flying in good weather during the daytime; nor was there any money to buy the necessary navigation equipment for them to work in poor visibility. They were also unreliable. Therefore, exercises with the army were arranged rather artificially, so that the training could continue without helicopters in case the latter were grounded.[111] Further difficulties ensued from the fact that many pilots and technicians of the SRT Force were not permanently assigned to helicopter units but were between postings in fixed-wing aircraft squadrons. According to the army, they "lack[ed] . . . experience of an Army in the field," and by the time they "began to think like soldiers," their time with helicopters had expired and a new batch of RAF personnel replaced them.[112] But the RAF also complained that infantry units were not assigned solely or even regularly to airmobile duties. Thus, despite keenness on both sides, relearning of basic details was a constant necessity and prevented training for more complex operations. As if this were not enough, detailed and uninterrupted developments were hampered by the further reduction in 1964 in BAOR's helicopters to cope with the simultaneous emergencies in Borneo and the Radfan; East of Suez once again became the focus.[113]

During this period the AAC did not display any notable interest either in air transport or airmobility. The disappointment that arose when control of JEHU passed to the RAF stemmed more from the frustration of the desire "to expand Army Aviation" by acquiring the RAF support helicopters rather than any inability to make dramatic breakthroughs in their use.[114] Since the AAC did not acquire these helicopters it was not surprising that, both in theory and practice, its involvement with air transport was limited. For example, one senior officer saw the future transport responsibilities of the AAC as "the carriage of patrols and reinforcements" and "emergency forward supply"[115]—essentially what it was already doing in the counterinsurgency conflicts with its light utility helicopters.

Counterinsurgency Ends—Airmobility Emerges

The withdrawal of Britain from the East and the more complete devotion to the NATO environment brought a major change for air transport, with sufficient helicopters becoming available to prove the concept in exercises. There was also an opportunity for a reassessment of helicopter practices as planners asked, "Has the British Army, for its role in Europe, got the right kind of helicopters, has it enough of them and is the system of operating them the most effective and efficient?"[116] Air transport was soon accepted in the new environment, with the helicopter's superior mobility being used to overcome congested roads caused by refugees. This advantage over surface transport was to be exploited mainly for logistic support—echoing the Royal Army Service Corps' enthusiasm for helicopters in the 1950s. It was also suggested that helicopters could provide transport for a "rapid reaction force," which would deal with both sabotage activities and enemy air landings behind friendly lines. This proposal ingeniously adapted the infantry-based security operations that had been so successful in counterinsurgency operations to the European battlefield.[117] Yet it also demonstrated the first signs of a concern that was to grow considerably in the next decade—that unmechanized light infantry were unlikely to achieve anything significant in an arena dominated by armor.

However, the Directorate of Air/Land Warfare was eager to develop a more advanced concept of airmobility.[118] Its main proponent was Major General Glyn Gilbert, first as the commandant of the School of Infantry and later as director of the Joint Warfare Establishment (JWE). He "was continuously involved in the development of airmobile operations between 1968 and 1974."[119] Gilbert's training and experience with air-oriented light infantry operations in the Parachute Regiment was a most appropriate background from which to develop airmobility. He was greatly assisted by his association with the forward-thinking advocate Colonel John Waddy, who had a similar light infantry background. From 1968 onward the School of Infantry attempted to determine, from the lessons learned East of Suez, the best procedures and tactics for the use of transport helicopters in Europe. Its work was made easier by the arrival in 1971 of the more powerful and reliable Aérospatiale SA.330 Puma, which overcame previous technical limitations.

The confirmation of the value of air transport within Europe came in 1972 from a major exercise, entitled Sky Warrior, which tested 5 Airportable Brigade Group, transported by helicopters in "missions . . . specifically related to possible rear area tasks."[120] Following this trial and in concert with the director of the AAC, General Roy Dixon, Gilbert and Waddy planned another exercise, Sky Sword, to examine airmobility in more detail. "Instead of planning a repeat of Sky Warrior . . . the JWE wrote a more advanced concept [for] . . . an airmobile all arms battlegroup."[121] This battlegroup was primarily designed for anti-tank defense and would be available to division or corps commanders to counter enemy armored penetrations. It was to be battalion sized and dedicated to airmobility to overcome the training problems noted above. Although signals troops and artillery observers would complement the infantry, the key component that made the unit "all arms" was an integral force of anti-tank helicopters. Not only would the anti-tank helicopter aid the infantry, but

> if the best value is to be obtained from ATHs [anti-tank helicopters] then, in the same way that the tank is accompanied by mechanized infantry, they must be supported by airmobile infantry organized, trained and equipped specifically to kill tanks and to work closely with the ATHs when necessary.[122]

This emphasis on joining armed helicopters and airmobile forces was, first, an intellectual advance since it recognized the need for combined arms without linking helicopter units to slower, ground-based forces. Second, it was a subtle attempt to win for airmobility some of the acceptance that the anti-tank helicopter had gained in establishing its own counterpenetration mission.

However, the progressive thinking by Gilbert, Waddy and the JWE was not matched by their superiors, who apparently did not understand "the difference between an air transported force and the airmobile battlegroup." Sky Sword was canceled, partly because of financial and resource constraints, but also because of opposition to airmobility from the Directorate of Military Operations. A smaller exercise, Nap Archer, was held instead in 1975 and, not being innovative, did nothing but confirm the army's existing views about air transport: "Interest in airmobility languished."[123] Some attention was paid to the air transportation of tactical reserves by helicopter, and the commander of 1 (BR) Corps had an infantry battalion available for this purpose. The RAF SH Force (Germany)—commanded by one of the RAF's most enthusiastic helicopter advocates, Wing Commander Sandy Hunter—played a considerable part in these developments, showing rather more interest than the army. However, this small unit within the RAF was unable to extend the army's thinking much, if at all, beyond air transport to the concept of airmobility that Gilbert had proposed.[124]

The obstacles that airmobility faced were numerous but by no means insurmountable. Gilbert was convinced that the opponents of helicopters were "the 'Do Nothing Brigade,' always a well recruited formation" and thought that if the British army had wanted to pursue airmobility, it could have no matter what the difficulties. His recollection may well be tinged with some anger that airmobility, which was clearly very important to him, was not taken further. Nevertheless the general contention was valid: airmobility suffered severely because there was little enthusiasm to overcome the practical problems and to resolve people's doubts about it.[125] For example, anti-tank helicopters faced similar problems but were able to overcome many of them because of the commitment of the AAC to their development.

The first practical difficulty with airmobility in the 1970s was the financial restriction imposed by the 1974 defense review.[126] There was already a shortage of helicopters—the Royal Navy had helped out during Sky Warrior—and now the RAF was unable to buy any more. Without these assets it was indeed a challenge to establish any sort of permanent airmobile force. Medium-lift helicopters to replace the Belvederes, which had been scrapped in 1969, were particularly necessary to the future of the airmobile force but, although the RAF was very interested in acquiring such aircraft, it had been continually thwarted by financial constraints.[127] As if the lack of money for additional assets was not enough, use of the existing helicopter force to develop airmobility was severely constrained by operations in Northern Ireland. Although the conflict taught the British much about using helicopters against terrorists, it was of no help in pursuing the concepts that Gilbert had outlined. Advocates had stressed that they were not calling for an immediate move to airmobility and were proposing an experimental, developmental force.

However, as Waddy recalled, the army did not even bother to experiment with it, assuming the concept was not viable.[128]

After 1974 airmobility also suffered because no one was available to provide the necessary leadership for its development. The division of helicopter ownership between the RAF and army was unhelpful in this regard. Those who worked with helicopters in the RAF—and by now there were a number of highly experienced officers whose career path lay in the helicopter force—were enthusiastic, but generally the RAF gave low priority to aerial transport operations. Moreover, RAF helicopter pilots had little influence with senior army decision makers on whom the RAF relied for decisions about what had to be done on this issue. The army was undecided about helicopters and so, as one RAF officer complained, "We are sometimes unsure of exactly what the Army does want."[129] Therefore, any momentum for airmobility had to come from the army. Gilbert's advocacy had been insufficient and his successors were much less inclined to pursue the issue. Even those who might have favored airmobility were not prepared to support it to the extent of transferring scarce resources from their own corps, where their main political responsibilities and organizational loyalties lay.

The only group to which this did not apply was the AAC, since helicopters were its responsibility, but it also gave relatively little support to airmobility. In 1970 one AAC officer, who had marked himself as a progressive by his support for the anti-tank helicopter, was somewhat dismissive of airmobility, concluding that "a little goes a long way in the Far East; but not in Europe." A report on Sky Warrior in the *Journal of Army Aviation* mentioned only that it was a very valuable training exercise for anti-tank helicopters.[130] These comments were not necessarily representative, for there was some interest in airmobility, but the overworked, understaffed AAC was more concerned with establishing its anti-tank helicopter force and was not prepared to take on yet another mission.[131] Even if the AAC had been actively considering airmobility, it was restricted because it did not own the transport helicopters and was rarely able to exert control over them in joint operations with the RAF. The split in ownership was clearly a major stumbling block. It meant that the interest shown by the AAC in airmobility focused on transferring the RAF helicopters to the corps, so that the latter could survive and expand as a separate entity. Instead of cooperating against the conservative elements in the armed forces, the RAF helicopter force and the AAC tended to argue with one another.

The responsibility to sustain the advances made in Gilbert's time lay most of all with the infantry because they were the users of the RAF support helicopters. In the 1950s and early 1960s, the helicopter was very important to them because of its utility in the seemingly unending counterinsurgency campaigns that Britain was fighting. However, due to the withdrawal from the East, this interest mostly evaporated, and the development of doctrine more suitable for Europe took priority. It even turned into opposition because of restricted defense spending and the pressing need to buy armored personnel carriers. Gilbert's interest in airmobility was atypical: influenced by his career with the nonmechanized infantry of the Parachute Regiment, he was not as strong wedded to ideas about mechanized combat as were most other senior infantry officers.[132]

Anti-tank helicopters were slowly being accepted at this time, with concerns for their survivability being resolutely addressed by helicopter enthusiasts. Yet there were no similar advocates for airmobility, and the issue of helicopter vulnerability remained a potent weapon for critics, especially given the cost involved with airmobility.[133] There were also grave doubts about the survivability of light airmobile infantry. Field Marshall Sir Michael Carver, then Chief of the General Staff, declared that he had "little faith in airmobile troops . . . or in the effectiveness of attacks by dismounted infantry."[134] Rejecting unmechanized infantry was a recognition of the massive armored strength of the Soviet Union and expressed the view that suitably protected infantry might well play a major part in future conflicts. Therefore, the resources that were available for the infantry went to strengthening its mechanized forces and not to airmobility.

A Change in the 1980s: Airmobility Finds Favor

Official interest in airmobility resurfaced in 1980 when an as yet unpublished study of the subject was conducted.[135] Reflecting this interest, the *British Army Review*—a journal that served as a forum for debating new ideas—published an article entitled "The Value of Their Mobility," in which the author deliberately reopened the debate on airmobility.[136] Reviving Gilbert's ideas, the author outlined an organization combining airmobile infantry and anti-tank helicopters that would act as a reserve against enemy armored breakthroughs and air assaults. Aware of possible criticisms, the article concluded by calling for the establishment of an experimental airmobile force to test the concept.

Obstacles remained, however. There was an incorrect belief that airmobile advocates were suggesting the helicopter could replace the tank. Resources were still extremely limited. The split of helicopter ownership between the RAF and army remained, with all its attendant problems. Thus, even the published responses that indicated approval of the article were pessimistic. Airmobility appeared to face a "minefield of financial constraint, political maintenance of the status quo and reluctance to rock the inter-service boat."[137] Certainly, these metaphorical minefields proved more of a barrier to the helicopters than real ones. Perhaps the best guide to the problems facing airmobility was an article by the CGS, General Bramall, early in 1982. Although he had been involved in airmobility trials in 1969 and was a keen advocate of anti-tank helicopters, he could do no more for the infantry than suggest a distinction between light mechanized and heavy mechanized forces; despite the fact that Bramall was explicitly calling for change and was promoting forward thinking, there was no mention of airmobile forces.[138] The reason was "the inevitable shortage of resources," which was particularly acute in 1980–82: airmobility was a distant third in the order of priority for vertical flight behind Lynx/TOW and the Harrier.[139]

The period 1982–83 saw a number of important developments that helped remove some of these obstacles, culminating in the announcement that 6 Brigade would conduct an extensive series of airmobility trials. First of all, the RAF began bringing the first of thirty-three Boeing Chinooks into service. The Chinook could carry a nominal fifty personnel (and over eighty if necessary) against the Puma's sixteen and had five times the latter's cargo capacity. Yet even this decision was not due to some sudden upsurge of enthusiasm for airmobility but because the defense budget in 1977 had been underspent and the RAF, which had long coveted the Chinook for logistics tasks, seized its opportunity.[140] Moreover, within months of the Chinook's arrival it was being used in action in the Falkland Islands conflict and thereby gave a dramatic example of what such helicopters could do—not least because only one Chinook was available after the remainder were destroyed before being unloaded from the transport vessel *Atlantic Conveyor*. As the official report stated, "[The Chinook] adds greatly to the mobility of our land forces. . . . [its] value was clearly demonstrated."[141] Although the Vietnam War had already shown the importance of such helicopters, its use in the Falklands was a much more immediate, and thus more telling, example.

Yet the arrival of the Chinook, even with its value proven in the South Atlantic, was not on its own sufficient to cause a change in thinking. The lesson of the Falklands war could well have been labeled irrelevant to the European situation and ignored; or the Chinook could have been withheld solely for logistics tasks. A key factor in the continuing development of airmobility was the support and encouragement given by senior British army officers, led by General Sir Nigel Bagnall. In July 1983 he was appointed to command BAOR and NORTHAG, and then in 1986 he became CGS. The establishment of 6 Brigade as a trial unit, within months of his appointment to command BAOR, as well as its continued existence for the next five years, was partly due to Bagnall's strong support. Moreover, his successor in 1 (BR) Corps, BAOR and NORTHAG, General Farndale, was even more keenly interested in helicopters, not only for direct anti-tank combat as noted above, but also for airmobility.

The birth of the trial airmobile unit, 6 Brigade, was not in itself a conclusive endorsement of airmobility. The brigade was an armored unit that, with a shortage of money for its reequipment with new tanks, was unable to take its place in Europe as a mechanized unit.[142] Although a cynical view of the trial would be that it was simply giving the brigade's two infantry battalions something to do, the decision to employ 6 Brigade in this way was clever opportunism in the quest to further develop airmobility. Any perceived threat to vested interests was concealed by the temporary nature of the airmobile force and the cost of the trial was reduced. It was a crucial first step on the path to a permanent airmobile brigade.

The brigade was composed of two battalions of infantry, equipped specifically for anti-tank combat, and a small squadron of Gazelle and Lynx/TOW helicopters; it was supported by engineers and air defense troops. Although supplied with light vehicles, the brigade was basically "boot mobile." The RAF provided the transport, although the unit also needed the support of numerous logistics vehicles. Designated as a corps reserve, the brigade's primary mission was to counter and delay enemy armored breakthroughs. In determining whether a major airmobile force could play a part in BAOR, the unit was also to establish the procedures and tactics that such a force might employ.[143] The brigade's efforts showed that an airmobile force had specific strengths and weaknesses. First, the increased allocation of ATGW to the brigade, as well as the close cooperation of even a small number of Lynx/TOW, made it an effective anti-armor force for its size. More important, the force had tremendous operational mobility: one brigade member stated that "we can deploy our fighting element quicker than any other ground based force."[144] Mobility meant far more than just speed since the unit's ability to respond rapidly gave it great flexibility.

> Committing [the commander's] reserve is arguably his most important decision. When these reserves are scarce their commitment at the right time becomes even more important. . . . [With airmobile reserves, he] can keep his options open longer.[145]

However, the potential success of 6 Brigade was hampered by a number of factors. The flexibility and speed that the use of helicopters generated was offset by the limited air support available: only one of the battalions was normally lifted for a particular operation, and even then it could not be carried in a single helicopter lift. As well, the "tortuous command and control system" required that requests for RAF helicopter support be directed to the corps headquarters. This system was a result of the lack of helicopters and the consequent need to centralize their control, and it greatly inhibited the ready reaction that ought to have been the hallmark of the brigade's operations.[146] There were also problems in achieving adequate cooperation between the RAF SH Force and 6 Brigade. While the respect and understanding between the services in the 1980s had outgrown the historical disputes of the 1960s and 1970s, the army nevertheless believed that "Support Helicopter pilots . . . have not been brought up to understand the environment of the ground battle."[147] While RAF officers would dispute this, there was no doubt that the need to work with the RAF was seen by many army officers as less satisfactory than if the AAC had been the sole helicopter operator.[148] It would be quite incorrect to blame the RAF SH Force for every problem that occurred, or to imagine that it was uninterested in developing its ability to support the army, but its interests lay more with the technical and procedural details of flying in close support of the army rather than the tactical and operational issues that had to be addressed if an innovative doctrine like airmobility was to develop.[149] As in the 1970s, RAF officers also had much less influence within the army in the battle to overcome skepticism about airmobility.

Although 6 Brigade's five-year trial was successful to some extent in proving the potential of airmobility, it was also important in increasing the level of awareness about helicopters and their possible uses, as well as stimulating much thinking and debate on the subject.[150] For example, the AAC now became far more involved. Although still seeing airmobility as a means to expand itself, the corps was more interested in the concept for its own sake than in past years. It recognized the potential of combining anti-tank Lynx squadrons with airmobile forces, thereby providing a degree of combined arms action, which improved the effectiveness of its armed helicopters.[151] The increased acceptance of airmobility was also due to a more positive reevaluation of nonmechanized infantry. To some extent this change made a virtue out of necessity since half of the regular British infantry would have to fight dismounted and there were insufficient funds to procure infantry combat vehicles for them all. Nevertheless, it was realized that there were times in modern warfare in which nonmechanized infantry would be very valuable, especially for fighting in the urban or forest areas of Western Europe.[152]

The most telling factor in the recognition and acceptance of airmobility in the 1980s resulted from the "great leap forward in operational thinking" within NORTHAG while it was commanded by Generals Bagnall and Farndale. Essentially NORTHAG developed as an operational unit in its own right, rather than as a name for a collection of individual, national corps. Moreover, Bagnall wanted to create "strong reserves," which would be used to ensure operational flexibility and to take the initiative, conducting counterstrokes and counterattacks.[153] The NORTHAG concept gave preeminence to armored reserves, but if armored units were to be withheld in this way, then airmobile forces were needed to replace them in the defensive task of blocking enemy penetrations. Furthermore, Bagnall's concept required maximum mobility "so that NATO's smaller forces can rapidly concentrate their firepower at critical points,"[154] thus favoring forces moving at helicopter rather than ground speed. At the same time that the British were reassessing their doctrine, General Ferdinand von Senger und Etterlin (then Commander-in-Chief, Allied Forces Central Europe—CINCAFCE) was arguing that, because of helicopters, there were "certain new dimensions of operational planning." While he applauded the current use of transport helicopters to give operational mobility (what he termed "airmobility"), von Senger thought that a better exploitation of the new dimension required "airmechanization," in which the "means of transport must become a means of combat."[155]

While it would be farfetched to assume that von Senger personally initiated the changes in NORTHAG, both he and the British generals were thinking along the same lines in stressing the role of the army group and operational art. The important point about von Senger's concept was that it combined his endorsement for army group operations with his enthusiasm for airmobility. He explicitly linked his proposed airmobile forces to the need for reserves with which the army group commander could directly influence the course of the battle, since the "wide radius of action [of airmechanized units] allows them to be landed at practically any point in the sector of an Army Group."[156] Senior British officers may not have had quite as much open enthusiasm as their German colleague, but clearly von Senger's views on airmobility played no little part in increasing their support for the idea.[157] In particular, the explicit connection made by von Senger between army group operations and airmobility helped to encourage interest in the latter in Britain at a time when the former was being vigorously pursued.

The most practical follow-up to the airmobility trial was the decision to establish a permanent airmobile brigade once 6 Brigade had finished the trial and been remechanized (Figure 3.4). The lack of open opposition to this move perhaps reflected most clearly the positive impact made by 6 Brigade.[158] However, putting theory into practice proved more difficult, and there were serious problems with the new airmobile formation, 24 Airmobile Brigade.[159] For example, despite the fact that it would be used in BAOR in time of war and supported by RAF helicopters in Germany, financial restraints meant a UK-based unit was chosen. It had to rely on the already over-stretched SH Force in England

Figure 3.4
24 Airmobile Brigade

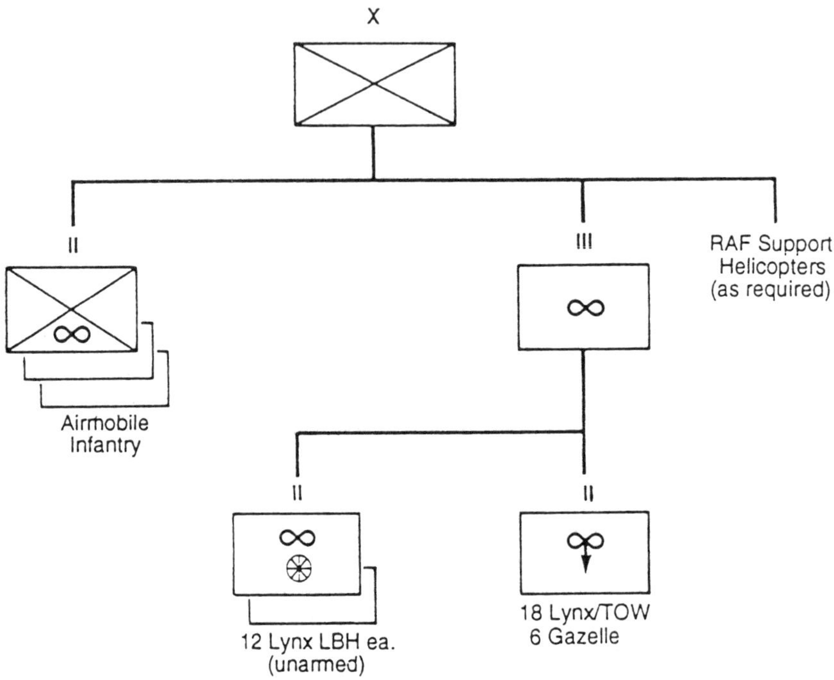

Total Helicopter Assets:

18 Lynx/TOW (anti-tank)
6 Gazelle (observation
24 Lynx LBH (utility)
up to 20 Puma (medium transport)
up to 20 Chinook (heavy transport)

Source: Interviews with British military personnel.

for much of its training, hindering the development of the close rapport between helicopters and infantry upon which successful airmobile operations depended.

There continues to be a problem with the small size of the SH Forces, and even if a new transport helicopter is acquired (as is currently planned), it will replace the Puma, not supplement it. Nor has there been any agreement about the size of the helicopter needed. The lack of helicopters meant that 24 Brigade did not have integral transport helicopters except for its small number of army Lynx and could not rely absolutely on the support of the SH Force in Germany because in wartime there would be many other demands made on the SH Force, not least for logistic work.[160] It was presumed that the majority of this work would be completed before the enemy broke through, and thus some helicopters would have been available to move 24 Brigade to counter such a penetration; whether the Warsaw Pact would have been so accommodating is less certain. With inadequate helicopter support, the brigade was forced to operate one of its three battalions in wheeled armored personnel carriers (APCs).[161] As with 6 Brigade, the new unit also had a large, roadbound logistics tail that could well be unable to keep up with the helicopter-borne elements of the force.[162] Both of these problems limited the unit's flexibility and rapid response. The less certain military environment after the collapse of the USSR only exacerbates the problems caused by these compromises, and how the brigade handles them remains to be seen.

The lack of helicopters emphasized the difficulty of putting theory into practice, and the continuing skepticism about airmobility suggested that until implemented, theoretical innovation would receive only guarded acceptance. The small number of helicopters also pointed to the problem that the British army faced in being caught by the political desire to maintain an indigenous helicopter industry (the Westland company) and the higher costs or inappropriate equipment purchases that that entailed. For example, the RAF and army would probably prefer to see more Chinook helicopters bought for the airmobile brigade, but the Westland-built EH–101 (like the Lynx, principally developed for naval use) is the more likely option in the future.

Another contentious matter, at least as far as many airmobility advocates were concerned, was that nothing was done in forming 24 Brigade to alter the existing division of responsibility for helicopter support between the RAF and AAC. The issue of helicopter ownership had been the source of considerable ill feeling for some members of the British armed forces and was still a highly political issue, despite the advances in interservice relations in the last twenty years.[163] The dispute was usually expressed in terms of whether or not RAF ownership of some helicopters had impeded innovation. But it was itself an impediment to gaining acceptance for airmobility within the army because doctrine and the divisive debate were seen as linked. Thus senior British officers were eager to defuse any tension over this matter, stressing that "colour of uniform is not vital... we should not make an issue of it."[164] Yet the AAC, which had always coveted its neighbor's helicopters, saw the 6 Brigade trial as providing the best evidence yet in support of its position.[165] However, the army's direct challenge to the division of ownership between the RAF and AAC in 1986 met with failure because, on economic grounds, the size and complexity of the Pumas and Chinooks meant it was better for the RAF to maintain ownership of the SH Force.[166]

In discussing this issue, it is important to distinguish between ownership and control, a point that some of the more extreme supporters of the AAC's position have failed to appreciate. In wartime the SH Forces were to be controlled by army commanders. Moreover, the 24 Brigade commander would have been given operational control of the helicopters if they were assigned to an airmobile mission.[167] As a result, some of the problems of RAF ownership would have been averted. Yet despite the improvements in command and control associated with 24 Brigade and the willingness of most RAF SH pilots to make the system work, the division of ownership was still a complicating factor. No matter how much control devolved to army commanders, joint army-RAF operations were inevitably more difficult. For example, in the 1960s RAF radios were incompatible with the army network. These and other problems, such as different approaches to navigation between army and RAF aviators—"the inhibiting wax of 'Jointery' "[168]—greatly affected the coordination of the two services. Although 24 Brigade and its support helicopters established an excellent working relationship, as good as any between RAF and army,[169] it is significant that the RAF had to change its command practices to achieve this result. As the very success of 24 Brigade implies, there were many problems in the past with army-RAF cooperation—problems that hindered Gilbert's attempts to convince senior army officers about airmobility.

While at an individual level there have been excellent relationships between army and RAF officers working with helicopters, at an institutional level the division of ownership caused problems because the majority of RAF officers saw supporting the army as less important than their other duties. It is notable that the deployment of 6 and 24 Brigades coincided with the development in the mid–1980s of the view in senior British military circles that modern warfare has become a joint land/air battle.[170] Also, by the 1980s the SH Force was staffed entirely by personnel committed to operating helicopters and not simply in between postings to other parts of the RAF.[171]

Most of all, RAF ownership of the transport helicopters affected not just the actual capability of the airmobile forces but also, and more important, the development of doctrine—especially where airmobility and the anti-tank helicopter force intersected. One airmobile enthusiast concluded, "It is the split of the helicopters between two services that prevents the AAC *becoming the focus* for battlefield helicopter and airmobility matters [emphasis added]."[172] There is no doubt that, on the whole, the RAF tried assiduously to meet the needs of the army for helicopter support and had increasing success in this regard in the 1980s. Many of these problems—such as limited assets and skeptical senior decision makers—would probably have been less serious had the army owned all the helicopters.

CONCLUSION

Although the British army was eager after the Second World War to develop helicopters for use in Europe, their first extensive experience with them came from the counterinsurgency campaigns of the 1950s and 1960s. Employed for troop transport and supply missions, as well as for liaison and command and control, helicopters proved to be essential to the successful prosecution of these campaigns. The helicopter's ability to hover and land in unprepared areas enabled the British troops to combat the large distances and difficult terrain in which they operated while its speed compensated for their numerical weakness.

When the British focused once more on Europe in the late 1960s, they sought to continue using helicopters. The army was eager to acquire helicopters armed with the new ATGW to form an air reserve to counter expected enemy armored breakthroughs. It feared that the RAF either could not or would not provide effective close air support in future conflicts and sought a replacement from within its own ranks. Other influences on this development were the desire to exploit ATGW to the fullest, the example of other countries' developments and the organizational imperative stemming from the AAC's desire to expand and play a more important role in the army. Thus in 1969 Britain acquired its first anti-tank helicopter: the Scout/SS–11 combination. However, further progress with armed helicopters, including the acquisition of more capable machines, was halted for a decade—mainly because of financial constraints, which in turn heightened the opposition of other army corps and the RAF, both of whom feared a transfer of resources to the AAC.

By the late 1980s, however, the anti-tank helicopter had become an accepted part of the British army and the AAC was receiving sufficient resources to implement its doctrines. Along with infantry and armor, army aviation was regarded as one of the three direct fire, "teeth" arms. The AAC had established itself as a competent member of the army despite its limited equipment but, more important, a number of senior army commanders became convinced of the helicopter's value and were in positions from which they could influence its further development. As a result, the range and importance of missions assigned to anti-tank helicopters began to expand. Most important, the helicopters changed from being an independent reserve force to an integral member of the all-arms battle. Furthermore, as a part of the offensive operational concept developed in the mid–1980s, their missions began to assume a much more aggressive character, and the British sought to use them to initiate combat rather than react as a last resort. Mobility was still paramount and, to preserve the helicopter's freedom of action, they were not used in direct, tactical combination with tanks and infantry at battalion level but were controlled by division or corps commanders.

These developments traced parallel paths: from dependence on other arms for resources and status, and concomitant battlefield subordination, through various stages toward strength in peacetime organization and independence and equality in combat. Much now depends on the actual arrival of the new helicopter and the extent of its capabilities, and on the coming reorganization of the British army. The AAC is still to become a fully integrated participant in the all-arms battle; even though it receives considerable funding priority, it still depends on helicopter advocates and supporters outside the Corps. Although the AAC became a much stronger institution in the 1980s and attracts highly qualified personnel for both officer and other rank positions, it has a very high percentage of the army's short-term commissions. Von Zugbach has suggested that the AAC tended to attract people who wished to learn to fly and then pursue a career outside the armed forces.[173] If this continues, then the Corps will struggle to produce high-ranking officers who identify solely with aviation; instead, outside officers such as Generals Gilbert, Ramsbotham and Farndale will have to carry the arguments in favor of helicopters in the decision-making process and to command aviation units on the battlefield in the way the AAC would prefer.

The future of attack helicopters is connected closely to that of airmobile forces. The offensive, maneuver-based concepts developed in the mid–1980s favored both types of helicopter operation, and the continuing trend in British thinking was to combine the two. As Farndale stated in 1988, the AAC has "vital roles to play in 24 Airmobile Brigade." Although there may well also be attempts by the AAC to isolate its attack helicopter activities from the airmobile brigade,[174] Grist's appointment to command the AAC after having been in charge of 24 Brigade would suggest that the opposite will occur. There will continue to be a desire in the AAC to dominate the theory and practice of all army helicopter operations and to exploit the potential combined arms advantage of joint airmobile and attack helicopter forces.

This potential has taken over two decades to emerge. Like anti-tank helicopters, airmobility also began developing after the 1960s with the application and extension of counterinsurgency experience to the potential European battlefield. It was proposed that a light infantry formation, with anti-tank weapons and transported by helicopter, could be used in concert with the anti-tank helicopter air reserve. Yet for a number of reasons, the practice and even the theory of airmobility were abandoned in 1974–75. At a time of severe financial restraint, which caused great concern about the future of existing institutions within the armed forces, there was no group of advocates willing to accept responsibility for developing the new concept of airmobility and promoting it within the armed forces. Without this intellectual and political leadership, airmobility was incapable of overcoming the many obstacles set in its path, the most important of which was the perception that light, airmobile infantry was out of place in modern armored warfare. Because of the lack of advocates, as well as a shortage of helicopters due to the lack of money, the airmobility concept did not develop with the same strength as anti-tank helicopter doctrines. It languished until 1983, when a trial began of an airmobile brigade. After five years of tests, and with the support of senior officers who saw the unit as a crucial defensive complement to offensive armored reserves, a permanent airmobile force was created.

The growth in airmobility was a tremendous advance from the point where progress had halted in 1974, but doubts remained and, remembering that it took anti-tank helicopters at least fifteen years to establish themselves within the British armed forces, there is some way yet to go.[175] Perhaps in the future developments will occur most readily in the NATO context. Before the collapse of the Soviet threat, NORTHAG was already planning to develop multinational airmobile forces.[176] Now, the British contribution to the new NATO defensive structure will be as the main component of a multinational Ready Reaction Corps, in which airmobile forces will play a significant part. Given the decreasing size of NATO's armies and the need for mobile units, which can cover large areas against many potential threats, helicopters would seem to offer much to this corps. Furthermore, the successes of NORTHAG and the 2d Allied Tactical Air Force (2ATAF) in rationalizing and combining the actions of a number of nations in the 1980s suggest that it may prove easier to acquire the necessary resources and recognition for airmobility in the face of multinational financial and political constraint than within the still-suspicious British armed forces.[177]

A number of constructive comparisons can be made between the development of helicopters for armed action and for airmobility. First, after the reduction in colonial commitments in the 1960s, both concepts struggled to gain recognition with Britain's plans in case of war in Europe. Perhaps the most significant difference between the two, however, was that armed helicopters had the support of senior officers who were concerned that the RAF was not going to provide adequate close air support. Airmobility, on the other hand, had no such support. Not only did airmobility lack the strong advocacy provided for anti-tank helicopters by the AAC, but the army's leaders were far less receptive to it because it was at odds with the development of mechanized infantry. While acceptance of the anti-tank helicopter did not mean implementation, and the AAC's combination of Lynx/TOW and the Gazelle observation helicopter was only introduced in the early 1980s, the decision to acquire the Scout/SS–11 in 1969 opened the door—albeit narrowly—for future developments. Airmobility was in contrast unable to gain admittance to the army's repertoire of missions and capabilities.

The army's more assured approach to anti-tank helicopters in the 1980s was as much because such weapons had become reasonably familiar in its activities as because of any radical changes in military planning. The introduction of the Lynx was significant mostly because it increased dramatically the numbers of helicopters and, thus, the rest of the army's exposure to them. The AAC vigorously promoted anti-tank helicopters and, as a result, army officers in the 1980s understood helicopters more and appreciated better the arguments for expanding the AAC's resources and responsibilities. This degree of familiarity also improved the actual implementation of helicopter doctrine by the divisional and corps commanders who were not aviators. Airmobility was not practiced much in exercises and officers were not familiar with moving troops by helicopter—except in the rather different manner used in Northern Ireland. Even though there were advocates of airmobility in the army, they had little opportunity to sell their ideas by practical demonstration. The division of helicopter ownership prevented both a political and intellectual focus on airmobility from developing within the army.

Another significant comparison can be made between the differing progress made by the anti-tank helicopter and airmobile infantry in the last twenty years: the former was a specific weapons system and the latter a concept, a way of thinking. Whereas anti-tank helicopters were tied to an existing idea—that of an aerial reserve force—and seen as a replacement for it, airmobility did not have this advantage and thus had to try and establish itself against the opposition of those who concentrated solely on improving current military activities.

These comparisons help to explain why, for most of the last twenty years, anti-tank helicopters prospered and airmobility did not—even though the two concepts were relatively alike in their aim and similar doubts were expressed about their vulnerability and cost effectiveness. Better advocacy for anti-tank helicopters, the fact that they did not threaten current army priorities, the effect of linking them to an existing military concept and the familiarity that developed during the 1970s and early 1980s all meant that, by 1991, airmobility was barely established, while the AAC was actively pursuing the assumption of an offensive, integrated role and the acquisition of a third-generation attack helicopter with which to perform it. Attack helicopters had become accepted and familiar, while airmobility still presented a challenge.

NOTES

1. Anthony Farrar-Hockley, *Opening Rounds—Lessons of Military History 1918–1988* (London: Andre Deutsch, 1988) 155.

2. *Design for Military Operations—The British Military Doctrine*, prepared under

the direction of the Chief of the General Staff, British Army (London: HMSO, 1989) 47–48.

3. Quoted in Wing Commander John R. Dowling, *RAF Helicopters—The First Twenty Years* (London: HMSO, 1987) 77.

4. Quoted in Dowling, *RAF Helicopters*, 77. Dowling's history was replete with discussion of the problems suffered by RAF helicopters. The full name of the helicopter was the Westland WS–55 Whirlwind Mk. 4.

5. Richard Clutterbuck, *The Long Long War* (London: Cassell, 1967) 157–58. Figures from Dowling, *RAF Helicopters*, 92.

6. Hickey, *Out of the Sky*, 201–5.

7. General Sir Walter Walker in *Air Support of Land and Maritime Operations*, report of a Royal United Services Institution Seminar (London: RUSI, 1968) 5.

8. Paul Beaver, *Today's Army Air Corps* (Wellingborough, Northants.: PSL, 1987) 24; and Walker in *Air Support of Land and Maritime Operations*, 4.

9. Air Chief Marshal Sir David Lee, *Eastward—A History of the RAF in the Far East 1945–72* (London: HMSO, 1984) 230. For air support in Borneo, see *Air Support of Land and Maritime Operations*, 4–5, 17–20, and Colonel David F. Mallam (AAC), correspondence (1989).

10. Walker in *Air Support of Land and Maritime Operations*, 4.

11. Air Vice Marshal Foxley-Norris quoted in John Everett-Heath, *British Military Helicopters* (London: Arms and Armour, 1986) 86. For the Near East campaigns see Dowling, *RAF Helicopters*, 134–37, 142 and 273ff, and Colonel F. Graham-Bell, "Army Aviation in Aden," *Journal of Army Aviation* [hereafter *JAA*] (1974): 51–58.

12. Dowling, *RAF Helicopters*, 112.

13. Various sources quoted above give evidence of the importance of Borneo, and they were confirmed by Field Marshal The Lord Bramall, correspondence (1989) and Mallam, correspondence.

14. Everett-Heath, *British Military Helicopters*, 36; and Brigadier Richard A. Norman-Walker, "The Formation of the AAC—1955 to 1957," *Army Air Corps Journal* [hereafter *AACJ*] (1982): 13. Norman-Walker was in charge of Army Aviation in the War Office during the period preceding the establishment of the AAC.

15. Colonel S.M.W. Hickey, "The Evolution of British Army Aviation," *JRUSI* 120.12 (1975): 19; and Norman-Walker, "The Formation of the AAC—1955 to 1957," 9–10. More correctly, the AAC was re-established, since during the Second World War, it had existed as the parent formation of the Glider Pilot Regiment, the Parachute Regiment and the Special Air Service.

16. See the anti-RAF sentiments expressed by the following Army officers: Hickey, *Out of the Sky*, 217; and Major General Sir Napier Crookenden in Wing Commander David L. Eley, "The Development and Use of Helicopters," *JRUSI* 112.6 (1967): 116.

17. Dowling, *RAF Helicopters*, 14.

18. Everett-Heath, *British Military Helicopters*, 208.

19. Brigadier P. E. Collins, correspondence (1989). (Collins was a senior Aviation commander in the late 1960s and early 1970s.)

20. "Auster" (a pseudonym for Brigadier Peter Mead), "More Aircraft For the Army: 'The Integration Scheme,' " *British Army Review* [hereafter *BAR*], 19

(1964): 6–13; and Peter Mead, *Soldiers in the Air* (London: Ian Allen, 1967) 197–206. Mead eventually commanded the AAC with the rank of Brigadier before retiring in 1964.

21. Everett-Heath, *British Military Helicopters*, 98, 116–119 and Colonel John L. Waddy, correspondence, (1989). Waddy was a long-standing helicopter advocate.

22. Dowling, *RAF Helicopters*, 408.

23. Ibid., 408–410 and Mead, *Soldiers in the Air*, 207–8.

24. Norman-Walker, "The Formation of the AAC—1955 to 1957," 14.

25. Major General F. D. King, "The Case for the Armed Helicopter in Europe," *BAR* 31 (1969): 36–37; and Collins, correspondence.

26. Crookenden and Lt. Colonel Somerville in *Air Support of Land and Maritime Operations*, 11, 20–21; and *The Role of the Helicopter in the Land Battle*, Report of a Royal United Services Institution Seminar (London: RUSI, 1972) 39. For the withdrawal of the Hunters, see Ministry of Defence, *Supplementary Statement on the Defence Estimates*, Cmnd.4521 (London: HMSO, 1970) 8.

27. Air Vice Marshal L. D. Mavor in *Air Support of Land and Maritime Operations*, 6.

28. Mead in *Air Support of Land and Maritime Operations*, 16; King, "The Case for the Armed Helicopter in Europe," 38–39; and Colonel Sutcliffe (later Brigadier, AAC) in *The Role of the Helicopter in the Land Battle*, 39. The British convention was ATGW instead of ATGM.

29. Crookenden in *The Role of the Helicopter in the Land Battle*, 53; and Mead, *Soldiers in the Air*, 158–59.

30. Kenneth Macksey, *A History of the Royal Armoured Corps* (Beaminster: Newtown Publications, 1983) 209; and Kenneth Macksey, *The Tanks 1945–1975* (London: Arms and Armour, 1979) 173–75.

31. For the social and political dominance of the British Army by the RAC, the Infantry and, to a lesser extent, the Royal Artillery, see R.G.L. von Zugbach, *Power and Prestige in the British Army* (Aldershot, Hants.: Avebury, 1988) 14–35 and 107. Von Zugbach was unable to analyze the AAC in more than passing detail because of its institutional youth: 143.

32. Crookenden and Hutton in Major General William M. Hutton (ex-Director-General of Fighting Vehicles), "The Future of Armour," *JRUSI* 113.6 (1968): 102.

33. See, for example, Colonel John Moss (Ret.) and Colonel John L. Waddy (Ret.), "Towards an Even More Effective Army," *BAR* 80 (1985): 7–8; Waddy, correspondence, and Colonel (Ret.) John Moss, correspondence (1989).

34. Macksey, *The Tanks*, 174–75; and Waddy, correspondence.

35. General R. Dare Wilson (Director, AAC) in *The Role of the Helicopter in the Land Battle*, 31 (Wilson had been the last DLAW), and "Flap Back" (a pseudonym for Brigadier Peter Mead), "The Next Helicopters," *JAA* (1967): 24.

36. Mead, *Soldiers in the Air*, 207 and Collins, correspondence.

37. King, "The Case for the Armed Helicopter in Europe," 40.

38. Wing Commander P.D.A. Austin in *The Role of the Helicopter in the Land Battle*, 38.

39. Major General Sir Napier Crookenden, "Aviation in the British Army

Today," *JRUSI* 110.9 (1965): 205; and Eley, "The Development and Use of Helicopters," 114.

40. Compare Major General R. Dare Wilson, "Facing Reality," *JAA* (1971): 9 with, for example, " 'Chairborne,' Whither Army Aviation?" *JAA* (1970): 16–19, in which the author strongly supported anti-tank action and referred to observation as a "sacred cow."

41. 28 Flight AAC, "Newsletter," *JAA* (1968): 59; and Moss, correspondence.

42. 652 Squadron AAC, "Newsletter," *JAA* (1970): 62.

43. Editorial, *JAA* (1971): 5 and Development Wing AAC, "Newsletter," *JAA* (1968): 58.

44. Lt. Colonel J. R. Shrimpton, "Framework for the Future," *JAA* (1973): 6.

45. Major General R.L.C. Dixon, "Progress Report 1975," *AACJ* (1975): 8.

46. Everett-Heath, *British Military Helicopters*, 210 and Major D. G. Dudley, Correspondence, *BAR* 53 (1976): 70.

47. Major General R.L.C. Dixon, "Progress Report 1976," *AACJ* (1976): 8.

48. Air Vice Marshal S.W.B. Menaul, "The Use of Air Power in Vietnam," *JRUSI* 116.6 (1971): 12. An example of this criticism was Air Marshal Sir Maurice Heath in Hutton, "The Future of Armour," 102.

49. Wing Commander John Price in *The Role of the Helicopter in the Land Battle*, 36; and Group Captain (Ret.) John Price, "The UK Helicopter Gap," *Defence* 11.9 (1980): 709–12.

50. "Talarius," "Survivability Not Vulnerability," *AACJ* (1975): 20; and Colonel S.M.W. Hickey, "The British Army and the Battlefield Aerial Vehicle. The Way Ahead," *JRUSI* 121.6 (1976): 7.

51. John L. Waddy, correspondence, *JRUSI* 125.9 (1980): 95. The issue was pursued in a long-running debate in the *JRUSI*: Charles Dick, "Tactical Air Power on the Battlefield," 123.2 (1978): 42–45; and Air Chief Marshal Christopher Foxley-Norris, "Tactical Air Power on the Battlefield A Reply" 67–69; Charles Dick, "A Reply to Criticism" 69–70; and Group Captain R. A. Mason, "Further Thoughts on Tactical Air Power on the Battlefield," 70–71, all 123.4 (1978).

52. Ministry of Defence, *Statement on the Defence Estimates*, Cmnd.5976 (London: HMSO, 1975) 1. See Julian Lider, *British Military Thought After World War Two* (Aldershot Hants.: Gower, 1985) 545, for the rivalry that the cuts generated.

53. Ministry of Defence, *Statement of the Defence Estimates*, Cmnd.6432 (London: HMSO, 1976) 24, Ministry of Defence, *Statement on the Defence Estimates*, Cmnd.6735 (London: HMSO, 1977) 39; and Major General W.N.J. Withall, "Preface," *AACJ* (1979): 5.

54. Dudley, correspondence, 70.

55. Air Vice Marshal Michael Armitage, "Air Power in the Central Region," 124.4 (1979): 36 and M. J. Armitage and R. A. Mason, *Air Power in the Nuclear Age* (Urbana and Chicago: University of Illinois Press, 1985) 265–66.

56. Major General J.D.W. Goodman quoted in Paul Beaver, "British Army Helicopter Needs in the 1990s," *Defence Helicopter World*, 5.2–3 (1986): 9; and S. O'Dwyer-Russell, "Eternal Triangle," *Armed Forces*, 3.2 (1984): 71–72.

57. Major General J. A. Ward-Booth, "Preface," *AACJ* (1978): 5. The lack of progress and continuing frustration with acquiring the Gazelle sight is well

documented in successive Directors' Prefaces to the *AACJ* in the late 1970s and early 1980s.

58. Sutcliffe in *The Role of the Helicopter in the Land Battle*, 39–40.

59. King, "The Case for the Armed Helicopter in Europe," 40.

60. Moss, correspondence; and Mead, "Army Aviation," 45.

61. Macksey, *The Tanks*, 176; and Colonel P. E. Collins (then Commander Aviation, BAOR), "Anti-Tank Helicopters," *JRUSI* 119.6 (1974): 45.

62. 28 Flight AAC, "Newsletter," 59; and Kenneth Macksey, "The Qualifications of Mobility," *BAR* 66 (1980): 46–47.

63. Collins, "Anti-Tank Helicopters," 45.

64. Dixon, "Progress Report 1975," 8 and Lt. Colonel A. E. Woodford, "Antitank Fireteam Tactics—The British View," *AD* 21.10 (1975): 4–16. This article—a report of a briefing given to the Americans—was based upon the doctrine published that year and the author's experience of AAC training and exercises.

65. Major L. C. Bond, "Helcat," *AACJ* (1979): 8.

66. Woodford, "Antitank Fireteam Tactics—The British View," 4–5, 12; and Bond, "Helcat," 8.

67. Lt. Colonel P. R. Ralph, "Tactical Deployment and Roles of the New Look Regiment," *JAA* (1973): 20; and Lt. Colonel Rod Canning, correspondence, *JRUSI* 124.6 (1979): 94.

68. Major General W.N.J. Withall, "Preface," *AACJ* (1982): 8.

69. General Sir Edwin Bramall, "British Land Forces: the Future," *JRUSI* 127.6 (1982): 18.

70. Major General W.N.J. Withall, "Preface," *AACJ* (1983): 7.

71. Commandant, Army Air Corps Centre [hereafter AACC], *Presentation at Director, Army Air Corps' Conference* (unpublished: AACC, 1982), 7; and Lt. Colonel Rod Canning, "The Role of the Armed Helicopter in NATO," *NATO's Fifteen Nations*, 26.spec. ed. 2 (1981): 103.

72. Major General J.D.W. Goodman, "The British Army Air Corps," *AACJ* (1987): 9; and "Arms Capacity, Endurance Termed Key to Military Sales," *AWST*, 14 January 1985: 81.

73. Paul Beaver, "Army Aviation's Potential Proven," *Defence Helicopter World*, 3.12 (1984): 26–27; Commandant AACC, *Presentation at DAAC's Conference*, 6; and Major General J.D.W. Goodman, "Preface," *AACJ* (1985): 6.

74. Withall quoted in "Army Aviation Gets Tough," *Flight International*, 17 June 1982: 140–141. See Withall "Preface" (1983), 6–7 for the link between more development in the AAC and the general army changes.

75. Bond, "Helcat," 8; and Lt. Colonel J. L. Dawson, "Progress with the Lynx," *AACJ* (1978): 8.

76. "British Anti-tank Helicopter Operations," *Defence Helicopter World* 3.9–11 (1984): 103; and Beaver, "Army Aviation's Potential Proven," 26.

77. See, for example, Ralph, "Tactical Deployment and Roles of the New Look Regiment," 20–24.

78. Withall, "Preface" (1982), 8.

79. Brigadier David Canterbury (Commander, Aviation BAOR) quoted in Paul Beaver, "A Critical Moment," *Defence Helicopter World* 5.6–7 (1986) 48. See also Goodman, "The British Army Air Corps," 9.

80. Air Marshal Sir Patrick Hine quoted in Alfred Price, *Air Battle—Central Europe* (London: Sidgwick and Jackson, 1986) 22.

81. "Rotary File," *Jane's Defence Weekly*, 20 August 1988: 325. See also Group Captain Tim Garden (previously Commander, SH Force (UK)), "The Air-Land Battle" in Air Vice Marshal R. A. Mason (ed.), *War in the Third Dimension* (London: Brassey's, 1986) 161.

82. Ed Hooton, "Air-to-Air Missiles," *Defence Helicopter World* 2.12 (1983): 34.

83. Goodman, "Preface" (1985), 6.

84. Canterbury quoted in Beaver, "A Critical Moment," 46.

85. Lt. General Sir John Learmont, "Military Theory and the Helicopter," *AACJ* (1989): 7–11; and General Sir Martin Farndale, "The Operational Level of Command," *JRUSI* 133.9 (1988): 28.

86. Learmont, "Military Theory and the Helicopter," 11.

87. *Design for Military Operations—The British Military Doctrine*, 47–48 and Lt. General Sir Peter Inge (Commander in Chief, BAOR), "Developments in the Land Battle," *JRUSI* 134.12 (1989): 13.

88. Brigadier C.R.S. Notley and Group Captain J.J.R. Cohu, "Armour/Anti-Armour: The Future," *JRUSI* 132.3 (1987): 19.

89. General Sir Martin Farndale, "The Role of Helicopters in the Central Region to the Year 2000," *JRUSI* supplement 134.12 (1989): 3; and General Sir Martin Farndale, "The Use of Helicopters on the European Battlefield," *Army Quarterly and Defence Journal* [hereafter *AQDJ*] 119.4 (1989): 139.

90. Most clearly revealed in Inge, "Developments in the Land Battle," 12 at a time when the process of deciding which armed helicopter to acquire was beginning to reach a climax.

91. See Colonel E. J. Everett-Heath, "The Helicopter in the Armoured Battle," *NATO's Fifteen Nations* 27.10–11 (1982): 58 for the earlier, tactics-based concept. The change was foreshadowed in General W.N.J. Withall, "The Aim Is to Survive," *Defence Helicopter World* 2:12 (1983): 13 and made explicit by Canterbury quoted in Beaver, "A Critical Moment," 46–48.

92. Ward-Booth, "Preface" (1978), 5.

93. Goodman, "Preface," *AACJ* (1987): 7.

94. For example Brigadier Peter Mead, "The Vital Roles," *AACJ* (1976): 56–61.

95. Farndale, "The Use of Helicopters on the European Battlefield," 139. Farndale was at this time Colonel-Commandant, AAC and Commander-in Chief, BAOR.

96. Major General L.F.H. Busk, "Preface," *AACJ* (1988): 5. As with the Americans in the 1980s, there was a constant exhortation for the AAC to be "soldiers first, fliers second."

97. See von Zugbach, *Power and Prestige in the British Army*, 8, 171 on the importance of teeth arm status. Being a direct-fire arm was even more vital.

98. General Sir Martin Farndale, "Farewell Message," *AACJ* (1988): 7.

99. General Sir Martin Farndale, "Exercise Reforger 87," *AQDJ* 118.1 (1988): 17, 19.

100. General Sir Brian Kenny (Commander, NORTHAG), "FOFA in the Northern Army Group," *International Defense Review* 23.2 (1990): 147; Lt. Gen-

eral Sir David Ramsbotham, "Battlefield Helicopter Operations and Airmobility," *AQDJ* 119.4 (1989): 143–148; and Learmont, "Military Theory and the Helicopter," 11.

101. "Grist attacks conservative thinking," *Defence Helicopter* 10.8–9 (1991): 43.

102. "Britain to Retain Number of Army Helicopter Units Despite Restructuring," *AWST* 29 July 1991: 62. For the RAF, House of Commons Defence Committee Session 1985–86, *The Defence Implications of the Future of Westland plc: Minutes of Evidence and Appendices*, HC169 (London: HMSO, 1986) 18 (hereafter *The Defence Implications . . . Minutes*).

103. Major General L.F.H. Busk, "Preface," *AACJ* (1989): 5.

104. "Britain to Retain Number of Army Helicopter Units Despite Restructuring," 62.

105. Goodman quoted in Beaver, "British Army Helicopter Needs in the 1990s," 9.

106. Learmont, "Military Theory and the Helicopter," 11.

107. Although never defined exactly, the term airmobility in official British Army usage was much more narrow than the American idea of "airmobility" which developed in the 1950s. Airmobility was more than air transport—the ad hoc movement of personnel and materiel by air—but only in that the troops involved were seen as dedicated to operating as light, airmobile forces, in close contact with the supporting helicopters. While critics claimed Britain's eventual deployment of an airmobile brigade was not "airmobility" because its methods were more "pedestrian" than that of, for example, U.S. air assault units, this work will retain the term "airmobility" as generally understood in the British army.

108. Crookenden, "Aviation in the British Army Today," 202–203.

109. Colonel S.M.W. Hickey, *The Army Air Corps Past and Present* (Middle Wallop: Army Air Corps Centre, 1986) 27–28 and Crookenden, "Aviation in the British Army Today," 203. Note the negative view of Dowling, *RAF Helicopters*, 114.

110. Dowling, *RAF Helicopters*, 398–400.

111. Ibid., 262–271; and Everett-Heath, *British Military Helicopters*, 210.

112. Mead, *Soldiers in the Air*, 209–210.

113. Crookenden, "Aviation in the British Army Today," 207 and Walker in *Air Support of Land and Maritime Operations*, 4–5.

114. Mead, "Army Aviation," 46.

115. "Flap Back" (pseud.), "The Next Helicopters," 25.

116. Air Vice Marshal Menaul in *Air Support of Land and Maritime Operations*, 2.

117. Air Chief Marshal Sir Augustus Walker and Crookenden in *The Role of the Helicopter in the Land Battle*, 43–44 and various comments on 44–45.

118. Wilson in ibid., 47.

119. Major General G.C.A. Gilbert, correspondence, *BAR* 67 (1981): 76. Major General Glyn Gilbert, correspondence (1989), Waddy, correspondence.

120. Gilbert, correspondence, 76; and Ministry of Defence, *Statement on the Defence Estimates*, Cmnd.5231 (London: HMSO, 1973) 17.

121. Gilbert, correspondence, 76; and Mallam, correspondence.

122. Major General G.C.A. Gilbert, "An Air-Mobile Force for the Future," *JRUSI* 119.12 (1974): 53.

123. Gilbert, correspondence, 76–77.

124. Air Vice Marshal Sandy Hunter, correspondence (1990); and Air Commodore Tim Garden, correspondence (1990).

125. Gilbert, correspondence, 77.

126. Ministry of Defence, *Statement on the Defence Estimates* (1975), 18.

127. Dowling, *RAF Helicopters*, 21, 400.

128. Waddy, correspondence.

129. Wing Commander Tony Salter, correspondence, *BAR* 66 (1980): 85–86; and Hunter and Garden, correspondence. Pro-helicopter army officers supported the RAF view that it lacked direction from the army: Mallam, correspondence.

130. "Chairborne" (pseud.), "Whither Army Aviation?" 19.

131. Lt. Colonel D. F. Mallam, "Airmobility—Linking in the Third Dimension," *AACJ* (1976): 80; and Mallam, correspondence.

132. "Opsimath" (pseud.), "In Praise of Infantry—Thirty Years On," *BAR* 52 (1976): 6–11 indicated that the use of mechanized infantry had become the major priority for senior Infantry officers. For opposition, see Colonel F. Graham-Bell, "The Army Goes Rotary," *JAA* (1967): 9; and Bramall and Mallam, correspondence.

133. Field Marshal The Lord Carver, correspondence (1989).

134. Field Marshal Sir Michael Carver, "An Address to the IISS," *Survival* 19.1–2 (1977): 36. One, admittedly biased, airmobility proponent described Carver as a "tanks man, through and through."

135. Lamont (then Minister for Defence Procurement) in *The Defence Implications . . . Minutes*, 292; and Major P.D.P. McQueen, correspondence, *BAR* 66 (1980): 83.

136. "Blackbird," "The Value of Their Mobility," 5–12. Blackbird was a pseudonym for Colonel David Mallam.

137. Major T. M. Deane, correspondence, *BAR* 66 (1980): 86. Other responses were Macksey, "The Qualifications of Mobility," 46–47; and McQueen, correspondence, 83–85.

138. Bramall, "British Land Forces: the Future," 20–21.

139. Bramall, correspondence. For the financial shortages of the early 1980s, see Ministry of Defence, *Statement on the Defence Estimates*, Cmnd.8212 (London: HMSO, 1981) 40.

140. Douglas W. Nelms, "HC Mk.1: New Chinook Helicopter for the RAF," *JRUSI* 125.6 (1980): 66.

141. Ministry of Defence, *The Falklands Campaign: The Lessons*, Cmnd.8758 (London: HMSO, 1982) 34; and Wing Commander G. A. Woolley, "The RAF's Helicopter Forces," *JRUSI* 134.3 (1989): 29.

142. Mallam, correspondence.

143. Captain B. W. Barry, "An Airmobile Battalion," *BAR* 79 (1985): 9, 13; and Lt. Colonel A.I.G. Kennedy, "Airmobility—A Practitioner's View," *BAR* 85 (1987): 17. Kennedy later commanded the first established airmobile unit, 24 Brigade.

144. Lt. Colonel Walker, quoted in Captain G.A.B. Grant, "Airmobility: A Trial of Strength," *BAR* 84 (1986): 58.

145. Kennedy, "Airmobility—A Practitioner's View," 17.

146. Ibid., 17–18.

147. Lt. Colonel D. C. Parker, correspondence, *BAR* 84 (1986): 80.

148. Colonel (Ret.) P. G. Fleming, correspondence *AQDJ* 118.4 (1988): 135.

149. Group Captain J. R. Day (Commander, SH Force (UK)), correspondence (1990).

150. Hall and Lamont in *The Defence Implications . . . Minutes*, 292–95. See also the Committee's conclusion: House of Commons, Defence Committee Session 1985–1986, *The Defence Implications of the Future of Westland plc*, HC518 (London: HMSO, 1986) xix (hereafter *The Defence Implications* [Report]).

151. Commandant AACC, *Presentation at DAAC's Conference*, 4; and Goodman quoted in Beaver, "British Army Helicopter Needs in the 1990s," 7.

152. Brigadier J. F. Rickett, "Employment of Non-Mechanized Infantry in 1 (BR) Corps," *JRUSI* 131.6 (1986) 27–32.

153. The first indication of change were lectures given by General Sir Nigel Bagnall and Air Marshal Sir Patrick Hine, Commander of 2nd Allied Tactical Air Force (2ATAF), "Concepts of Land/Air Operations in the Central Region: I and II," *JRUSI* 129.9 (1984): 59–66.

154. General Sir Nigel Bagnall, "Airmobile Operations in Northern Army Group," *NATO's Sixteen Nations* 33.10 (1988): 75.

155. General Ferdinand M. von Senger und Etterlin, "New Operational Dimensions," *JRUSI* 128.6 (1983): 11, 12. See also "The Air-mobile Divisions: Operational Reserves for NATO," *JRUSI* 132.3 (1987): 23–30.

156. Von Senger und Etterlin, "New Operational Dimensions," 15.

157. See, for example, comments by Goodman in Beaver, "British Army Helicopter Needs in the 1990s," 9.

158. T. Sainsbury (Under-Secretary for Defence Procurement), "The British Army: 1988," *AQDJ* 118.4 (1988): 140.

159. These problems were best expressed in Paul Beaver, "Realities of UK airmobility plans," *Jane's Defence Weekly*, 13 February 1988: 258.

160. *The Defence Implications* [Report], xi; and Woolley, "The RAF's Helicopter Forces," 30.

161. "UK forms airmobile brigade," *Flight International* 6 February 1988: 15; and editorial comment, *AQDJ* 118.4 (1988): 136.

162. Kennedy, "Airmobility—A Practitioner's View," 18.

163. As attested to by the tone and content of correspondence and numerous interviews with serving and retired officers.

164. Farndale, "Farewell Message," 7.

165. Note the optimistic, though subtly worded, remarks in Goodman, "Preface," *AACJ* (1984): 8.

166. Goodman, "Preface" (1987), 7–8; and "Decision over UK support helicopters," *Jane's Defence Weekly*, 30 January 1988: 162.

167. Kennedy, "Airmobility—A Practitioner's View," 19.

168. Collins, correspondence.

169. Group Captain John R. Day, "The Contribution of Support Helicopters to Airmobile Operations," *JRUSI* supplement 134.12 (1989): 18.

170. Bagnall and Hine, "Concepts of Land/Air Operations in the Central

Region: I and II," 59, 63; and *Design for Military Operations—The British Military Doctrine*, 40.

171. Woolley, "The RAF's Helicopter Forces," 29.

172. Major B. W. Barry, correspondence, *BAR* 82 (1986): 76.

173. Von Zugbach, *Power and Prestige in the British Army*, 143.

174. See Learmont, "Military Theory and the Helicopter," 11, for an implicit approach along these lines. Compare with Farndale, "Farewell Message," 7.

175. Lt. General Sir David Ramsbotham, "British Developments," *JRUSI* supplement 134.12 (1989): 7; and Inge, "Developments in the Land Battle," 11–12.

176. Von Senger und Etterlin, "New Operational Dimensions," 13, 15; and Brigadier R. D. Grist, "Airmobile Forces in Central Europe," *JRUSI* 133.3 (1988): 41–48.

177. Ramsbotham, "British Developments," 9.

4

"A Tale of Two Helicopter Forces"—Developments in West Germany and France

After initially using helicopters for transport and light aviation tasks, West Germany began developing armed helicopters from the late 1960s onward. These helicopters were intended to operate in close coordination with front-line units acting as a highly mobile anti-tank force at the tactical level. Helicopters were also used with infantry brigades to provide fast-reacting airmobile reserves at the operational level. Until recently, these two forces were kept separate, but the West German army began to add anti-tank helicopters to the airmobile units, creating a much stronger mobile reserve—in the process reviving an idea that had first been suggested by German helicopter advocates some twenty years before.

Developments in France were greatly accelerated by the need for helicopter-based mobility in the Algerian War. After this campaign, the French army had nearly six hundred helicopters and sought to formulate doctrine for their use in conventional warfare. At first they were deployed piecemeal, subordinated directly to divisions. As anti-tank helicopters were introduced in the mid–1960s, the possibilities of using helicopters in a more independent fashion were recognized and, although the process of actual change was slow, by 1977 most of the army's helicopters were centralized under the control of its three corps. The importance of helicopters was further exemplified by the creation in 1985 of an army-level airmobile division with 240 helicopters whose task was not to support or to combine with other ground forces but, in fact, to lead France's new *force d'action rapide* (FAR) in case of a war in Europe.

In both France and West Germany, the late 1980s saw military helicopters come of age: their roles and organizations were more firmly established and their contribution to each army was more highly valued and accepted. In both cases, the developments throughout the last four decades came about because of growing recognition of the value of the mobility advantage of helicopters over armored vehicles. The defense planning of France and West Germany was affected by many of the same factors, and their helicopter forces were relatively similar in size and capability, but the way in which each country applied helicopter mobility to its situation differed. In particular, the French army accorded the helicopter a more active and aggressive role, primarily because of its peacetime strategic circumstances. On the other hand, West Germany's defensive orientation and desire to prevent any incursion into its territory led to doctrines in which most helicopters were integrated into the first line of defenses.

Because of more limited access to source material, this chapter does not go into as much detail as the previous three. What is interesting about France and West Germany, and what compensates for the lack of detail, are the comparisons that can be drawn between them. Therefore, after examining them both, the chapter will conclude by comparing the two.

WEST GERMANY

Development of the Anti-tank Helicopter

It was only when armed helicopters were introduced that the *Heeresflieger* came to be regarded as an important part of the army. In 1956, the first *Heeresflieger* (Army Aviation) units were established, using British Scout and Sycamore helicopters for observation, liaison and light logistic work. These duties remained a secondary task for the *Heeresflieger* into the 1980s, with squadrons of Aérospatiale SA.318 Alouette IIs and then Messerschmitt-Bölkow-Blohm (MBB) BO–105Ms being assigned to each division and corps.[1] In 1960 an Alouette II was tested as a firing platform for the SS–11 anti-tank missile and throughout the decade West German aerospace companies and the armed forces continued to study suitable technology and employment tactics. There was considerable discussion about the missions helicopters could perform, with the *Luftwaffe* and the *Heeresflieger* supporting the respective merits of fixed-wing aircraft and helicopters for anti-tank and other forms of close air support.[2] Debate intensified after 1967 when NATO abandoned their tripwire strategy, with its reliance on nuclear weapons, in favor of flexible response, which emphasized conventional defense.[3] Given the preponderance of Warsaw Pact armored forces facing West Germany and the country's defensive orientation, it was not surprising that the main requirement of the new strategy was to improve the country's anti-tank defenses.

The *Heeresflieger* began developing an innovative and wide-ranging doctrine for the employment of helicopters in these new conditions. It was inspired by the U.S. airmobility concept and its application in Vietnam,[4] but sought to modify it to suit the NATO environment. The *Heeresflieger* wanted to acquire a large number of unsophisticated, cheap helicopters armed mainly with ATGMs. They would form the basis of a new airmobile organization, combining reconnaissance, fire support, anti-tank and transport helicopters with regular and anti-tank infantry in numerous units of approximately company strength, each with about twelve helicopters. These units were to have been the army's main reserve, exploiting the helicopter's superior mobility to respond to armored breakthroughs or air landings.[5]

However, these units were never formed, not least because the *Luftwaffe* was opposed to armed helicopters and wanted instead to modernize its own force by buying the expensive Tornado fighter-bomber. All funds for aircraft and helicopters came from the same part of the defense budget and although the commander of the *Heeresflieger* at the time, Brigadier Hans Dreibing, optimistically asserted that this simply meant there was "healthy rivalry" between the two flying forces, in the 1970s helicopters received a low priority in comparison to *Luftwaffe* procurement.[6] The situation was exacerbated by the technical impossibility of producing the inexpensive, simple helicopters on which the *Heeresflieger* concept was based.[7] Equally, the *Heeresflieger*'s relatively poor status in comparison to the armor, artillery and infantry branches made it difficult to gain acceptance within the army for the innovative plan. Because of the army's lack of experience with helicopters, it was hard to establish convincingly the actual value of the new units. Reservations mainly centered on the assumed vulnerability of helicopters, since the offensive tactical action envisioned in the *Heeresflieger*'s concept greatly increased the risk of confronting enemy air defenses. Yet armed helicopters were not rejected entirely, because they offered some advantages that the army was eager to exploit, and Ansbach had shown that, with the right tactics, they could survive. West Germany's evaluation of the anti-tank helicopter, which culminated in its participation in the Ansbach trials of 1972, gave more positive results on survivability, but only if the helicopters remained concealed, using NOE flight as much as possible, and engaged the enemy outside the range of its air defense weapons.[8]

Noticeably, it was only after Ansbach that anti-tank helicopters began to figure prominently in West German defense planning. Instead of forming the independent airmobile units with their offensive orientation, the army decided to pursue the more limited concept of an anti-tank helicopter force that would utilize defensive ambush tactics.[9] This force was to be integrated with the rest of the army, an approach that accorded with current doctrine and so was more easily accepted than the *Heeresflieger*'s independent plan.[10] Finally, because a limited anti-tank force was less expensive, it did not conflict with existing priorities in defense spending—primarily the acquisition of the Tornado and the Leopard tank—which were based on more proven ideas and supported by more powerful groups within the armed forces.

Helicopters were particularly attractive because their mobility offered a partial solution to the problem caused by West Germany's lack of operational depth and its resulting commitment to "forward defence."[11] This strategy was politically necessary since it guaranteed that no part of West Germany would be given up willingly to an enemy and that every possible effort would be made to recapture lost territory; yet it made defense planning much more complicated, particularly for an army that had traditionally favored defending in depth. While tanks were regarded as the most important part of the defensive system, it was considered that "anti-tank helicopters are ten times as fast as tanks and they can quickly establish, reinforce, or shift points of main anti-tank effort."[12] Thus armed helicopters were able to compensate for the lack of operational depth because of their ability to react quickly, re-deploying rapidly across West Germany's shallow defensive zone.

Although these relatively scarce resources were concentrated into three regiments, one per corps, with a smaller unit allotted to the independent division in Schleswig-Holstein, these regiments were not, however, reserves to be employed by the corps commanders. They were an administrative organization that simply allowed a corps commander to retain control over the helicopters; in combat, the PAH–1s (*Panzerabwehr Hubschrauber*, or anti-tank helicopter) were to be assigned downward through divisions to the front-line brigades. It was recognized that, because of their mobility, anti-tank helicopters might perhaps have to operate as a corps reserve in a "situation of last resort when . . . there are no other quick reaction forces available."[13] However, this possibility was thought to be very remote. The normal and most practiced method of operation involved the regular attachment of PAH–1 platoons to brigades and the establishment of a close understanding between them.[14](Figure 4.1)

Since the PAH–1s were assigned to brigades, they were able to engage the enemy as part of the main battle, thereby attempting to prevent any breakthrough from occurring, rather than being relied on to counter such penetrations after they had happened. Close integration with front-line units decreased reaction times and distances from the intended target, which compensated for the relatively poor endurance of the PAH–1. The association with the brigades also promoted better coordination between the anti-tank helicopters and other forces and created "an interlinkage of all anti-tank weapons, forming one single system."[15]

Photo 4.1
BO 105 PAH–1

The *Heeresflieger*'s principal armed helicopter. The six HOT anti-tank missile tubes are clearly visible. (Photograph: Eurocopter. Used by permission of Eurocopter and *Heeresfliegerwaffenschule*.)

Figure 4.1
***Heeresflieger* Organizational Structure**

Heeres flieger Aviation Command

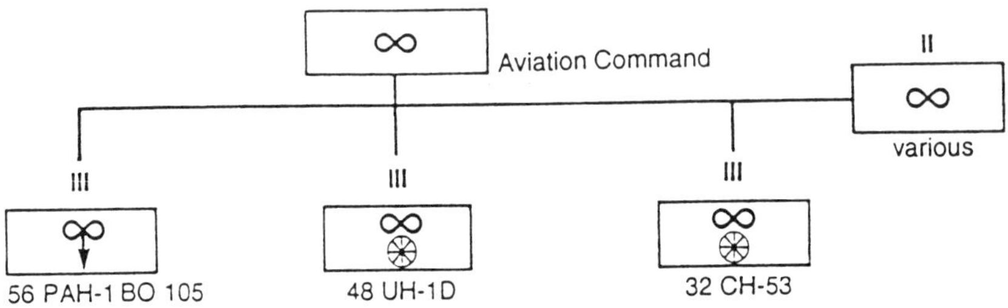

Source: *World Airpower Journal*, 1.1 (1990):140.

Integration meant that the West Germans did not require observation helicopters to find targets for the anti-tank units, since the brigade or division's own reconnaissance and intelligence-gathering assets were to fulfill that function.[16] While operationally integrated with other ground forces, anti-tank helicopters were to fight separately at the tactical level. Indeed, even though the normal helicopter detachment given to a brigade was no larger than a platoon of seven PAH–1s, this unit was considered to be virtually an independent command. The brigade would generally have used them in a situation when other forces could not respond to an enemy threat, either because they were not mobile enough or because all the brigade's forces were otherwise committed.[17]

The choice of which helicopter to meet the anti-tank requirement was determined mainly by cost and time constraints and, as one senior *Heeresflieger* officer remarked, this meant that "compromises" were made to the detriment of combat capability.[18] MBB had proposed initially to build, in cooperation with the Italian firm Agusta, the BO 115, a new helicopter that would have been specifically designed for armed combat. However, the cost and fears that it would take too long to develop meant that this offer was rejected. Instead, a modified version of the existing BO 105 was selected and brought into service in 1980. This helicopter—which in fact proved to be harder to modify than first thought (particularly in relation to engine performance and maintenance requirements)[19]—was never intended to be more than a short-term solution; by 1974, the requirements for a new helicopter, the PAH–2, were already being defined with the aim of acquiring it by the mid–1980s.[20]

However, the PAH–2 suffered numerous delays and near cancellations. Therefore, although the tactics that evolved during the 1970s to govern the PAH–1's combat employment remained relatively stable,[21] the helicopter itself underwent a number of modernizations to extend its useful life until the arrival of the PAH–2. The first improvement addressed the threat posed to the PAH–1 by enemy helicopters. For many years it was thought that the considerable agility of the PAH–1 and the presence of ground-based air defenses would prevent enemy helicopters from attacking it successfully; furthermore, there was some thought that the Luftwaffe's Alpha Jets might attack enemy helicopters,[22] even though it was assumed that enemy fixed-wing aircraft did not threaten the PAH–1. This attitude probably resulted as much from a realization that the resources needed to protect West Germany's helicopters more directly were not available as from any innate confidence in the army's air defenses.[23]

By 1984–85, the increasing numbers of enemy armed helicopters made the problem too serious to ignore, not just for the *Heeresflieger* but for the army as a whole.[24] While ground anti-aircraft units remained the primary defense, the *Heeresflieger* decided to equip some fifty observation BO 105Ms with air-to-air missiles and designate them *Begleitschutz Hubschrauber* (BSH, or escort/protection helicopter). The new helicopters were designed "to deal with enemy helicopters that have managed to slip through the defence network of the other systems and thus jeopardise the helicopter operations of friendly forces."[25] Similarly, it was planned to improve the intrinsic survivability of the PAH–1 by fitting radar warning receivers and infrared radiation suppression devices to assist the helicopter to escape detection. In the 1970s it had been thought that anti-tank helicopters ought to be as simple as possible and that they did not need such added safety features.[26] In an example of the way in which actual capability can lag well behind theoretical requirements, the funds for the BSH project and the fitting of survival equipment only recently became available.[27]

The last development concerned the PAH–1's capacity for night fighting. The need for this capability had been recognized when the PAH–1 was being developed but had been omitted, first because night vision devices were both expensive and relatively unproven and second because the BO 105 was insufficiently powerful to take the extra weight of the equipment. In the late 1980s, with the rest of the army acquiring a night-fighting capability and the PAH–2 many years from completion, modifications began on the PAH–1 to provide the necessary additional lift to allow for night-vision equipment and night sights. Transport helicopters were also equipped for flight with night-vision goggles, giving them more chance of using the cover of darkness to move about the battlefield without detection.[28]

The PAH–2

The modifications to the PAH–1 were necessary because of the numerous problems that beset the PAH–2 project. It was a joint development between France and West Germany, initiated in the mid-1970s, which aimed to produce one variant, the PAH–2, for the Germans and two variants, HAC–3 and HAP, for the French (see below). Saving money was the main reason for this approach since it was more expensive to design and build two separate helicopters, especially when only two hundred machines were needed by each country.[29] Yet the financial advantages were offset when the difficulties of resolving the distinct military and industrial views of the two partners proved almost insurmountable. For example, the French decided that having the pilot in the nose of the helicopter was better, while the West Germans wanted the gunner stationed there. As well, in one of the few examples in the development of helicopters where industrial concerns were not determined strictly by the requirements of a country's armed forces, Aérospatiale was interested in designing a lighter, cheaper helicopter more suited to its well-established export market; in contrast, MBB and the West German army concentrated solely on the requirements for fighting a major war in Europe. Finally, the French were adamant that the helicopter be equipped with European-designed avionics because of their desire to demonstrate political and economic independence, whereas the West Germans were eager to import an existing system from America.[30]

These differences of opinion soon delayed the helicopter's design and, although the companies and the military eventually worked out a compromise, in the meantime the price had risen so much because of the postponements that the economic feasibility of the project was called into question.[31] The PAH–2 faced a very uncertain future for a number of years but was eventually saved because of military and political necessity. First, the continuing emphasis that combat helicopters were receiving in each country meant that neither army could afford not to develop a new helicopter. Given that any new helicopter almost certainly had to be produced indigenously for political reasons, the financial benefit of joint development, design and production remained. More important, France and West Germany's defense ministers were determined to display real evidence of their countries' new-found cooperation on defense matters, helping to overcome concern in both countries about the high cost of the new helicopter.[32] In concert with the Franco-German Brigade, the PAH–2 project was a valuable symbol for French involvement in Europe's defense and West Germany's acceptance of that involvement, at a time when the INF treaty negotiations were making Europe nervous about America's commitment to its defense. Recently, the contracting size of the helicopter market has forced Aérospatiale and MBB's helicopter divisions into a complete merger, suggesting that military design and acquisition is moving in the same direction as other political and economic structures in the increasingly common European community.

When the PAH–2, now known as the Tiger, finally arrives, it will be a considerable advance on the PAH–1, not least because it was designed specifically as a weapon system and not adapted from a civilian design. It will be more reliable, have better endurance and be more resistant to damage; it will carry eight newer and more effective ATGMs as well as its own air defense missiles. Moreover, the provision of a mast-mounted sight, as on the American OH–58D AHIP, will enhance its survivability by allowing the helicopter to remain behind cover except for the few seconds needed to fire its missiles.[33]

Airmobility

While the *Heeresflieger* was introducing its anti-tank helicopters, a gradual change also took place in its transport activities. The two hundred or so Bell/Dornier UH–1D helicopters bought in the late 1960s for support tasks were complemented in the early 1970s by the purchase over one hundred Sikorsky/VFW CH–53 helicopters as part of the renewed emphasis on conventional defense, greatly enhancing the *Heeresflieger*'s capabilities.[34] These helicopter transport regiments, also assigned to West Germany's three corps, continued to be used for liaison, medical evacuation and moving supplies, but a more important task began to emerge in the late 1970s as they began to provide airmobility for the corps' *Luftlande*, or parachute brigades.

An army-wide reorganization in 1979 made these brigades the most important corps reserves to be used either to counter enemy armored penetrations or to reinforce other units in the corps.[35] Previously, each corps had an armored regiment as its main reserve, but these units were converted to front-line brigades,[36] and the constraints of the forward defense doctrine decreased the ability of corps commanders to withhold such brigades from the main battle. In the reorganization, two airmobile anti-tank battalions equipped with TOW ATGMs and light vehicles were added to each *Luftlande* brigade to augment their existing two paratroop battalions. As with the PAH–1s, it was the speed of aerial movement that made these units suited to this role: although lightly equipped and thus unable to engage in sustained combat, the *Luftlande* brigades were able to react much more quickly than armored units to any enemy incursions past the defensive front line.

While the brigade retained the capability to deploy by parachute, it became normal to use helicopters following the addition of the new battalions. Helicopters were able to extract and relocate the force once

Photo 4.2
HAC/PAH–2 Tiger

First prototype of the French and West German armies' latest, purpose-built attack helicopter. Note the mast-mounted sight. (Photograph: Eurocopter. Used by permission.)

it had deployed, thus helping to overcome the inherent immobility of paratroops after they had landed. Helicopters were also more effective for rapidly deploying engineer equipment to create minefields and other barriers. Finally, while not invulnerable, there was less risk of aerial interception when using low-flying helicopters than there was during a parachute drop. Thus, by the mid–1980s, it was usual for a considerable portion of a corps' transport helicopters—particularly the CH–53s—to be assigned to supporting its *Luftlande* brigade.[37]

The developments and the introduction of the PAH–1 meant that, by the mid–1980s, helicopters were an integral part of West Germany's defense planning. Yet, although it was not inconceivable for the PAH–1 units and the *Luftlande* brigades to have operated together, the two were not closely linked. The common method of employment was for the corps commander to send the anti-tank helicopters forward to support the armored brigades, while retaining the *Luftlande* brigade as the last reserve unit, thus precluding practical cooperation between the two. However, this approach, which utilized helicopters in a rather fragmented fashion, began to change in 1988. Using the term coined by one of their most senior officers, Field Marshal von Senger und Etterlin, the West German army began combining its corps level *Luftlande* and *Heeresflieger* assets into five smaller airmechanized brigades.

Airmechanization

As noted in chapter three, von Senger was convinced that helicopters could only be exploited fully if, instead of "scatter[ing] the new forces among the old," they were organized and used en masse.[38] He and others saw airmobile or airmechanized units as the solution to NATO's most pressing problem: its lack of operational reserves, controlled by senior commanders, which were sufficiently mobile to react in time against a fully mechanized enemy. While von Senger's ideas were more of a general theory than a specific blueprint for reform, they were clearly responsible for more than just the name of the new brigades.

As in the reorganization of 1979, the primary purpose of airmechanization was to strengthen the combat power of the corps' reserves by adding anti-tank helicopters to the existing airmobile infantry, thus creating a much more powerful unit. The reserves were designed to

> Block advancing or penetrating enemy forces; control terrain threatened by the enemy or prevent the enemy from seizing it; constitute a quick and flexible anti-armour barrier; threaten the aggressor's flanks; and fight enemy elements such as enemy air landings effectively in their own rear.[39]

Moreover, because of their speed and range of operations, they "enable[d] the commander to keep or recover his battle initiative by launching highly mobile surprise attacks over wide areas."[40] The ability of the airmechanized units to operate independently of other forces, although in coordination with them, ensured that the speed and reaction capability of such helicopter-based forces could be exploited even more fully than before.[41] As one of the likely commanders of the new divisions stated, "In the long term airmobility of land forces will lead to airmechanized forces able to conduct combined arms fighting in accordance with its own particular air-to-ground and air-to-air rules."[42] Only by creating separate units was it possible to exploit properly the helicopter's superior mobility and range, while at the same time allow them to coordinate their actions with ground-based forces.

Until this time, helicopters had generally been regarded as playing a supporting role, subordinate to the main combat arms of infantry and armor. In contrast, the *Heeresflieger* had considered itself to be comparable to these arms in stature and independence, if not in number or power. Therefore, it was eager to pursue airmechanization because of the chance to establish the importance of helicopters more widely throughout the army.[43] Equally, the new formation was a way of ensuring that financial and personnel resources would continue to be devoted to helicopters and to the *Heeresflieger*. Thus there were organizational and bureaucratic pressures that reinforced the purely military reasons for creating the new brigades.

It will be many years before airmechanization is fully implemented, and changes in Germany after reunification will presumably lead to even more radical organizational upheavals. The pace of its introduction is yet to be determined and will depend on political and financial considerations. The new organization will rely on existing helicopter resources until at least the next century and the introduction of the PAH–2, a new utility helicopter and the modernization of the CH–53.[44] Nevertheless, even without implementation, airmechanization was a fundamental doctrinal development. While anti-tank helicopters had come to play an important part in West Germany's defense planning, the full exploitation of the helicopter's speed—always seen as its most important characteristic—was hampered by the perception that the PAH–1 units had to be used in support of existing arms.[45] Then, faced by an enemy that was growing more capable of projecting combat power past the forward defense that West Germany hoped to maintain, a highly mobile reserve became even more necessary than before. As far as senior officers were concerned, in responding to attacks by forces that were themselves fully mechanized, the mobility of the helicopter was essential.[46] In the uncertainty of the post-USSR military situation, where unpredictable threats might emerge, the importance of the mobility differential can only increase and with it the attention paid to integrated, large helicopter formations.

FRANCE

Developments in the 1950s and 1960s

France began using military helicopters during its postcolonial counterinsurgency campaigns, at first in small numbers for casualty evacuation in Indochina and then in Algeria for a much greater range of missions. The latter war was fought against a guerrilla enemy over a large and often mountainous area. These conditions demanded great mobility and flexibility and, to meet this need, over six hundred helicopters were eventually deployed. They performed a variety of tasks but were mainly used to move troops in response to enemy attacks and in offensive operations against known guerrilla strongholds. As one observer stated, "In Algeria, the helicopter is the manoeuvre itself."[47] It was the first time that they were used on such a large scale, and the extent of France's commitment can be compared with the major British effort in Borneo, which was supported by no more than eighty helicopters. Helicopters proved to be the only means by which a small number of French troops could combat its elusive enemy. The Algerian War played a very important role in establishing the helicopter in the French army.[48] After assessing its experiences, the army realized that the helicopter could play an important part in more conventional warfare. But while movement of troops had been crucial in North Africa, the use of armed helicopters had also become commonplace when detachments of airmobile infantry began coming under enemy attack during and after landing. Helicopters were fitted with rockets and machine guns to provide readily available fire support. The war also provided an opportunity for France to test helicopter-fired guided missiles in combat, thus developing further its pioneering work of the 1950s.[49] And, in the shift back to Europe, it was the armed and ATGM-carrying helicopter that became the new focus.

The Algerian War was also important because it led to the acquisition of a large number of helicopters that were then available for application to conventional warfare. Such a fleet of helicopters would have been very difficult to obtain but for the needs of the fighting in Algeria, since financial restrictions and bureaucratic opposition would have been much greater in peacetime. It also brought crucial structural and organizational changes. Although helicopters from all three services had assisted the ground forces in the Algerian War, in the early 1960s the army gained exclusive control of helicopters supporting the land battle, leading to the formation of *aviation légère de l'armée de terre* (ALAT, or Light Aviation of the Army). After Algeria, each division received a *groupe aviation légère division* (GALDIV) with a mix of liaison and observation (the Alouette II), anti-tank (the Aérospatiale SA.316 Alouette III with SS–11 missiles) and transport helicopters (initially the American CH–21, replaced in 1969 by the Puma).[50] This regular organization and the army's exclusive ownership of helicopters, when allied with the combat experience gained in Algeria, laid the foundation for future developments.

The 1970s: The First *Regiment d'Hélicoptères de Combat*

Yet the first major doctrinal change eventually resulted in the emergence of a very different arrangement of ALAT's helicopters within the army than had been set up after 1962. The majority of helicopters were concentrated at corps level, thus removing support from divisions and resulting in a more independent role for ALAT. Previously, helicopters had "to serve all branches of the Army" and were restricted to "auxillary or support missions," operating in close association with ground forces and in reaction to their needs.[51] However, in the late 1960s, ALAT began to develop the concept of preplanned "autonomous combat missions," combining observation, fire and maneuver, with helicopters being employed in situations where ground troops were in fact not being used. This idea was designed to take advantage of the helicopter's superior mobility and to ensure that helicopters were not employed piecemeal.[52] At the same time, ALAT was concentrating on destroying enemy armored forces, and there was continuing interest in how anti-tank helicopter units might be used more effectively. With the support of successive chiefs of staff, ALAT conducted extensive trials that focused mainly on grouping anti-tank helicopters in larger units that then operated independently.[53] General Maurice Cannet, a key figure in French helicopter developments who eventually commanded ALAT in the late 1970s, summed up the idea: "At the end of the 1960s, I was one of those trying to convince people of a new concept, namely, that a formation of about 50 helicopters could block an [enemy] armoured division."[54]

Thus, by the late 1960s, ALAT had become convinced that the decentralization of aviation assets throughout the army was incompatible with the perceived capabilities of the combat helicopter. In 1969 units similar to the GALDIVs were formed at corps level (GALCA—*groupe aviation légère corps d'armée*), and by 1975 helicopter squadrons from the two different groups were sometimes assembled temporarily into larger formations. Yet the detailed planning and coordination required to make airmobile missions successful was inhibited by these ad hoc arrangements; in particular, the use of ground troops in cooperation with the helicopters was very difficult because no infantry were integrated with the helicopter units.[55] A new organization was required to put these doctrines into practice. However, as can be inferred from Cannet's remark, there was opposition because of doubts about the helicopter's viability in modern warfare.[56] There were also financial constraints, both

Figure 4.2
***Regiment D'Hélicoptères De Combat* (1977)**

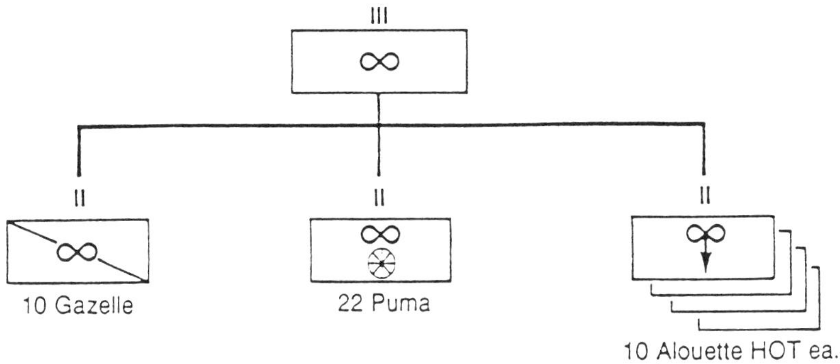

Source: Cannet, "Aviation in the French Army," 1–2.

because of the decision to limit conventional defense spending in favor of the nuclear *force de frappe* and the economic crisis in 1974–75.[57] Although a lack of funds affected the entire army, the ability of a small supporting arm such as ALAT to bid for resources against more established sections was particularly limited.[58]

However, in 1977, the changes that Cannet and ALAT were seeking began to be implemented. While the acquisition of more capable SA.341F Gazelle and Puma helicopters gave ALAT a demonstrably more effective force, making its theories more acceptable and helping to overcome previous opposition, more important factors were at work that were outside the control of helicopter enthusiasts. The financial restrictions were relaxed and more money was made available for conventional defense preparations. Furthermore, the whole army was reorganized, leading to an increased emphasis on corps operations and a reduction in the size of divisions. Consequently, it was logical to concentrate army helicopters at corps level. Moreover, the number of divisions more than doubled and, given the army's "relatively small aviation resources," maintenance of the old GALDIV units in each division was no longer feasible without even greater dispersion of helicopters throughout the army.[59] As a result, each divisional aviation unit was reduced to just ten liaison and observation helicopters, and corps aviation assets were dramatically strengthened. Each corps had one *groupe d'hélicoptères légère* (GHL, or light helicopter group) of thirty liaison and observation helicopters and, more importantly, two *regiment d'hélicoptères de combat* (RHC, or combat helicopter regiment—Figure 4.2) of seventy-two helicopters, combining "intelligence gathering, firing and movement" and "operating in *coordinated action within the same organisation* [emphasis added]."[60]

The reorganization confirmed that ALAT was a combat arm capable of independent action. As noted above, helicopter units had previously been devoted mainly to supporting other arms. Now they would work "in front of divisions, on their flanks or between divisional boundaries."[61] To ensure that the RHCs were not distracted from their independent missions, the other corps helicopter unit, the GHL, was tasked specifically to conduct liaison and light logistic work in direct support of the corps' other troops.[62] Thus, at a time when most armies were stressing the need for combat helicopters to work closely with ground forces, ALAT was preparing for more independent operations. The helicopter units were still linked to ground-based forces, but the tactical methods of the RHCs were such that it was the results of armored vehicle and helicopter actions that were to be combined, not the actions themselves.[63]

Drawing to some extent on their experiences in Algeria, the French emphasized the value of combined ground and air action within the RHC.[64] This meant that the RHC's Puma helicopters were accorded considerable importance, being termed "maneuver" or even "combat" helicopters. The Pumas were also crucial because they gave the RHC a resupply capability equal in mobility to the anti-tank helicopters, thus enabling the unit to sustain itself in combat more effectively.[65] The new organization facilitated the close training and coordination required by airmobile units. Since most of the difficulties with the old makeshift method had resulted from the lack of infantry directly attached to the helicopter units, the RHCs were now "colocated with the corps infantry regiment."[66] Although the ground troops were still not completely integrated with the helicopters, the arrangement was better than before, and for all practical purposes they were regarded as a combined unit.

While the RHC was an advance in theory, it did not immediately lead to an equal improvement in actual capability. Despite the purchase of Puma and Gazelle helicopters in the early 1970s, and even with the continued use of its obsolescent Alouette IIs, in 1978 ALAT still needed 110 extra helicopters to meet the requirements of its new organization. Even by the mid–1980s the situation had not improved. Only two out of six RHCs were at full strength, with a third RHC at 67 percent of its establishment. Of the three other RHCs, two were below half strength and equipped with a varied collection of helicopters and one was to be formed from ALAT's training unit in time of war. The crucial weakness was in anti-tank helicopters since only half of the 158 HOT anti-tank missile-equipped SA.342M Gazelles necessary to meet the requirements of the 1977 reorganization had been acquired; nor had plans to equip the SA.341F Gazelle with HOT progressed very far.[67]

Nevertheless, because of the considerable advances in employment doctrine that it represented, the formation of the RHCs was a substantial step forward for ALAT. The need for more anti-tank helicopters was recognized and, in time, sufficient numbers were acquired, boosting

ALAT's total aircraft inventory by some 20 percent. ALAT's personnel strength also rose from 3,700 to about 6,500, with double the number of officers, thus improving its organizational strength.[68] Because these officers were required to serve in other branches, the army as a whole became more experienced and more familiar with helicopters. Above all else, ALAT's helicopters became for the first time "a significant combat element on the modern battlefield."[69]

The 1980s: From the RHC to *4e Division Aéromobile*

The RHCs had no sooner became an established, if not yet full strength, part of the army when ALAT underwent another major reorganization. After the creation of the FAR in 1983, in 1985 a number of helicopter regiments were concentrated to form the *4e division aéromobile* (4e DAM) with 240 helicopters. The new unit was a further expression of the movement toward powerful, independent helicopter units that had begun with the RHC; indeed, 4e DAM was based on the experience gained with the earlier organization.[70] Once again, like the RHCs, the very gradual buildup of helicopter assets within ALAT meant that the new organization did not achieve full strength for some years.

While members of ALAT were in no doubt about the importance of helicopters in modern war, the creation of 4^e^ DAM did not result from their advocacy alone but was linked directly to the establishment of the FAR. Following many examples of the importance of being able to respond quickly and decisively to emergencies, the FAR was set up to allow France to demonstrate by military means its political authority and determination in a crisis.[71] It was composed of 4e DAM and another new division, *6e division légère blindée* (6e DLB, or light armored division), as well as *27e division alpine*, *9e division d'infanterie de marine* (9e DIMa) and *11e division parachutiste*.[72] The use of strategically mobile troops such as marines and paratroopers was not new in the French army; they already constituted an intervention force for use in former colonies and France's overseas territories. What was so novel about the *force d'action rapide* was the intention to employ it within Europe, and it was for this reason that 4^e^ DAM was set up. According to a senior member of ALAT, General Betrand de Lacroix de Vaubois, creating the new division was "a political act,"[73] based on the strategic possibilities of the helicopter's ability to move rapidly over long distances and yet still have sufficient combat power to fight in a major conventional war. Indeed, to highlight the great difference in mobility between 4^e^ DAM and the rest of the army, the commander of the FAR likened the latter to the normal railway system, while 4^e^ DAM was, in comparison, the TGV or very fast train.[74] It cannot be overemphasized that, although 4^e^ DAM and its supporting unit 6^e^ DLB were discussed as part of the whole concept of the FAR and its many possible uses, the predominant assumption was that these two units were intended specifically for a major European war. In particular, the repeated references to 4e DAM being the spearhead of the FAR can only be understood in reference to Europe: in any intervention in a major European conflict, the FAR would have to rely first of all upon 4e DAM, supported primarily by 6e DLB.[75]

Thus, 4e DAM was an "*innovation stratégique capitale*," as Defense Minister Charles Hernu stated in 1984.[76] It gave France additional flexibility and political freedom of action with which to implement its declared strategy of deterrence. It was considered that a rapid, conventional response in support of NATO during a crisis in Europe might well be sufficient to deter any threatened aggression; even the possibility of such a reaction was a deterrent.[77] This expansion of deterrence from purely nuclear force to include conventional means was part of France's efforts to make more contribution to the defense of West Germany (the latter was never comfortable about being defended by France's short-range nuclear weapons). It was also a response to the growing threat of a conventional Warsaw Pact attack on NATO, against which it might prove politically impossible to use a traditional nuclear deterrent.[78]

The importance of the helicopter in this new strategic approach was that very rapid action was needed to deter an enemy before or just after it had attacked. Only a unit such as 4e DAM, which combined firepower and mobility, would be able to intervene quickly and effectively. Even if deterrence failed, such mobility combined with the emphasis on antitank weapons would make the division an ideal counter to the major threat that France's forces would face: the penetration of NATO's front line by Soviet operational maneuver groups and other deep battle forces.[79] Thus it was their superior mobility that led one aviator to describe helicopters as "*la carte maîtresse*" or trump card of future conflict. It was especially important because of the growth of urban areas and the likelihood of refugees blocking roads, which meant "mechanized units on their own will not have the necessary capacity for rapid reaction or intervention."[80] ALAT's helicopters were indeed "*forces d'action rapide*," possessing sufficient speed to concentrate and then disperse to avoid being attacked, to react quickly in a defensive emergency and to accelerate the tempo of operations during the offensive.[81]

The extent of the faith placed in helicopters and perhaps the persuasiveness of their supporters can be judged from the fact that, when the FAR was announced amid much emphasis on 4e DAM, the concepts behind the new division had not even been tested to any great extent. Following the 1983 defense plan that established the FAR, intensive studies were commenced into how an airmobile division might best be organized and employed. The culmination of this activity was a major field exercise in September 1984 known as *Damoclès*, in which a prototype

Figure 4.3
4e Division Aéromobile

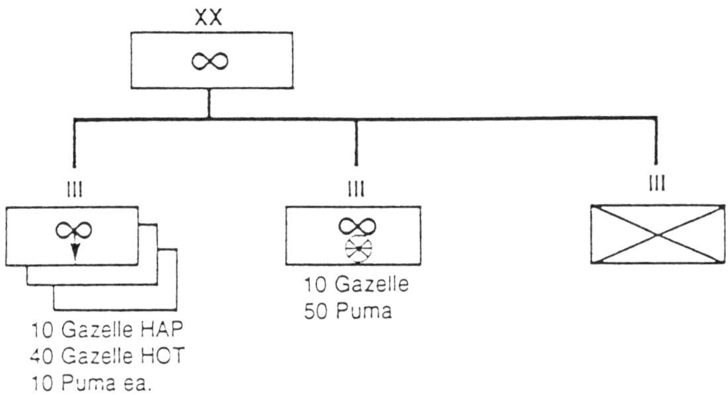

Source: Various articles, including Bonnet, "French Army Aviation," 5.

of 4e DAM was tested in its primary mission of independently countering an enemy armored thrust.[82] The trial was regarded as a complete success—although, given the emphasis placed on 4[e] DAM by the commanders of both the FAR and the army as a whole, it must be wondered whether any other result could have been expected.

4[e] DAM was formally created in July 1985 (Figure 4.3). It consisted of three RHCs, which, with sixty helicopters each, were slightly smaller than before; each of the three corps was also allotted a similarly sized RHC. The reduction in numbers per regiment allowed the creation within 4e DAM of a new *regiment d'hélicoptères de commandement, de manoeuvre et de soutien* (RHCMS, or command, maneuver and supply regiment), again with sixty helicopters, of which fifty were Pumas. 4e DAM also had its own infantry unit, the *regiment combat aéromobile*, as well as integral logistics support.[83] 6e DLB, with a mix of mechanized infantry and light armor, was seen as "a force which greatly complements the airmobile division."[84] It was recognized that 4e DAM's ability to occupy terrain against an enemy attack—even with its own infantry regiment—was limited. Designed to strike swiftly and repeatedly, its method of operation relied not on holding ground but on the mobility of its helicopters and the weight of anti-tank fire it could deliver. With 4[e] DAM either delaying the enemy force or stopping it completely, the ground-based units would be given time to follow up this attack, and 6[e] DLB was specifically created and assigned to the FAR to ensure that such forces would be available as soon as possible.[85]

The reorganization did not bring about any great change in the tactical employment of France's combat helicopters, either in the corps RHCs or in 4[e] DAM itself. The concept of combining anti-tank, maneuver and observation helicopters within a single organization was still very important, as was the coordinated action between helicopters and airmobile infantry. Although operating under different levels of command, both units were designed to exploit the flexibility and mobility of helicopters without the constraints of direct interaction with ground-based forces. Yet, although the previous use of RHCs at corps level was the antecedent to their employment within 4^{e} DAM, there were significant differences between the operational doctrines of 4^{e} DAM and the individual regiments. 4^{e} DAM was a more distinct expression of the idea that combat helicopters were capable of operating independently as a maneuver unit.[86] It was much larger than any one RHC and, more important, with its own supply units and additional Pumas to sustain the anti-tank helicopters, was less dependent on other formations for logistic support. 4^{e} DAM's infantry regiment was integrated, supported by more maneuver helicopters and designed to cooperate very closely with the helicopters, supporting and extending their anti-tank punch.[87] Finally, the combination of three RHCs gave 4^{e} DAM far greater flexibility and more endurance in action than any individual RHC, since the division could use its three RHCs to maintain a continuous engagement.

These differences both reflected the desire for and gave effect to a level of operational freedom for 4^{e} DAM quite unlike the limited tactical independence of the RHC in each corps. As a consequence, the unit's relationship with ground forces was very different. Not only would 4^{e} DAM have sometimes fought on its own, but even when other forces such as 6^{e} DLB were in support it would have led the fighting, setting the tempo rather than responding to the pace of ground-based combat. Furthermore, as 4^{e} DAM was an army-level unit and not attached to a corps, the area in which it was able to operate would have been much greater. Whereas a corps-level helicopter regiment might move to engage an enemy under one hundred kilometers away, it was assumed that 4^{e} DAM was able to move within a radius of at least four hundred kilometers and up to six hundred kilometers within twenty-four hours and then redeploy up to 160 kilometers the next day.[88]

As might be expected, there was some doubt as to whether 4^{e} DAM, which was based on helicopters instead of the more accepted armored vehicles, was capable of fulfilling the tasks that it had been set. But most criticism was not in fact inspired by military conservatism but by political conservatism. The changes in France's nuclear strategy and attitudes toward NATO, implicit in the formation of the FAR, were a distinct move away from the original, fiercely independent defense concepts of General de Gaulle. Thus strong attacks on 4^{e} DAM were more a defense of France's traditional strategy than a realistic assessment of its operational capability.[89] Military criticisms also were not directly related to the combat ability of helicopters but to issues such as logistics, command and control, and intelligence gathering, which were more easily answered by supporters who argued that they would be overcome with further study and practice as well as the acquisition of new equipment.[90]

Many of the answers to these concerns required cooperation and training with the West German armed forces, and as early as 1983 arrangements were being made between France and West Germany to help the FAR in this respect.[91] Of more importance were joint exercises, such as Brave Sparrow in 1987, which specifically tested the deployment of 4^e^ DAM and 6^e^ DLB to West Germany. The extent of cooperation can be judged from the fact that, during the latter exercise, French forces were placed under command of a West German general.[92] Other problems were solved with the acquisition of additional equipment. For example, the problem of how 4e DAM might gather target information in enough time to exploit its mobility was addressed by the acquisition of Pumas equipped with the long range *Orchidée* surveillance radar.[93] Other criticism focused on the dangers posed by a possible lack of air cover. However, as Brave Sparrow indicated, "The risks of detection . . . turned out to be lower than expected," and the airmobile force was at least as capable of overcoming enemy air interdiction as a ground force. The exercise also answered critics who thought that the division lacked combat power, since 4^e^ DAM was judged to have destroyed 150 tanks for the loss of only fifteen helicopters on the first day alone.[94] While 6^e^ DLB was hard pressed to arrive within the necessary time required to support the airmobile unit, the same problem of slow reaction would have applied even more to other armored forces. At least 4^e^ DAM was itself capable of deploying rapidly and could perhaps have been assisted by other NATO ground forces.[95]

The creation of 4^e^ DAM was a most important step forward for ALAT but, because of the shortage of helicopters, it came at the expense of the corps' regiments. These RHCs were initially reduced to under forty helicopters, with their capacity to carry airmobile infantry as well as to resupply themselves being halved to provide the Pumas for 4^e^ DAM's fourth helicopter regiment. Furthermore, the lack of Gazelle helicopters meant that the obsolescent Alouette III, with the obsolete SS–11 ATGM, had to fulfill the anti-tank requirement. However, the French acquired more Gazelle-HOT helicopters as part of the buildup that followed 1983 reorganization, so that by 1989 all anti-tank helicopter squadrons were fully equipped with Gazelles. Priority continued to be given to 4e DAM, as it received the first *Athos* night-sights for its observation Gazelles and thirty Aérospatiale SA.332 SuperPuma helicopters; similar discrimination will no doubt occur when the first Tiger, France's new armed helicopter, arrives.[96]

As indicated by the concentration of the most capable helicopter assets into 4e DAM, the new division was very highly regarded in the French army. The process was similar to that carried out in the first formation of the RHC in the 1970s. At that time the French had sought to enhance the effectiveness and power of their combat helicopters by placing them in a specific and more independent unit at corps level, at the expense of lower formations (Table 4.1). 4ᵉ DAM was a further consolidation at a higher level, with yet another increase in independence. Thus ALAT obtained, for the first time, a maneuver force of equal stature to the mechanized divisions that comprised the bulk of the army.[97] The new organization established that helicopters had a crucial combat role at the operational level of warfare and, by removing a considerable portion of them from supporting the ground forces, enabled ALAT to develop its new independence more freely.[98] In 1982 a senior ALAT officer had commented that in the future, "Because of the plans of the high command, as well as technological progress, [the helicopter]... will take its rightful place, slowly but surely, as part of the ground manoeuvre force."[99] By 1991 the helicopter had indeed become integral to the ground maneuver force, so much so that it was redefining the notion of what constituted ground action.

Technical Developments

The changes in 1977 and 1985 emphasized the combat role of ALAT and led to attempts to improve the helicopter as a combat weapon system. There were three main areas of concern: the provision of night-vision equipment, increased survivability and the expansion of armed action to include "*appui et protection*" (fire support and protection) against enemy helicopters. There was, first, much concern over the inability of helicopters to operate at night and in adverse weather, not least because the Warsaw Pact armies were capable of fighting twenty-four hours a day. In the 1970s there was little chance that a change in capability could result from this concern since resources were limited. Yet even by the 1980s little had changed. In 1984 ALAT's commander stressed that the achievement of night and all-weather capability by 1990 was still "the number one priority."[100] Only with the formation of 4ᵉ DAM in 1985 was action taken to alleviate the problem since night vision was essential to allow the division's large number of helicopters to exploit the cover of darkness and move "with discretion and security."[101]

The French also sought to keep their helicopters in combat on an increasingly dangerous battlefield by improving their survivability. In the 1960s and 1970s the use of nap of the earth flight to reduce the chances of detection was thought to be the best way of reducing helicopter vulnerability. The emphasis on "discretion and deception" remained important,[102] and France's new anti-tank helicopter, the *hélicoptère antichar–3* (HAC–3), will be equipped with a mast mounted

Table 4.1
French Helicopter Numbers and Organization

Total helicopter numbers, distributed by *year* and *type*, according the *level* of organisation at which they were deployed. Percentages indicate the proportion of total assets assigned at each time to that particular level.

Year	Type	Level of Organisation		
		Division	*Corps*	*Army*
1969	Observation	120	28	
	Armed	60	30	
	Manoeuvre	60	40	
	TOTAL	240 (70%)	98 (30%)	
1977	Observation	160	150	
	Armed		240	
	Manoeuvre		132	
	TOTAL	160 (25%)	522 (75%)	
1985	Observation	160	120	40
	Armed		120	120
	Manoeuvre		30	80
	TOTAL	160 (25%)	270 (43%)	240 (32%)

Note: All data refer to nominal establishments.
Source: Various references cited in this chapter.

sight so it will rarely need to leave the cover afforded at NOE altitudes.[103] However, in 1980 a more intense search for survivability began because of an assessment that new threats "endanger the relative impunity that the helicopter in tactical flight (NOE) has enjoyed until now."[104] Although there were some modifications to existing helicopters,[105] the main results will be seen when the HAC–3/HAP is deployed. The French believed that only by designing helicopters from the start to be survivable was it possible to reduce vulnerability significantly. Thus the new helicopter will be more resistant to damage, more powerful to enhance its NOE flight capability and will have much smaller radar, infrared, visual and audible signatures.

ALAT also developed the ability of its helicopter force to provide fire support and to protect itself from enemy helicopters. While fixed-wing aircraft were not considered to be particularly dangerous, in the late 1970s ALAT became increasingly concerned about the "*prolifération . . . alarmante*" of armed enemy helicopters and the threat they posed.[106] As ALAT was thinking in terms of combat both outside and inside Europe, the first sales of the Soviet Mi–24 Hind to African and Middle Eastern countries exacerbated the problem.[107] Both transport and anti-tank helicopters required protection if they were to accomplish their missions, and ground-based air defenses were unsuitable since they were unable to match the helicopters' mobility. Thus ALAT moved swiftly to acquire its own capability. Since the Gazelle was unable to carry self-defense missiles except at the expense of its anti-tank armament, unarmed observation helicopters were converted for air-to-air combat, and by the mid–1980s each RHC was allotted one squadron of ten Gazelles equipped with 20 mm cannon and designated *hélicoptères d'appui et protection* (HAP, or support and protection helicopters).

While other countries were concerned mainly with arming helicopters for aerial self-defense, ALAT was prepared to use its HAP units offensively so as to counter "enemy helicopter attacks penetrating our lines."[108] It considered creating a "special reserve of anti-helicopter helicopters, which would be made available to the commander of the FAR or the 1st Army" in the same way that a reserve of anti-tank helicopters might be held back. This approach would appear to be heavily influenced by the likely confrontation between French forces, led by 4^{e} DAM, moving east and Soviet OMGs moving west in a future war in Europe. As noted above, the OMG would be strongly supported by helicopters seeking to disrupt possible counterattacks. This threat provided the impetus for ALAT to take more than just passive precautions. The introduction of the gun-armed Gazelles not only reflected concerns over air-to-air combat, but also a desire to expand the range of ground targets that ALAT's helicopters might attack. Although the Puma was often fitted with a 20 mm cannon or rockets, particularly when deployed overseas, the arrangement was not satisfactory since it was a large helicopter and not designed for the task. Since 4^{e} DAM was more than likely to be isolated from ground forces for at least some of the time and would thus require its own fire support,[109] the matter became even more pressing in the late 1980s and drove development of the HAP version of the PAH–2/HAC–3 helicopter. It will combine both missiles and cannon for air-to-air combat with rockets for ground attack.[110]

Although some modifications to ALAT's existing force were made, notably in respect of night vision, the developments outlined above probably had more effect on shaping the design of France's next combat helicopter, the HAC–3/HAP. Indeed, its deployment next century will bring together many of the concepts that ALAT has been formulating since the 1970s and represents the next stage in a gradual process of development that was forced upon ALAT because of financial restrictions.[111] As part of this evolutionary process, ALAT moved steadily but surely over the last two decades from being dedicated to supporting others to playing a leading role in the French army. The helicopter was once thought of in terms of a lone aircraft evacuating the wounded or bringing in urgent supplies,[112] but by 1989 the Chief of Staff, General Maurice Schmitt, was declaring, "The new main battle tank [the Leclerc] and the new combat helicopter . . . constitute the two major trumps of a modern army."[113]

CONCLUSION

In the past thirty to forty years, France and West Germany have developed helicopter forces, both armed and airmobile, primarily for use in anti-tank combat in Europe. Although it was thought that these forces were limited by their lack of endurance and combat staying power, they were seen as a critical complement to mechanized units because of their superior mobility. Both countries preferred to operate helicopters and armored forces separately rather than with close tactical cooperation. This method ensured that the helicopters' mobility advantage was not lost by reducing them to the speed at which ground units operated.

With respect to each country's equipment, the Gazelle and BO 105 armed helicopters were relatively similar in terms of power and survivability, with the West Germans having an advantage because of their large CH–53 transports. In the 1980s, both ALAT and the *Heeresflieger* embarked on modification programs to improve survivability, night and poor-weather operating capability and air-to-air combat capacity. In the future, the purchase of the Tiger and a common light transport helicopter will further reinforce the similarities. However, all these developments, as well as the procurement of new helicopters throughout the last three decades, were restricted by a lack of money. Not only were France and West Germany's defense budgets tight, but the ability of helicopter proponents to gain priority at the expense of other arms was limited, especially in West Germany in the 1960s and 1970s.

However, there were also differences, and the most striking was France's involvement in the Algerian War. While West Germany was able to learn second-hand from this and other counterinsurgency campaigns, it never had direct combat exposure to helicopters. Therefore, the majority of German officers gained little practical experience, and as a result the army as a whole was less convinced of the advantages stressed by the few helicopter advocates. While ALAT's innovations still met with opposition, the French army acquired from its experiences in Algeria a basic understanding and acceptance of helicopters that no amount of peacetime exercise could have provided. This experience was certainly not sufficient to ensure the development of new helicopter doctrines, but it was an important precursor to the successful introduction of these innovations. France's campaign in Algeria also coincided with existing developments of helicopter-launched guided missiles and accelerated the testing and proving of this weapon system. In West Germany twenty years passed between the first test firing of a helicopter ATGM and the fielding of an anti-tank helicopter; in France it took only half that time. France was similarly advanced in terms of doctrine. The airmobile and anti-tank helicopter groups of the 1960s were similar to those proposed unsuccessfully by the *Heeresflieger* in 1970–71. The combination of RHCs and infantry at corps level in 1977 was, given the different sizes of the two armies, roughly equal to the airmechanization plans announced a decade later in West Germany.

It is interesting to note that the acceptance of new ideas for helicopters by the West German army became more pronounced once the *Heeresflieger* had introduced its first armed helicopters. The PAH–1 was not completely deployed until 1984; within four years, airmechanization was being approved. Developments in France were similar, with the widespread introduction of anti-tank helicopters in the late 1960s leading to the considerable doctrinal innovation eventually embodied in the formation of RHCs. The success of armed helicopters in Algeria was itself an important foundation for this next development. For helicopters to be accepted as an important and worthy part of modern warfare and for their advocates to become more powerful, it had to be clear that, like other leading army systems, helicopters were able to destroy enemy forces directly. Thus it would appear that a significant factor in shaping the degree of official acceptance and implementation of proposed helicopter doctrines was a lack of practical experience in the initial stages of development. The introduction of armed helicopters and a successful combination of airmobile infantry and transport helicopters began the process by which the army as a whole became more informed about helicopters: since the *Heeresflieger* was not strong enough to carry through further innovations alone, broad acceptance was vital (Table 4.2).

Perhaps the most fundamental distinction between French and West German developments stemmed from each country's differing strategic circumstances. Because of the strictures of forward defense, West Germany's posture was essentially reactive and required an emphasis on preventing breakthroughs with rapid lateral movement behind the front to reinforce threatened sections of the line. France's strategy necessitated rapid movement out of France and into West Germany to forestall any enemy incursion into French territory. Moreover, once they had arrived in West Germany, French forces would probably have been given the mission of counterattacking enemy units, while most West German units would have remained on the defensive. Thus, as helicopters became a more important part of the French army from the late 1960s onward, they were given an increasingly independent and aggressive role to match the needs of France's operational and strategic doctrines. Whereas West German helicopter units were seen as a rapidly moving reinforce-

Table 4.2
Comparison of French and German Helicopter Numbers

French and German helicopter acquisitions and deployments in comparison to armoured weapons systems in 1980 and 1989.

	France			W.Germany		
	1980	1989	% inc.	1980	1989	% inc.
Armed Helicopters	90	240	167%	0	210	na
Total Helicopters	540	680	26%	540	740	37%
Tanks (Primary)[1]	1100	1350	23%	2500	4100	64%
Total AFV	6000	7000	15%	9000	9900	10%

	France		W.Germany	
	1980	1989	1980	1989
Armed Helicopters as a %age of Tanks	8%	18%	0%	5%
Total Helicopters as a %age of AFV	9%	10%	6%	8%

[1] France: AMX–30 and Germany: Leopard I and II.
Source: *The Military Balance* (London: IISS, 1980, 1989).

ment for existing defenses or as a last ditch reserve, ALAT's doctrine emphasized operating in front of and on the flanks of mechanized units to protect their advance, using the helicopter's speed to cover large areas with limited resources. 4[e] DAM opened up even more possibilities for independent action since it was, in effect, the spearhead for the entire French army. A strict comparison of the numbers and capabilities of the helicopters in each army in 1989 might suggest that France and West Germany attached roughly the same importance to the military helicopter. However, when the number of helicopters is compared to other weapons systems, it becomes clear that the French army was in fact far more committed to helicopters. Because of its experiences in Algeria, its specific strategic needs and, it would seem, a much greater willingness to innovate, France was far ahead both organizationally and in terms of the acceptance of helicopters by the army as a whole.

NOTES

1. Paul Beaver (based on an interview with the commanding officer of the *Heeresflieger*, Brigadier Kurt-Josef Veeser), "Advancing on All Fronts," *Defence*

Helicopter World 5.5–6 (1986): 16; and Everett-Heath, *British Military Helicopters*, 38, 46.

2. Messerschmitt-Bölkow-Blohm (MBB), *PAH–1* (videotape recording: 1979); and Donald E. Fink, "New Versions of BO 105 in Development," *AWST*, 1 May 1972: 45.

3. *The Security of the Federal Republic of Germany and the Development of the Federal Armed Forces*, Defense White Paper (Bonn: Federal Ministry of Defense, 1971–72), 13, 18–19, 142.

4. Brigadier Hans E. Dreibing (Commanding Officer, *Heeresflieger*), "Views on Airmobility," *Aviation Digest* 21.3 (1975): 10. See also "Europe Reviewing Combat Helicopter Role," *AWST*, 2 August 1969: 250.

5. "Army Aviation Stresses Helicopter Units," *AWST*, 24 April 1972: 49 (based on an interview with Brigadier Dreibing).

6. "Political, Economic Shifts Hamper NATO," *AWST*, 19 March 1973: 21; and "Helicopter Capabilities Sparking Strong Market in Western Europe," *AWST*, 28 May 1973: 181.

7. Richard Simpkin, "An Airmechanized Force for the 90s," *Armor* 90.7–8 (1981): 55.

8. MBB, *PAH–1*. See also Mark Lambert, "Germany's PAH–1 in Service," *International Defense Review* [hereafter *IDR*] 14.9 (1981): 1172. The U.S. tests were known to the Germans as the Katterbach trials.

9. Compare *The Security of the Federal Republic of Germany and the Development of the Federal Armed Forces* (1971–72), 142–43 with Rudi Meller, "Federal Germany's Defense Potential, Part 1: The Armed Forces," *IDR* 5.4 (1974): 171.

10. "Tactical Dialogues 2—Germany," an interview with General J. A. von Kielmansegg, *JRUSI* 118.9 (1973): 37.

11. *The Security of the Federal Republic of Germany and the Development of the Federal Armed Forces*, Defense White Paper (Bonn: Federal Ministry of Defense, 1979) 126.

12. Ibid., 166.

13. Lt. General M. Glanz (Chief of Staff, West German army), "Defeating Enemy Armour," *NATO's Sixteen Nations* 28 spec. ed. 1 (1983): 46; and Paul Beaver (based on interviews with officers of the *Heeresflieger*), "*Schwarm Führer*," *Defence Helicopter World* 5.5–6 (1986): 10–13.

14. The extent of training for this method of operation made it clear that it was the norm. See "West Germany Developing Anti-Tank Helicopter Force," *AWST*, 17 July 1978: 57 (based on an interview with Brigadier Dreibing).

15. *The Security of the Federal Republic of Germany and the Development of the Federal Armed Forces* (1979), 166.

16. Lambert, "Germany's PAH–1 in Service," 1172; and "West Germany Developing Anti-Tank Helicopter Force," 57.

17. Beaver, "*Schwarm Führer*," 13.

18. Brigadier H. Tiedgen (Commanding Officer, *Heeresflieger*), quoted in Paul Beaver, "German Army Aviation in the Next Decade," *Defence Helicopter World* 3.9–11 (1984): 15. See also Wolfgang Flume, "Letter from Bonn," *Military Technology* 11.3 (1987): 16.

19. Peter Pletschacher, "German Military Helicopter Programs," *IDR* 19.5 (1986): 643.

20. MBB, *PAH–1*; and *The Situation and Development of the Federal Armed Forces*, Defense white paper (Bonn: Federal Ministry of Defense, 1985) 356.

21. Comparing MBB, *PAH–1*, a videotape recorded in 1979, with a *Heeresflieger* demonstration the author witnessed in 1988, it was clear that little had changed.

22. Mark Lambert, "Helicopters Prepare for Air Combat," *Interavia* May 1985: 445.

23. G. Charnbost and K. G. Benz, "The Franco-German Combat Helicopter—Reconciling Contradictory Requirements," *IDR* 17.5 (1984): 577; and MBB, *PAH–1*.

24. *The Situation and Development of the Federal Armed Forces*, 198.

25. Brigadier General Csoboth (Command Officer, *Heeresflieger*), "Air-to-air Fighting with Helicopters I," *IDR* 22.2 (1989): 153–54.

26. "West Germany Developing Anti-Tank Helicopter Force," 60; and Beaver, "German Army Aviation in the Next Decade," 12–13.

27. "MBB to Update PAH–1," *Flight International*, 7 May 1988: 10; and "Hoverview," *Defence Helicopter World* 5.8–9 (1986): 85.

28. *The Situation and Development of the Federal Armed Forces*, 196; and "MBB to Update PAH–1," 10.

29. Michael Feazel, "Government Approval of European Helicopter Tied to Commonality," *AWST*, 3 February 1986: 69.

30. Charnbost and Benz, "The Franco-German Combat Helicopter—Reconciling Contradictory Requirements," 575–78. See also "Franco-German PAH–2 Project Stalled," *AWST*, 10 November 1980: 48.

31. Michael Feazel, "Germans, French to Study Substitute for PAH–2 Helicopter," *AWST*, 20 October 1986: 38–39 and "Cath Programme Threatened," *Flight International*, 20 June 1987: 21.

32. W. Schütze, *France and Germany: Co-operation and Conflict in Defence and Security* (Geneva: PSIS occasional paper No. 2, GIIS, 1987) 46–47; and "An Inside View of the French Army," an interview with General Maurice Schmitt (Chief of Staff, French army), *Military Technology* 11.8 (1987): 34.

33. "Eurocopter Announces PAH–2 Workshare," *Flight International*, 14 May 1988: 14.

34. *The Security of the Federal Republic of Germany and the Development of the Federal Armed Forces* (1971–72), 143; and "Army Aviation Stresses Helicopter Units," 51.

35. *The Security of the Federal Republic of Germany and the Development of the Federal Armed Forces* (1979), 155, 164; and Glanz, "Defeating Enemy Armour," 46.

36. Meller, "Federal Germany's Defense Potential, Part 1: The Armed Forces," 169.

37. Peter Macdonald, "*Fallschirmjaeger*," *Armed Forces* 5.11 (1986): 518; and *Heeresflieger* demonstration (1988).

38. See chapter 3 for discussion of von Senger und Etterlin. Von Senger's advocacy of airmechanized forces was continued by General Hans-Henning von Sandrart (CINCAFCE): Guy Willis and Charles Dick, "Beyond the Corps Battle in NATO's Central Region," interview with General Hans-Henning von Sandrart, *IDR* 22.2 (1989): 170.

39. Major General Georg Bernhardt, (Commander, 1st *Fallschirmjäger* Division) "The West German View," *JRUSI*, supplement 134.4 (1989): 16.
40. Csoboth, "Air-to-air Fighting with Helicopters I," 154.
41. Wolfgang Flume, "Letter from Bonn," *Military Technology* 11.9 (1987): 16.
42. Bernhardt, "The West German View," 16.
43. Beaver, "German Army Aviation in the Next Decade," 12; and *The Situation and Development of the Federal Armed Forces*, 191.
44. "Hoverview," 85. See also H. Schulte, "Cost-cutting Will Hit W. German Army," *Jane's Defence Weekly*, 5 March 1988: 379.
45. Generalmajor Georg Bernhardt, "Airmobile Formations: The Federal Republic of Germany," *IDR* 23.4 (1990): 380.
46. "Beyond the Corps Battle in NATO's Central Region," 170.
47. Quoted in Hickey, *Out of the Sky*, 201; and Major Victor J. Croizat, "Helicopter Warfare in Algeria," *Marine Corps Gazette* 71.8 (1987): 23–25.
48. "Aviation in the French Army," *Aviation Digest* 18.9 (1972): 28–30.
49. Hickey, *Out of the Sky*, 201.
50. Lt. Colonel P. M. Bonnet, "French Army Aviation," *Aviation Digest* 31.10 (1985): 2; and Headquarters (HQ) ALAT, "French Airmobility," *Aviation Digest* 21.3 (1975): 17.
51. "Aviation in the French Army," 31.
52. Colonel M. Lartigue, "*L'ALAT: arme du chef?*" *Armées d'Aujourd'Hui* [hereafter *AAH*] 12 (1976): 61. The author had been involved in studies of airmobility in the early 1970s.
53. General Maurice Cannet (Commanding Officer, ALAT), "*L'aéromobilité: le sens d'une évolution*," *AAH* 35 (1978): 44.
54. General (Ret.) Maurice Cannet, "Air-to-air Fighting with Helicopters," *Interavia*, September 1988: 874.
55. HQ ALAT, "French Airmobility," 15, 17.
56. Lartigue, "*L'ALAT: arme du chef?*" 60–61, 73. See also Cannet, "*L'aéromobilité: le sens d'une évolution*," 44.
57. G. Longeret, "*L'armée de terre de 1960 à 1988*," *Défense Nationale* 44.4 (1988): 25; and, for the general effects of the cuts, Daniel S. Yost, *France and Conventional Defense in Central Europe* (Boulder, CO: Westview, 1985) 21.
58. For example, it had been decided in 1967 to acquire Lynx helicopters to replace the Alouette III, but in 1969 the order was cancelled on financial grounds: Everett-Heath, *British Military Helicopters*, 167.
59. Colonel Claude F. Regis, "French Army Aviation," *Aviation Digest* 24.12 (1978): 5; and Cannet, "*L'aéromobilité: le sens d'une évolution*," 44.
60. General Maurice Cannet, "Aviation in the French Army," *Aviation Digest* 26.8 (1980): 1ff.
61. Regis, "French Army Aviation," 5.
62. Cannet, "Aviation in the French Army," 1.
63. Ibid., 3; and Colonel Georges Baffeleuf, "*J'ai commandé un R.H.C.*," *AAH* 35 (1978): 50. Baffeleuf was to be the commanding officer of ALAT in the mid–1980s.
64. Regis, "French Army Aviation," 5; and Cannet, "*L'aéromobilité: le sens d'une évolution*," 44.

65. General Henri Navereau (Commanding Officer, ALAT) quoted in Paul Beaver, "The Battlefield Helicopter—The French View," *Defence Helicopter World* 3.3–5 (1984): 9; and Colonel Xavier de Reviers de Mauny, "*La carte maîtresse: l'aéromobilité*," *AAH* 68 (1982): 29.

66. Regis, "French Army Aviation," 5. Note that French regiments were roughly equivalent to battalions.

67. An excellent analysis of ALAT's helicopter numbers was provided by Paul Jackson, "ALAT Regroups," *Armed Forces* 5.3 (1986): 137–39.

68. Helicopter and personnel figures from *The Military Balance 1976–77* (London: IISS, 1976) 21, and *The Military Balance 1978–79* (London: IISS, 1978) 23.

69. Regis, "French Army Aviation," 5.

70. Paul Beaver (based on an interview with General Georges Baffeleuf, Commanding Officer, ALAT), "The French ALAT in the 1990s," *Defence Helicopter World* 4.6–8 (1985): 8.

71. "*La réorganisation de l'armée de terre*," an interview with General René Imbot (Chief of Staff, French army), *AAH* 82 (1983): 6; and "*La force d'action rapide*," an interview with General Guy Forray (Commanding Officer, FAR) *AAH* 87 (1984): 20.

72. Schütze, *France and Germany: Co-operation and Conflict in Defence and Security*, 20; and "*La réorganisation de l'armée de terre*," 6.

73. Gerard Turbé, "The French Army's Light Air Arm," interview with General Bernard de Lacroix de Vaubois (Commanding Officer, ALAT) *IDR* 21.8 (1988): 949.

74. "*La force d'action rapide*" (1984), 22. One of the important proponents of the concept was General Georges Fricaud-Chagnaud, "*L'Armée de Terre face à ses missions en Europe*," *Défense Nationale* 39.5 (1983): 35–44.

75. Lt. Colonel Charles-Henri de Monchy, "*Le Bouclier Aéromobile*," *AAH* 96 (1984): 41; and "*La force d'action rapide*," *AAH* 97 (1985): 26.

76. Henru quoted in de Monchy, "*Le Bouclier Aéromobile*," 41; and in "La FAR," *AAH* 96 (1984): 43.

77. "*La réorganisation de l'armée de terre*," 4.

78. Schütze, *France and Germany: Co-operation and Conflict in Defence and Security*, 20–21. The reorganization of France's tactical nuclear forces, which removed them from direct support of *1ᵉʳ armée*, emphasized the move away from a reliance on nuclear deterrence: see "*La réorganisation de l'armée de terre*," 7.

79. General René Imbot, "The French Rapid Action Force," *NATO's Sixteen Nations* 28, spec. ed. 1 (1983): 39.

80. de Reviers de Mauny, "*La carte maîtresse: l'aéromobilité*," 28–29.

81. Lt. Colonel Claude Battlo, "*Hélicoptères en appui aérien*," *AAH* 82 (1983): 42; and Defense Minister Hernu commenting in "La FAR," 43.

82. Imbot, "The French Rapid Action Force," 38–39. For *Damoclès* see Bonnet, "French Army Aviation," 6; and Jacques de la Hersière, "*L'Exercice «Damoclès»*," *Défense Nationale* 40.11 (1984): 167–69.

83. Bonnet, "French Army Aviation," 5; and "*La force d'action rapide*" (1985), 26.

84. "*La force d'action rapide*" (1984), 22.

85. "*La Force d'action rapide*" (1985), 26; and Bonnet, "French Army Aviation," 6. 9e DIMa is being converted to an organization similar to that of 6e DLB.

86. Bonnet, "French Army Aviation," 4; and "*La force d'action rapide*" (1984), 22.

87. Bonnet, "French Army Aviation," 5; and General H. Préaud (Commander, FAR), "Airmobile Formations: France," *IDR* 23.4 (1990): 378.

88. Compare the tone of the articles describing the RHC in 1978, such as Baffeleuf, "*J'ai commandé un R.H.C.*," 51 with Bonnet, "French Army Aviation," 5–6; and "*La force d'action rapide*" (1984), 22.

89. For example, see the comments of a senior retired officer, General Valentin, quoted in Yost, *France and Conventional Defense in Central Europe*, 93. Most of Valentin's criticisms were capable of being answered in any case. See ibid., 87–103, for a complete discussion of the impact of the FAR on French nuclear strategic thinking.

90. G. Vincent, "*Réflexions á bâtons rompus sur l'emploi de la force d'action rapide*," *Défense Nationale* 40.11 (1984): 166; and Bonnet, "French Army Aviation," 2, 7.

91. Imbot, "The French Rapid Action Force," 39.

92. Jacques Isnard, "Major Exercise for France, W. Germany," *Jane's Defence Weekly*, 26 September 1987: 661; and Keith E. Wilson, "Albatross X: Tri-national Aviation Exercise," *Aviation Digest* 33.1 (1987): 34.

93. Bonnet, "French Army Aviation," 6.

94. "The French Army's Light Air Arm," 950–51; and General Guy Forray, "French Ground Forces in Europe," *NATO's Sixteen Nations* 33.10 (1988): 28.

95. For criticisms see Gerard Turbé, "France's Rapid Deployment Forces," *IDR* 20.8 (1987): 1024; and Gerard Turbé, "Testing Franco-German Military Co-operation," *IDR* 20.11 (1987): 1458.

96. Jackson, "ALAT Regroups," 139; and "The French Army's Light Air Arm," 949.

97. Bonnet, "French Army Aviation," 4; and de Monchy, "*Le Bouclier Aéromobile*," 41.

98. Baffeleuf, reported in Beaver, "The French ALAT in the 1990s," 8.

99. Lt. Colonel A. Martini, "*Combat du futur*," *AAH* 68 (1982): 38.

100. General Navereau (Commanding Officer, ALAT), quoted in Beaver, "The Battlefield Helicopter—The French View," 8; and Colonel Bernard de Lacroix de Vaubois," "*La modernisation des equipements de l'Alat*," *AAH* 58 (1981): 14.

101. Beaver, "The French ALAT in the 1990s," 8.

102. Lartigue, "*L'ALAT: arme du chef*?" 60; and Beaver, "The French ALAT in the 1990s," 8. See also Colonel Bernard de Lacroix de Vaubois, "*Les ennemis de l'alat*," *AAH* 68 (1982): 32–35, for an assessment of the many dangers facing ALAT on the future battlefield.

103. "The French Army's Light Air Arm," 951; and "Franco-German PAH-2 Project Stalled," 48.

104. Cannet, "Aviation in the French Army," 3.

105. Martini, "*Combat du futur*," 38.

106. Battlo, "*Hélicoptères en appui aérien*," 42.

107. "The French Army's Light Air Arm," 950; Battlo, "*Hélicoptères en appui aérien*," 42–43; and Cannet, "Air-to-air Fighting with Helicopters," 874.

108. "The French Army's Light Air Arm," 949–50.
109. Imbot, "The French Rapid Action Force," 38.
110. Cannet, "Air-to-air Fighting with Helicopters," 874.
111. Cannet, "Aviation in the French Army," 4.
112. Cannet, "*L'aéromobilité: le sens d'une évolution*," 44.
113. "An Inside View of the French Army," 31.

5

"A Rotary-Wing Revolution?"—Helicopters and Air-Land Warfare

In searching for some general conclusion to the specific developments outlined in chapters 1 to 4, it became clear that there was a common thread running through the fabric of each country's experiences. This fundamental theme concerned the relationship between air and land warfare and, more precisely, between aerial and ground-based means of conducting land combat.[1] The concept might usefully be summarized as "air-land warfare." Although America's Airland Battle and Britain's land/air warfare doctrines were obviously part of the concept, the term "air-land warfare" is more generally applicable and, in fact, can refer to developments throughout the twentieth century. This chapter argues that, as an innovation, military helicopters were defined and used within the context of air-land warfare; at the same time, they have transformed air-land warfare because of their technical characteristics and the specific doctrines associated with them. Indeed, helicopters helped to make air-land warfare more important and more effective now than at any other time.

Thus the helicopter's most pronounced effect on modern warfare occurred not in competition with ground vehicles but at the intersection between air and land combat. If the helicopter's development has seemed unduly slow over the past forty-five years, it has not been because of the innate opposition of armored warfare proponents but more because the structure of armed forces threw tanks and helicopters into an organizational conflict instead of doctrinal cooperation.

AIR-LAND WARFARE

Defining Air-Land Warfare

Like "doctrine," air-land warfare seems to have acquired a variety of meanings in different military establishments, and so it is important that the present definition be clearly understood. To some extent, all land warfare has been air-land from the time when balloons and then aircraft were employed by armies to help them achieve their objectives. However, the phrase "air-land warfare" is more properly and more usefully defined in narrower terms. It can be understood to mean the intimate and regular interaction of air and ground forces in pursuit of an aim directly related to the land battle. This requires very close integration, implying effective unity of command over the ground and air elements. More important, it implies that doctrine makers do in fact recognize the need to harness these differing means of combat to a single end. It is perhaps best explained by analogy with the relationship of marines and naval vessels. When used in ship-to-ship combat in the nineteenth century, these soldiers were fully integrated into the naval combat process; when the ships simply transported them for an amphibious assault, then they were taking part in a joint action. In the case of air and land forces, the term "integration" certainly could not be applied to strategic bombing, nor even to interdiction. The latter was, again, a joint action in which the results of the ground and air operations were coordinated but not actually integrated. Indeed air-land warfare, while not limited to the tactical arena, seemed until recently to arise only at this level.

There were three fundamental aspects of air-land warfare that determined the degree of integration. First, the objectives of the combination of air and ground forces had to be directly related to land combat. Air power was not solely nor primarily useful for this purpose, but air-land warfare exploited the special characteristics of aircraft (speed, flexibility and so on) purely with reference to land battles. Moreover, concepts of air-land warfare were essentially constructed from a ground force perspective.[2] The relative status of the aerial and ground-based components was also significant. If they were of widely unequal status, then integration was much harder to achieve, with aircraft being used merely to assist land warfare. Status refers both to the specific employment of air power on the battlefield and to the way in which its importance was conceived by senior officers. Organization was the third critical aspect. Because of political and practical differences, air and ground forces grew increasingly separate both in fighting formations and in the structures of ownership and control. Aviation was also separated to overcome doctrinal constraints on the employment of helicopters because ground-oriented commanders tended to use them less according to the doctrines developed by aviation proponents. Yet air-land warfare required a particularly close combination—usually possible only when the air and ground forces were controlled by the same parent organization.

Air-Land Warfare in the Twentieth Century

Despite the rapid emergence of air-to-air combat and the slower development of strategic bombing, the First World War was a good example of a major conflict in which aircraft were mainly used for specifically land-oriented purposes—observation and ground attack.[3] Close connections between land and air were enhanced because aircraft were, mainly, operated by armies and because many pilots had some experience with and training in ground fighting. Yet unit organizations still tended to separate aircraft from the land forces they were supporting. Assumptions about the unique and even mystical nature of flying began to emerge, as did the requirement for aerial warfare, laying the foundations for the often bitter political disputes that occurred between air forces and other services after the war.

Furthermore, the status of aircraft was entirely different from that of ground troops: not so much unequal as not even comparable. At best, aircraft were seen as a peripheral supporting force that was certainly not a decisive factor in achieving victory. Newly invented war machines were not highly regarded, and most officers placed their faith in massive artillery bombardments backed by the courage and endurance of the infantry. Thus, like tanks, aircraft were rarely seen as central to combat activity. These negative perceptions of technology were compounded by the inadequacy of aircraft at that time. Simply keeping airplanes aloft was a battle for pilots; accurately aiming and delivering an effective weight of ordnance was very difficult indeed. Only in artillery spotting and photographic reconnaissance were they particularly and routinely effective, which further suggested their subordinate position in the hierarchy of military effectiveness.

After the First World War, air and land power moved further away from each other in many countries. Air forces were formed with varying degrees of independence from armies and, in extreme cases, aircraft were seen as competing with ground forces because they offered the chance to prevent a repetition of the bloody, drawn-out fighting that had characterized the war. The most potent theories about air power, although not actual practices, concerned strategic bombing and, belatedly, air defense—neither activity requiring much integration with land forces. Even in the few countries where cooperation was stressed, it took many years to develop techniques and technology that satisfied the minimum requirements of air-land warfare. Indeed, in the Second World War, only Germany and the Soviet Union really demonstrated an understanding of air-land warfare. The *Luftwaffe*, mainly composed of officers transferred from the army, cooperated relatively closely with German ground forces—most notably in France in 1940. The Red Army's *Shturmovik* aircraft, flown by an air force essentially subordinate to the ground forces, proved to be reasonably effective in attacking enemy tank and vehicle concentrations. The Western allies, with the exception of the actions of the desert air forces, did not develop close air support to the same degree until after the Normandy landings and, even then, reliable cooperation was rare.

This level of air-land warfare was probably the best that could have been achieved at the time and was certainly regarded as valuable by those involved. Yet the level of integration was nevertheless limited. Air force pilots often lacked training in cooperating with armies; fighter-bombers were usually called in to assist the land battle only on specific occasions and were not routinely assigned to ground units. Liaison and target marking procedures were neither reliable nor widely understood, hampering communication between air and ground. Moreover, the organizational routines of both air and land forces, generally focusing on their own specific activities, tended to impede the satisfactory resolution of these deficiencies. While the Red Army assigned some aircraft to the direct command of armies and corps acting as mobile groups in the latter stages of the war, on the whole air and ground units remained separate. The status of the air component was also very different from that of ground units. They were essentially a supporting arm like artillery but less readily available and certainly not included in planning to the same extent. Ground attack aircraft were mostly used as a rapid reaction force to assist ground troops in dealing with unexpectedly fierce enemy attacks.

The airborne operation, employing parachutes and gliders to deliver personnel and equipment at a vital location, was another form of air-land warfare. It had been developed mainly by the Soviets and Germans before 1939 and was first used in action during the Second World War. It demonstrated much more organizational cohesion between the land and air elements than close air support and was specifically oriented to ground objectives, but was greatly limited by technology. Although specialized attacks, as at Eben Emael,[4] were very effective, generally airborne operations by all countries in the Second World War were either outright failures or only marginally successful. The troops had limited mobility after landing, and it proved very difficult to coordinate the airborne operation with the ground-based component.[5] Nevertheless, the enthusiasm with which most armies developed an airborne capability and tried to use it in action was testament to the interest in exploiting aerial movement within the context of land combat.

If the level of air-land integration was relatively satisfactory in 1945, considerable pressures arose after the war that threatened to, and in some cases did, break down the interaction of air and land forces. The development of atomic weapons gave air forces the technology with which to pursue the concept of strategic air warfare prevalent in the 1920s and 1930s. The very need for land, and hence air-land, forces was called into question. Financial constraints saw air forces deemphasize those duties, such as cooperation with armies, that they regarded as less relevant to their main mission. With their increasingly high performance, aircraft were thought to be better suited to interdiction rather than close air support. Paratroop operations became less and less viable. Only the demands of many counterinsurgency campaigns made air and land forces cooperate more than they would have if left in peace. Where helicopters were not involved, these actions did not extend integration past the limited form that had emerged during the Second World War; moreover, the depth of integration in these conflicts simply highlighted its shallowness in preparations for high-intensity warfare.

Yet the importance of applying the fundamental characteristic of air power–its superior mobility over ground forces—purely in the context of land combat did not diminish and, if anything, increased. Guerrilla conflicts, in which aircraft overcame the problems of harsh terrain and small force-to-space ratios, demonstrated the particular value of helicopters and, despite the problems of applying lessons from one form of fighting to another, suggested that air-land combat might also have great value in conventional warfare. Air power was seen as essential for moving troops and ordnance speedily over wide areas, over battlefields obliterated by nuclear weapons, over difficult terrain and over the enemy's main line of resistance into its vulnerable rear. Some military thinkers began to imagine a more intense form of air-land cooperation, especially as they gained practical experience with helicopters. The expectation arose, most notably among paratroop officers in the United States, that air-land warfare would entail much more integration than in previous years. Thus, although the level of actual or expected interaction between air and ground forces diminished in many cases during the 1950s, 1960s and 1970s, it was not for lack of trying on the part of some sections of various armies and air forces. The final emergence of articulated doctrines for air-land warfare in the 1980s was not so much a radical change as the culmination and eventual expression of the work of the previous three decades—a putting into practice of what had previously been theoretical.

The helicopter began to emerge as a viable piece of military equipment during this period; its obvious technical advantages meant that it became a central part of the search for a greater air-land warfare capability. But it would be wrong to assume that the growth of air-land warfare was the reason for the development of helicopters or vice versa. The process of innovation and development was more complex. The potential for integrating air and ground forces preceded helicopters and, equally, it was not until the late 1980s that specific technological developments made practicable any explicit connection between helicopters and air-land warfare. Nevertheless, the two were subtly linked, with many of the arguments and imperatives used to support helicopter development stemming from the desire of the armies to employ aerial combat power. Innovative or experimental technology often proved impractical but did spur further theoretical advances. Moreover, air-land warfare set the context for the development of the helicopter, providing the concept on which acceptance of helicopter innovation was first founded. In other words, the missions to which helicopters were first assigned and even the way in which they were conceived in military terms stemmed from a desire to use helicopters, rather than fixed-wing aircraft, as the aerial component in the land battle. Yet, as helicopters were used more widely, air-land warfare changed to better exploit their capabilities.

THE HELICOPTER'S FUNDAMENTAL TECHNICAL CHARACTERISTICS

As will be examined in chapter 6, doctrine and technology interact. While air-land warfare was the military's theoretical basis for specific helicopter doctrines, there was also an underlying foundation of technical characteristics.[6] Not surprisingly, these characteristics favored the use of helicopters as aerial combat vehicles integrated into land combat. The helicopter's main defining characteristic was its ability, while airborne, to remain stationary relative to the ground. The helicopter was thus able to land in an area not much larger than itself and one that could be cleared easily and quickly in most circumstances. Even if unable to land, or if it was not tactically advisable to do so, the helicopter was still able to interact with the ground while hovering above it. Moreover, by remaining stationary or moving at very slow speeds, the helicopter gained benefit from terrain obstacles—cover from fire and observation—without being in any way hindered by them.

The reason the helicopter was able to ignore these obstacles provides the next determining factor, one that the helicopter shared with all aircraft—the ability to fly. In comparison to most fixed-wing aircraft of similar vintages, the helicopter was rather ordinary at this task; it was, however, altogether more effective at covering a distance than the average tank or truck. First, flying was quicker than surface movement in absolute terms, even with the need for time-consuming preflight checks and so on. More important, the problem posed to ground forces by terrain irregularities, often exacerbated by deliberate or accidental combat damage, meant that helicopters were an order of magnitude faster than wheeled or tracked vehicles.[7]

Flying brought certain disadvantages, however. The first of these was that much of the helicopter's available energy had to be spent in keeping itself in the air. Under extreme climatic conditions, as armies operating in the tropics or at high altitudes found, the portion of the helicopter's power usually available to lift cargo or weapons was almost entirely consumed in getting the aircraft itself off the ground. Until the introduction of more efficient turbines to replace piston engines in the 1960s, when technology began to move ahead of actual requirements, load limitations were a considerable hurdle.

Another problem with flying was complexity. There is no doubt that, in terms of maintenance, modern helicopters are less intricate than their predecessors and that, comparatively, armored vehicles have increased in complexity, especially in terms of their electronic equipment. However, for much of the last forty-five years the helicopter was the most complicated item of military equipment used for tactical land combat. In some ways, air-land warfare in the 1990s emerges as much from the distinct similarity of maintenance on the computer-dominated tank and armed helicopter.

Flying also consumed more fuel and spare parts than ground movement, so that the helicopter paid for its independence from terrain by increased dependence on supply lines. The use of cargo and utility helicopters to provide a logistic service of equal mobility to the airmobile and attack helicopter formations they were supporting helped to alleviate this problem, but the final solution lies with different ways of operating and supporting rather than massively increasing available logistic support.

The third crucial aspect was helicopter vulnerability or, to be more precise, its potential for vulnerability, most of which was the direct consequence of flying. Because of the need to maximize the power-to-weight ratio, helicopters were unable to carry much armor. Moreover, their dynamic systems—especially the rotors—were both very susceptible to damage and almost impossible to protect. Nor did helicopters possess high speed to assist in evading attack. Finally, if some damage was caused to a helicopter's engine or rotor blades while in flight, gravity usually finished the job for the attacker.

This potential vulnerability was, however, offset considerably by the first characteristic of helicopters: their ability to use terrain and fly in the nap of the earth, thereby seeking to avoid detection and to fly below the minimum engagement altitude for many surface-to-air missiles. Moreover, by combining their activities with other weapons—especially artillery to suppress enemy air defenses—and by using technology to counteract the tracking systems of anti-aircraft weapons, helicopters were able to achieve a measure of survivability consistent with other combat arms. For some years this use of supporting arms to assist helicopters was held back because helicopters were themselves supporters and could not call on support in the same way that infantry and armor regularly did to overcome their limitations. Only when helicopters became like infantry and armor—when they obtained maneuver status—could vulnerability be decreased even further.

The need to overcome their specific vulnerabilities played a major role in the way in which the military applications of helicopters developed. Perceptions of vulnerability also meant that helicopters were less easily accepted by soldiers accustomed to the protection afforded by armored vehicles. The culture of tanks and armored personnel carriers that had developed in most armies out of the Second World War, and that remains dominant, did nothing to assist those arguing in favor of such apparently fragile machines as helicopters; in the air forces, the culture of sleek, fast jets also worked against the slow, ungainly rotary-wing aircraft.

The final technological factor that defined the way in which the military helicopter evolved was its limited range and payload in comparison to fixed-wing transport or attack aircraft. Although some helicopters were able to ferry themselves over long distances with the aid of additional fuel tanks, once in actual combat, weight limitations again meant that most helicopter tasks were tactical in terms of their geographic scope; however, if those specific tasks were carried out at the appropriate place and moment, they could have significance at the operational level of war. These limitations on range again emphasize the closeness of helicopters to ground warfare and their distance from most air operations.

Some writers have likened the helicopter to a flying truck or flying tank, implying it is a land vehicle that just happens to have an aerial capability. Others have claimed categorically that it is part of air power and is simply an unusual form of aircraft. A few, such as British military thinker Robert Simpkin, have described it more accurately as occupying a "demi-monde" somewhere between the two.[8] Because the helicopter had a much closer relationship with the ground than most aircraft, it was very useful in land combat; what differentiated it from other ground weapons was its ability to fly. It was this characteristic that made it so attractive for air-land warfare: it was an aerial combat system, but its particular style of flight meant that it was best employed in concert with ground forces in the land battle. But, like most characters from the demi-monde, helicopters were regarded with considerable suspicion by the establishment.

HELICOPTERS AND AIR-LAND WARFARE

As noted at the start of this chapter, air-land warfare sought to exploit aerial combat power—either for movement, reconnaissance or fire—so as to achieve land battle goals more easily. It required more integration than joint action, and the degree of integration was reflected in the relative status of the air and ground components and how they were organized together—both in action and administratively—to achieve the necessary combination. The connection between helicopters and air-land warfare can be examined first through the missions and tasks that helicopters performed, revealing the land battle orientation that constituted the first critical aspect of integration; second through the issue of status, showing how helicopters recently obtained much more equality with ground forces; and finally through helicopter organizations, in which the integration of air and land actually occurred.

Missions: Reconnaissance, Movement and Fire

Helicopters were first of all seen as a replacement or complement for the many light, fixed-wing aircraft that had been used during the Second World War for aerial artillery observation, occasionally for casualty evacuation and for carrying important passengers around the battlefield. These light aircraft were able to operate from small, basic airstrips located very close to the front lines; however, helicopters offered the added advantage of requiring no airfield at all, which was particularly important during fighting in mountainous or jungle terrain. Their ability to hover meant helicopters were a better observation platform and also that they could interact with ground troops—accurately dropping supplies, for example—without actually having to land. Although the helicopter's much greater mechanical complexity offset these advantages to some extent, it was imagined that this problem would soon be resolved by advances in technology, perhaps indicating some of the faith in technology that has characterized much of warfare since the Industrial Revolution.

After the deployment of helicopters for liaison and observation, the possibilities of using them to move troops and supplies over the battlefield became clear, especially when more experience was gained in trials and in combat in Korea and Malaya. Although ideas soon outstripped both the technical capabilities of current models and the actual numbers available, it was generally accepted that helicopters had a crucial role to play in air transport. They were clearly able to overcome terrain restrictions that would limit ground vehicles, while not needing to use parachute delivery or an established airbase like fixed-wing aircraft. The helicopter's speed ensured that, even if only small loads of supplies or troops were carried, they were more likely to arrive at the right place and at the right time than if the loads moved by land.

In the 1950s, the American and Soviet armies concentrated mainly on preparing for nuclear war. Helicopters seemed to offer considerable advantages on a nuclear-devastated battlefield in terms of speed and mobility, especially in affording transport and resupply to units that would have to operate more widely dispersed than ever before. The Soviets used helicopters to provide short-range airborne forces of company or battalion strength that would land up to fifty kilometers ahead of advancing armored units. While the Americans considered the sky cavalry concept of large infantry units, entirely supported by helicopters, their actual deployments mirrored those of the Soviets, with each division having a helicopter unit capable of moving an infantry company or conducting limited resupply missions. However, in the 1960s, the U.S. Army did establish an air assault division. It had a broader operational role and all of the formation's activities—command and control, reconnaissance, firepower, logistics and maneuver—were carried out mainly by helicopters. While Soviet developments came to fruition in the late 1960s, progress in America was diverted to the low-intensity conflict in Vietnam and thus toward a very different set of priorities.

Although Britain and France also saw the value of helicopters for transport activities in nuclear warfare, their main experiences with helicopters also concerned these different priorities. The counterinsurgencies in which they were engaged were almost exclusively fought in extremely difficult terrain and with a very low force-to-space ratio. Troops had to travel rapidly over mountains or jungles if they were going to have any hope of bringing the enemy to battle. While it can be argued that the wars in Algeria, Malaya, Borneo and so on were decided mainly at the political level, in strict military terms neither France nor Britain could have fought as successfully as they did without helicopters.

Helicopters seemed particularly attractive when compared to parachute delivery. Undoubtedly, fixed-wing aircraft and paratroops had an advantage because of their strategic range. However, helicopters were better suited to tactical and even operational missions since they were based forward and the troops using them needed less training or special equipment. Helicopters could land units almost regardless of terrain, deploy them immediately ready for combat, provide mobility after landing and remain in the vicinity to retrieve them. Paratroop units were operationally and tactically inflexible because of the extensive preparations needed to launch them to their targets and their poor mobility once landed.[9] In a rapidly changing combat situation, helicopters offered a much more maneuverable and flexible means of carrying out air landings.

By the 1960s widespread development of armed helicopters had begun. At first, there was ad hoc arming of transport or observation helicopters to enable them to perform their missions more effectively; this step was often taken because of the needs of combat. It replayed the way in which aircraft had been armed in the First World War. A far more crucial development concerned the dedicated attack helicopters, whose task at that time was to provide aerial fire support, sometimes referred to as close air support, but more properly thought of as "flying artillery." Only the United States took this option in the 1960s. The Soviet Union preferred instead a more formal version of the combined transport-armed helicopter, the original Hind-A. France, Britain and West Germany pursued a more limited option by developing a simple anti-tank helicopter, reflecting their financial problems and their doubts about the helicopter's viability. By the 1970s America and the USSR were also focusing their attention on anti-tank combat, but both retained the capability and the willingness to use armed helicopters against a wide array of targets.

The development of anti-tank helicopters was greatly assisted by the almost simultaneous appearance of the ATGM. It was an accurate, long-range weapon that helped overcome criticism that helicopters were so unsteady they could never deliver accurate fire and that they would be too vulnerable to anti-aircraft fire. Since the missile notionally had the specific function of destroying tanks, its connection with helicopters helped allay the concerns of those who feared that the armed helicopter would supersede the close air support aircraft. Moreover, because ATGMs seemed to be so powerful in the late 1960s and early 1970s, it was thought that the missiles would be essential to any defensive operation and that only the mobility of helicopters would allow commanders to deploy, transfer and redeploy them at the necessary speed.

Helicopters were generally much better suited to the aerial fire support mission than fixed-wing aircraft. They were able to operate and remain in the vicinity of the frontline for many hours, either by landing and conserving fuel or by a rotation of helicopter subunits back and forth to refuelling points. These bases, unlike those required for even the least complex jet aircraft, were very simple; moreover, they were small and could be moved regularly and easily, thus decreasing the likelihood that enemy attacks might destroy or disrupt them. The cost effectiveness of the helicopter, in comparison to fighter-bombers, was also an important advantage for the former. Although helicopters have become complex and costly items of military equipment, for much of the last forty-five years their development was driven by the perception that they offered a cheaper alternative to the fixed-wing aircraft that might perform comparable missions.

Helicopters were also thought to be capable of delivering fire support more accurately than fast jets and, because of their forward location, they could provide more support more regularly and respond more quickly. Unlike aircraft, the helicopter's slower approach allowed it to serve as both the spotting and the firing platform at the same time, thus reducing the difficulties inherent in using separate observers to adjust the fire of close air support aircraft.

In the rush to concentrate on anti-tank warfare and direct-fire anti-tank helicopters, the more traditional uses of helicopters declined somewhat during the 1970s. Observation was more likely to be used to assist anti-tank helicopters rather than artillery, but the development of sophisticated targeting and guidance methods compensated for this change. More worryingly for many helicopter advocates, air transport was relegated to second place behind armed action and was rarely seen in combination with it; in particular, the integrated airmobility concept pioneered by the Americans fell from favor. A number of reasons contributed to this state of affairs. Financial stringency did not allow more than one main direction to helicopter developments in some countries; there were organizational and political impediments; and a reasonable fear existed that light, airmobile infantry were as unsuited to modern armored warfare as the anti-tank helicopter was appropriate.

Air transport continued to be important in some respects, particularly if operations were to be conducted in difficult terrain or were amphibious in nature; its use for emergency logistic support to mechanized units was also appreciated. Nor was airmobility entirely forgotten, with the American air assault division continuing to exist and other U.S. light infantry divisions retaining an interest. In France a small infantry unit—admittedly with a specific anti-tank function—was attached to corps helicopter regiments in 1977. Interestingly, the country with the least practical experience in airmobile operations, the USSR, maintained and improved the air-landing capability it had first introduced in the 1950s in line with its particular interest in deep, offensive operations. It was as if the Western countries' involvement in low-intensity wars had somehow inhibited their thinking about the use of light airmobile forces in conventional conflicts. Additionally, the USSR's more offensive doctrinal stance may have made air landings more attractive than in defensively oriented NATO countries, because many of the problems of light infantry related to the capacity to defend against armored attack. Their main advantages were, of course, in their capacity to attack where the enemy was weakest—in its rear areas.

The 1980s saw more importance attached to armed helicopters, with greater recognition that they could go beyond anti-tank combat. It was widely accepted that helicopters could be used to advantage against a variety of targets, even in the enemy's rear areas. It was realized that the helicopters might have to perform air-to-air combat. Airmobility received much more acceptance. Furthermore, integrated operations between attack helicopters and airmobile infantry became prominent in the development of American aviation brigades, Soviet air assault forces,

Table 5.1
Increase in Attack Helicopters 1970s–1990s

	Early 1970s	Early 1980s	Early 1990s
USA			
Attack Helos	900	1,000	1,600
Total Helos	10,000	8,000	8,300
Armed:Total	9%	12%	19%
USSR			
Attack Helos	0	580	1,700
Total Helos	2,500	3,400	4,500
Armed:Total	0%	17%	38%
UK			
Attack Helos	24	24	80
Total Helos	420	370	380
Armed:Total	5%	7%	21%
France			
Attack Helos	60	90	240
Total Helos	450	560	680
Armed:Total	13%	15%	41%
W.Germany			
Attack Helos	0	0	200
Total Helos	510	540	750
Armed:Total	0%	0%	27%

Note: All figures approximate.
Source: *The Military Balance* (London: IISS, 1970–1990).

the French *force d'action rapide* and the West German airmechanized units (Table 5.1).

Status of Air and Land Components

The relative status of the ground and air forces was an important indicator of the nature and degree of integration within air-land warfare. Part of the problem with air-land warfare in the Second World War had been that aircraft were thought to be capable only of supporting ground combat units. Helicopters also began their military careers in a supporting role, and the assumption throughout the period from the 1950s to the 1970s was that aerial combat power was merely an adjunct to land power. Thus the extent of integration was limited, similar to the relegation of the tank to a supporting role by Britain and France in the 1930s. However, in the 1980s it was accepted that helicopters were capable of more than supporting other forces because of their ability to maneuver on the battlefield in comparable style, although with different techniques, to armored units. Maneuver, in this sense, did not simply mean "move" but implied the capacity to lead combat actions and determine the course of the battle.

The eventual emergence of helicopters as an accepted maneuver arm was partly due to political pressure from helicopter advocates and the organizational needs of aviation groups. Better technology also played its part, making the helicopter more like other arms by increasing its reliability and sustainability, reducing its logistic dependence and introducing all-weather capabilities. Not only did technological changes improve actual combat capacity, but it also changed the way that ground soldiers perceived the helicopter, making them more likely to accept it. For example, the less frail and more menacing that helicopters looked, the more they seemed capable of dominating a battle in the same way that tanks were supposed to do. In addition, the development of maneuver status stemmed from the fact that, unlike other aircraft, the helicopter could utilize the ground for cover and interact with it. Most of all, it was realized that the helicopter's key asset, its speed, was not being fully exploited. In reacting to the needs of ground forces while performing a supporting mission, the aerial component of air-land warfare was always constraining its speed and flexibility by operating at the pace of ground movement. It meant that the potential uses of helicopters were poorly understood. For example, the U.S. Army found that, when treated as close air support aircraft, its attack helicopters were left out of battle planning and then employed as a last resort: they were, essentially, peripheral to land combat. No matter how decisive the aerial intervention might have been, the pace and form of the battle had been set by land units. By maneuvering, helicopters were instead able to determine when and where their associated ground-based forces were to fight.

It must not be thought that sole, or even primary, responsibility for maneuver in air-land warfare passed to helicopters because of this change. On the contrary, armored units continued to be the main maneuver force in terms of numbers and overall combat power, and the very nature of air-land warfare required a variety of different maneuver arms. However, the components of air-land warfare became far more equal, thereby allowing either to lead the battle and set the tempo for the other, whereas this had never been the case before. In counterinsurgency operations, helicopters were never truly maneuver forces (for all their importance as a supporting arm), because the style of combat was intensively infantry based. It was when helicopters began to be deployed for more intense forms of warfare that the possibility of maneuver emerged. The small numbers of helicopters (relative to armored vehicles) might in fact be an advantage here: given the increasingly—and in some armies near total—mechanization of ground units, the existence of a maneuver arm capable of moving much faster than the rest of the forces engaged in land combat will probably mean that aerial maneuver, while less frequent, might be more decisive than land-based maneuver.[10] Thus the helicopter-based aviation units of the future might act in the same way as the relative handful of armored divisions that so successfully dominated operations in the Second World War.

Organization

Even more vital than status was organization, revealing the way in which the integration of aerial and ground-based land combat power developed after the introduction of the helicopter. It concerned not only the arrangement of helicopter and ground units at both tactical and operational levels, but also the broader issue of ownership and control of helicopters within and between armies and air forces. Indeed, one of the greatest hurdles to integration was the almost universal assignment of responsibility for air power to a country's air force—in effect dividing the components of air-land warfare. The creation of separate air forces occurred mainly in the interwar period. This separation was justifiable, especially for the strategic or operational use of air power. However, because air forces regarded the exercise of air power as the basic reason for their existence, doctrinal dissonance arose with ground forces, particularly in Britain, France and America, over the extent to which air forces should integrate with armies.

A separate air force was not necessarily incapable of close combination with an army, as the *Luftwaffe* showed in the Second World War. Some of its success was due to the fact that most of its officers were originally from the army and its training and equipment were mainly designed for air-land warfare tasks. However, separate organizations were much more likely to go in the opposite direction, leading to less cooperation—either because of their members' attitudes or more basic deficiencies. For example, in the 1970s the radios in RAF support helicopters were unable to operate as part of army communication nets—a fact that hampered attempts to develop a workable doctrine for airmobile operations. More important, officers in air forces were rarely trained in the same manner as ground commanders, making it very hard for the two to understand each other's viewpoint. Even when both sides wanted to cooperate, simple matters such as differing terminology exacerbated communications problems and made it too easy for negative assumptions and hostile attitudes to develop.

During the last forty-five years in France, West Germany, the United States, but especially Britain, the air forces have impeded the development of army aviation, fearing that air power was in some way being compromised. Opposition was mainly expressed in these terms: either helicopters were unsuited to the task, or the army was unable to employ them according to the tenets of air power, or the money spent on them was better expended on more fixed-wing aircraft. In the USSR, the nature of the relationship between the VVS and the ground forces gave the latter much more power over the former than in Western armed forces; yet even in the Soviet army, the structural problems caused by dissimilar training and equipment remained unresolved.

Because of this air force opposition, aviation structures were established either inside armies or closely linked with them. Helicopters facilitated this process in some countries because the air forces were not particularly interested in spending money on what they saw as expensive, complex and inefficient aircraft; the army, however, saw helicopters as a form of ground vehicle that could fly. If an army wished to acquire helicopters at its own expense, and in doing so took up some of the tactical air support burden, then an air force was relatively content. Yet helicopters became a part of many armies early in their development when they were not carrying out tasks that the air forces claimed exclusively as their own—for example, armed action. Then, when such tasks began to be performed, the air forces were less able to prevent this expansion of army aviation—though they were quite happy to try and defeat any such expansion.

Forming an air component with specific responsibility to the land forces did not solve all the problems of integrating air and land systems. Rivalries developed between those advocating the supremacy of helicopters and those supporting armored vehicles. Often the problem was that advocates for either side constructed their case in terms of either air or land warfare; for example, one British helicopter enthusiast once suggested to a tank officer: "Wouldn't you like [to have] a Scout helicopter armed with SS–11 missiles instead of a Chieftain [tank]?"[11] The strength of sentiment on each side was caused partly by the reduced defense budgets that at one time or another affected all major armed forces, but also by a genuine, although misguided, belief from those involved that either tanks or helicopters, but not both, were predominant. As Simpkin concluded, the key issue was the "interface between the armored vehicle and the helicopter . . . [and] armies are peculiarly bad at coping with interfaces, let alone at identifying future ones."[12] The whole point of air-land warfare was that it required doctrinal cooperation and not conflict.

Yet, in the long term, the development of a group within the armed forces with responsibility for helicopters and other forms of aviation specifically intended for air-land warfare gave the helicopter an official advocate without which it would not have succeeded. The organizational imperatives, which were a consequence (and a cause) of the creation of an Army Air Corps or an Aviation Branch, lent further weight to campaigns for increased recognition of the helicopter's value. In other words, each progressive victory by such an organization was not seen as sufficient in itself but simply as laying the basis for the next step, the next innovation. Perhaps most important of all, however, such a group had the potential to develop aerial combat power in the context and with the objectives of land combat firmly in mind: in other words, to promote air-land warfare. While air forces were excellent proponents of air power, they were organizationally, as well as politically, unable to develop doctrines by which aerial means could be used solely for the purposes of land combat.

One of the most important aspects of air-land warfare was the actual physical combination of aerial and ground forces into combat units. A variety of different organizations was used, some more successfully than others, during the last four decades. At one end of the scale was the U.S. air assault division, which had over four hundred helicopters and was under the control of armies or army groups. Even though the infantry within this formation would actually fight on the ground, they were so closely associated with helicopters that, in relation to other divisions, they had become part of the aerial component of air-land warfare. France's *4^e^ division*, with 240 helicopters, was comparable in its independence and, with only a small infantry component, was even more clearly an aerial combat unit. West Germany's airmechanized divisions, should they be formed, will be similar to these two large formations. Attack helicopter battalions in US corps' combat aviation brigades also had an independent role.

Other types of organization were tried. They included using small numbers of helicopters at the tactical level, such as the arrangement of anti-tank helicopters in support of American battalions and German brigades; these helicopters were held at divisional level but assigned downwards for combat. British and French divisions also had helicopter units, but they tended to be used at divisional level. Britain recently began integrating attack helicopters within the ground battle but, as with France and West Germany, helicopters and armored forces were seen as coordinating their actions and not actually combining. In contrast, U.S. and Soviet divisional helicopter units were to be used in close tactical combination within battalions and regiments.

Despite their widely different forms, all of these organizations sought to resolve the same problem: how to integrate the air component into air-land warfare without losing its freedom of movement, while at the same time meeting the expectations of ground commanders. The organizations had to ensure that helicopters did not become so independent of land forces that the integration required by air-land warfare was impossible, and at the same time they had to prevent helicopters from becoming so closely linked that their unique mobility advantage over ground vehicles was lost.

Another consideration that determined these organizations was a desire to achieve combined arms, in which helicopters were to benefit from the association with other arms because their particular vulnerabilities were compensated for by the others' strengths. Equally, those arms would gain from the helicopter's specific advantages. Combined arms also had a political dimension because inclusion in a combined arms team seemed to imply recognition of the combat value of helicopters. It was often thought that combined arms required close integration, the most extreme example being the suggestion by some Americans in the 1970s of a company composed of helicopter, tank and infantry platoons. Yet combined arms was more subtle than that. For example, the American air assault division sought to include the necessary pieces of a combined arms team, but all based on helicopter movement. Thus there were two possible solutions: either add helicopters to existing arms or create a new combination based solely on aerial force.

The advantage of the latter was obvious. Such airmobile organizations (like France's 4^{e} DAM) allowed much greater freedom of movement and increased the flexibility and effectiveness of helicopters. Another benefit was that the unit's training and above all its mental approach were based on aerial maneuver of ground forces, which was greatly preferable to ad hoc organization. It was partly for this reason that the Soviets, who had used the helicopter for many years to reduce their dependence on special airborne troops, later created specific air assault units. The form of combined arms was also affected by the relative status of the various components. In the 1960s and 1970s, pure attack helicopter units were most likely to be found acting as support within a combined arms team based on armored forces. Recently, however, they began to be used as the sole maneuver arm, combining with artillery and other support services but not with armor or infantry. In this case, many of the combat benefits of combined arms remained but without the limitations on helicopter mobility.

Although no two countries were entirely alike, by the end of the 1980s these organizational issues were resolving themselves differently at tactical and operational levels. Whereas before there had usually been either too little or too much integration (as exemplified by the anti-tank helicopter doctrines in the 1970s of the British and Americans, respectively), in the 1980s a balance emerged between the two. The combination became one of effects rather than methods, since armored ground forces and helicopters operated in very different ways. First, helicopters formed tactical combined arms teams with armored infantry and tanks as well as supporting arms such as artillery. At the operational level airmobile formations based on helicopters provided their own form of tactical combined arms, while at the same time combining externally with the rest of the army. In addition, attack helicopter units were able to act as the sole maneuver force within a combined arms team.

At both levels, the aim was to achieve the integration necessary to promote air-land warfare and the mobility advantage of helicopters over ground vehicles was the key factor. The degree of closeness of integration and the organizational level at which it occurred depended on the range at which the helicopters were to operate. Over short distances, as found in tactical engagements (and presuming the terrain allowed ground forces to operate relatively easily), the difference in mobility between ground and air forces was slight, although still important; therefore, tying the two closely together exploited the helicopter's speed but did not restrict it too much. Over the longer distances—upward of several hundred kilometers—associated with operational movement, integrating a ground and aerial force tactically would have encumbered the latter too much, negating its mobility advantage. Thus the combined arms of large airmobile units (including attack helicopter formations) had to be based entirely on aerial means, and their integration with the rest of the army was operational in nature.

Mobility and time and space considerations are becoming even more important to the world's major powers. In the period between 1945–91, the close confines of Northwest Europe, with its high concentration of combat power, dominated theoretical military thinking in the countries most concerned with helicopter development. As the Persian Gulf War showed, these preparations were adaptable to a strange situation. But future planning is likely to concentrate on these different areas where time and space factors are greater and the ratio of force to space is smaller. The reduction in military forces will only exacerbate this development. In the future, or so army aviation proponents suggest, helicopters will come into their own by offering the capacity to cover large areas with small forces in a way that mechanized units simply cannot.

CONCLUSION

The helicopter's particular technical characteristics made it an air-land vehicle; moreover, as it was being introduced, armies were searching for new means of conducting air-land warfare. This confluence of ideas and technology led to helicopter doctrines being developed under the aegis of existing air-land warfare concepts. However, the helicopter was different from the aircraft previously used and, from within this established conceptual framework, the helicopter fundamentally altered the nature of air-land warfare. Through its associated doctrines, it redefined the relationship between air and ground combat power.

First of all, the development of the helicopter allowed aerial and land-based combat means to become much more integrated. By the 1980s, doctrines were mainly based on the assumed equality or near equality of helicopters and armored vehicles, and organizations had evolved to allow close and regular battlefield interaction of the two. Helicopters also led to an expansion of the scope of air-land warfare. With the exception of some relatively unsuccessful paratroop operations, it had been essentially tactical in scope. With the deployment of large brigade or divisional-sized airmobile formations—usually self-sufficient in terms of fire, movement and reconnaissance helicopters—there arose the capacity for aerial and land-based components to interact at an operational level. Although often containing ground-based infantry, the closeness of association with transport and other helicopters meant that the whole air assault formation became an aerial combat unit, integrating its actions with other ground-based divisions under the command of an army or army group.

This change arose from the combination of factors. Technical characteristics, especially as they matured in the 1970s and 1980s, were important in allowing helicopters to interact with the ground, using terrain and so on, without being limited by it. Armies increasingly began to need a more integrated form of air-land warfare because the growing mechanization of armies had eliminated the mobility differential, reducing armies to equal tempo. Therefore, helicopters appeared as a means of being able to increase the pace of combat, above that of the majority of the enemy's forces, and at the right time and place. In defense, this was seen mainly as offsetting the attacker's advantage of choosing when and where to strike; in offense, the advantage was expressed in terms of the speed with which combat power could be transferred into the defender's rear areas.

The new form of air-land warfare also arose because helicopters helped to unravel the confusion between means and objectives that had beset army-air force relations over fixed-wing aircraft. Many military personnel in both services were unable to see that, because aerial means might well be vitally important in the context of land combat, at least some of them had to be controlled by and integrated within the organization responsible for that combat. Flight—a means—became linked too closely with the objectives of air war. Helicopters also suffered from this confusion to some extent, with some air force officers (notably in Britain) construing them to be part of air power first and foremost. However, the relative uselessness of the helicopter for anything other than surface-oriented combat was too obvious for such perceptions to last very long or to carry much weight in the decision-making process.

Table 5.2
Attack Helicopter Numbers Compared to Tanks

	1980	1990	%age increase
USA			
Armed Helos	900	1,600	66%
Main Tanks	8,000	11,600	44%
AH: MBTs	8%	14%	
USSR			
Armed Helos	580	1,700	190%
Main Tanks	20,000	24,000	20%
AH: MBTs	1%	7%	
UK			
Armed Helos	24	80	233%
Main Tanks	900	1,300	44%
AH: MBTs	3%	6%	
France			
Armed Helos	90	240	167%
Main Tanks	1,100	1,350	23%
AH: MBTs	12%	18%	
W.Germany			
Armed Helos	0	200	0
Main Tanks	2,500	4,100	64%
AH: MBTs	0	5%	

Note: All figures approximate. MBTs are USA: M–60, M–1; USSR: T–62, –64, –72, –80; UK: Chieftain, Challenger; France: AMX–30; W. Germany: Leopard I, II.
Source: *The Military Balance* (London: IISS, 1970–90); checked against other references cited above.

The development of air-land warfare was not immediate and was as much the result of numerous small incremental changes as of one large incident of innovation. It emerged slowly, as armies struggled to understand new ideas and equipment without simply using them after existing fashions. However, by 1992, armies had begun to find the doctrines—organizations, concepts and training methods—that enabled the aerial freedom and flexibility provided by helicopters to be utilized without the constraints of ground action but, at the same time, without being remote from the specific purposes of land combat. Moreover, recent combat experience had dramatically shown the value of helicopters in providing an aerial maneuver force within the context of land warfare.

Although some advocates might see helicopters as causing a "rotary-wing revolution,"[13] the term "revolution" is too value laden to be used profitably in assessing the basic and fundamental changes that have come about. As was the intention of the more extreme supporters of this revolution, the term implies that the helicopter is replacing the tank as the most crucial land weapon system, whereas current helicopter doctrines are, in fact, dependent on the continuing presence of armored forces (Table 5.2).

The helicopter is a part of the mechanized revolution that began in the early twentieth century and not the beginning of a new era in warfare. If there has been a fundamental change because of military helicopters, it occurred at the intersection of air and land combat. Until the emergence of the helicopter, its technical improvement and the maturation of associated doctrines, armies had only a tenuous grasp of the use of air power within the land battle. The helicopter provided an aerial combat system that ideally complemented armored forces. Indeed, intellectual mastery of the military helicopter was most evident when soldiers stopped seeing its relationship with established arms as a competition and began to develop means of integrating the new with the old.

NOTES

1. Scibiorek, "Helicopters," 2; and Reznichenko, *Tactics* (1987).
2. The use in the 1980s of terms such as Airland Battle seems to be driven almost entirely by armies. While developed in consultation with air forces, they do not appear to reflect air force thinking.
3. A good summary of the development of close air support can be found in B. Greenhous, "Aircraft versus Armor: Cambrai to Yom Kippur," Tim Travers and Chris Archer (ed.), *Men at War: Politics, Technology and Innovation in the Twentieth Century* (Chicago: Precedent, 1982), 93–118.
4. Gavin, *Air Assault*, 21–32.
5. See, for example, the analysis in Richard E. Simpkin, *Anti-tank—An Air-mechanized Response to Armoured Threats in the 90s* (Oxford: Brassey's, 1982), 188.
6. See also Farooq Hussain, Ian Kemp and Philip McCarty, *The Future of the Military Helicopter*, (London: RUSI, 1986), 64–66.
7. Richard E. Simpkin, *Race to the Swift* (London: Brassey's 1985), 120–21.
8. Richard E. Simpkin, *Tank Warfare* (London: Brassey's, 1979) 173.
9. Only the Soviets developed adequate armored vehicles for their airborne troops, and even they saw the benefits of using helicopters for many air-landing missions.
10. von Senger and Etterlin, "New Operational Dimensions," 12.
11. Macksey, *The Tanks 1945–1975*, 175.
12. Richard E. Simpkin, "Flying Tanks? A Tactical-technical Analysis of the 'Main Battle Air Vehicle' Concept," *Military Technology* 8.8 (1984): 80.
13. Simpkin, *Race to the Swift*, 117.

6

Deciding on Innovation—Helicopters and the Decision-making Process

Multitudes, multitudes in the valley of decision

Joel 14.

The only way in which the innovations discussed in the preceding chapters could be implemented, and become an accepted part of army life, was through the doctrinal decision-making process. And, indeed, most analyses of doctrinal or, more usually, technological innovation have focused on decisionmaking. But most of the answers that have emerged from these analyses appear simple—even simplistic—when compared to the complex reality they seek to represent. Graham Allison's *Essence of Decision* established the basic conceptual paradigm of the interplay between "rational" actors, bureaucratic politics and organizational routines, which has been most influential in these analyses. But Allison was rather more circumspect than some of his successors. He noted that one paradigm did not reflect the complexity of the subject and, by acting as a "conceptual lens," filtering and focusing the light that a scholar hopes to throw on a subject, models in fact distorted the picture.[1] And, as James Kurth warned, the danger in this proliferation of models was that any one of them, on its own, could be construed as being a sufficient, even exhaustive, explanation.[2] Although analysts have been more cautious about single models than Kurth imagined they might be, nevertheless, they usually concluded that one or another model, this factor or another, had more explanatory power than the rest. In some cases scholars constructed a direct and explicit competition between the models.[3] In others, the contest was between analysts—each favoring a certain theory.[4] Even where many factors were considered together, conclusions were often drawn that established them in a hierarchy of explanation.[5]

Matthew Evangelista argued that the crucial factor in determining this approach was the desire of most analysts to influence the policy process they were studying. He contended that analysts wanted to find a "parsimonious explanation" because "a theory as complicated as the reality it seeks to explain [is] an undesirable consequence if one wants to use theory as a guide to policy."[6] The widespread occurrence in the literature of phrases such as "policy implications" and explicit comments about the authors' intentions to prescribe new policies is strong evidence in favor of this argument. Indeed, Evangelista was himself engaged on the same quest and, once again, his conclusions tended towards a simpler explanation over a more complex reality.[7] In Kurth's approach, various determinants of decisionmaking are discussed in this chapter, with two key assumptions.[8] First, that all factors have equal value in explaining decisionmaking and that no one factor or group of factors necessarily had more influence than the others; and second, that while these factors are discussed in turn, each is intimately connected with the others and no factor nor even a group of factors can be used to describe adequately all the nuances and ramifications of a particular decision. As some scholars have recognized: "The arbitrary isolation of various determinants tends to ignore the importance of their dynamic inter-relationship."[9] In this case, "dynamic" means that the component elements of decisionmaking were always realigning themselves and interacting in different ways. Not only did some factors become more or less important over time, but decisionmaking was ongoing, with the emergence of new doctrines in development affecting implementation of existing ones and vice versa.

It would appear that, for many scholars, complex description was too untidy, and so the search for austere conclusions also stemmed from what E. H. Carr termed, rather ironically, the "professional compulsion" to try and determine the "ultimate cause" of events.[10] In contrast, this chapter seeks only to present a picture of the influences on doctrine. None of these should be thought of as causing doctrine but, rather, as describing the way in which it developed. Two basic conclusions will emerge here. First, the interaction of factors was crucial and was more than just a cumulative process in which factors weighed up for and against a particular decision. Interaction meant that each factor only affected doctrine within the context established by the other determinants. The second conclusion is that the impact of individuals cannot be ignored: not because they were rational actors, taking control of the forces around them, but because many factors had an impact on doctrine only after they had been affected by individuals' subjective perceptions.

THE MECHANICS OF HELICOPTER DOCTRINE DECISIONMAKING

Before looking more closely at the many factors that affected decisionmaking about helicopter doctrine, we must be clear about what exactly they were affecting. All those making decisions about doctrine were seeking to establish how and why to use particular weapons—helicopters—so as to create or improve military power. It should also be remembered that doctrine had three distinct aspects: ideas, or doctrines in development; officially sanctioned doctrines; and accepted and implemented doctrine. The hierarchical nature of armed forces, as well as the division and delegation of responsibility within those forces, ensured that these aspects required decisions to be made at different levels, and decisions at one level had ramifications for the others.

As was made clear in the preceding chapters, certain organizations were given responsibility for helicopters and helicopter doctrines within their respective military hierarchies. However, full official sanction for most of these proposed doctrines came at a higher level. The exact level depended mainly upon the impact of the changes, the amount of resources required and repercussions within and between the various armed services. A decision to change the techniques for communication between helicopters—a minor matter—would probably have been sanctioned solely within the helicopter organization, with only pro forma ratification from higher up. To establish a new task for helicopters—for example, the anti-tank mission—needed approval from the most senior levels of the military and, because of the financial implications, civilian bureaucrats as well. Actually implementing doctrines shifted the decisionmaking back down through the hierarchy to the operational commanders who controlled helicopter units (even though they often came from other branches of the army). This process increased the number of decisionmakers involved, particularly nonaviators. Implementation depended on the willingness of these subordinate commanders, the authority inherent in the doctrinal decision, the practicality of that doctrine and the availability of necessary resources.[11] Where doctrinal changes were minor, they could normally be implemented within the helicopter organization itself. Since there was more awareness about helicopters among these decision makers and because the leader of the helicopter organization was directly connected with the decisions, this process was much easier than for more major innovations.

In general terms decisions were not made quickly, even for quite minor changes: it took France's ALAT four years to initiate, develop and introduce its protection and fire support helicopter doctrine. The length of time increased the number of factors affecting the decision-making process and meant that these factors did not remain constant. In the mid–1960s, the Armoured Corps supported the AAC's development of anti-tank helicopters in Britain but, by the 1970s, its interest and thus its important political support had waned. Such time spans were caused by the innovative nature of most helicopter doctrines and the reluctance of the armed forces to accept innovation; the increasing complexity of modern warfare and weapons; and the cautious nature of decisionmaking. Armed forces obviously sought well thought out decisions purely for military reasons—an importance reflected in the growing professionalism and sophistication of the armed forces,[12] but the cost of acquiring modern weapons or even of changing doctrines made budget-conscious soldiers and civilians still more careful. Moreover, technologies were developing so slowly that any doctrinal decisions linked to them could not proceed quickly.[13]

The issue of cost, and consequent political desires for increased accountability in defense spending, meant that decisions not only had to be right but had to be seen to be made as objectively and rationally as possible. It was simply not acceptable for money to be spent as if by chance or on a whim. The last thirty years in particular have seen increasing scrutiny of defense decisionmaking in the Western world, usually by people critical of the military. The Soviet Union's Marxist-Leninist approach to war also required an outwardly objective method. The rational approach, however, did more to conceal rather than actually eliminate the impact of subjective factors on decisions. These factors were present because decisions involved individuals who brought to the process their own experiences, opinions, personal aims and assumptions. Moreover, the organizations in which these people worked imposed their own irrational constraints and routines.

The determinants at the heart of doctrinal decisionmaking can be arranged in five groups: technology; military planning and threat analysis; experience; organizational process; and bureaucratic politics. They provided a background for doctrines; set the parameters in which doctrinal innovations were considered; and influenced the people making decisions and the structure in which they worked. Nor were these factors isolated inputs into doctrine, at various times either conflicting or resonating with each other. Further, a factor's influence depended on its perceived proximity to the essence of the decision being made, so that in the Soviet Union experience from Afghanistan was, on the whole, remote from established doctrines and had relatively little impact. Since they affected both the decisions and the decision-making process, the following discussion is concerned with how these factors determined the outcome of the decision-making process and also how they represented the nature of that process. And, ultimately, while some factors were more important than others in terms of specific outputs from the decision-making process, it is not possible to isolate any one factor as either a necessary or a sufficient reason why a decision was made in the way that it was.

TECHNOLOGY

First of all, what was the relationship between doctrinal and technological innovation? Since helicopter doctrine was based on the technical innovation of rotary-wing flight, technology was clearly essential, but the relationship between the two was very complex, changing in particular when technology moved from a state of the potential to the actual as part of attempts to meet specific requirements of doctrine. In the same way that doctrine had a number of aspects, there were various facets of technology. The base of applied scientific knowledge provided numerous potential technological solutions. Its intersection with doctrine resulted in commodities,[14] which were the differing equipment models based on scientific knowledge and suggested by defense manufacturers in response to the demands of doctrine. Although going beyond knowledge, commodities were still not actual technology, since they consisted mainly of designs or prototypes. Commodities then passed through a process of competition, selection and modification after which one of them was acquired and introduced into service as a weapon. This sequence, defining technology in its progress from potential military importance to an actual capability, was rather similar to the progression of doctrinal ideas into implemented doctrines, and the three aspects of technology had most impact respectively at these three levels of decisionmaking.

Immediately, this analysis of technology reveals the impact of subjective perceptions of technology on decisionmaking. Because of the host of "anthropological, psychological and cultural factors" working on decision makers,[15] and because technology and doctrine intersected primarily in the minds of decision makers, they interpreted technology in the light of existing ideas and imagined what it could do if harnessed to a proposed doctrine. One example was the redefinition of helicopter vulnerability. After periods in which the viability of helicopters had often been questioned, as if by magic helicopters were seen as more survivable. A technological limitation was reduced—not because of an actual change in technology, but because of a change in perception both of helicopters and the air defenses they were to face. If anything, these dangers were increasing.

The evolution of technology from potential to actual involved a great many people in various representative groups. Naturally perceptions of technology depended on the observer's standpoint, so that helicopter manufacturers, advocates, senior decision makers, unit commanders and others all tended to view the same technology—be it a weapon or potential weapon—in different ways. For example, those members of the military not closely involved with helicopter development tended to view the technology mainly in terms of its cost, rather than its effectiveness—thereby predisposing them to reject helicopter doctrines. As costs decreased or effectiveness was more clearly demonstrated, their perceptions changed. Most analyses of the relationship between doctrine and technology have not made an overt distinction between differing aspects of their subjects, and as a result there has been some confusion about what exactly was being studied. The concept of the military-industrial complex has been a particular source of problems. In this model of decisionmaking, it was held that rational decisions were being made about military equipment not in the interests of national security, but rather for commercial profit; and that technology was itself determining weapons acquisition, making the process "immune from political judgment."[16] Neither position is entirely unfounded, but if they are to be argued effectively, it must be realized that the technology with which scientists and innovators were concerned was not the same as that of manufacturers, which in turn was not that required by the armed forces. These distinctions arose because each group assigned different values and perceptions to their particular aspect of technology. Further, the helicopter's technical characteristics only gained any real meaning through such mental processes.

Knowledge

Knowledge was the foundation of applied science from which specific weapons were developed. Since technological innovations first appear as new knowledge, it might be wondered whether doctrinal innovations preceded technological ones or vice versa. While it is common among analysts to try and make this distinction,[17] most of the debate is irrelevant except for looking at the sort of simply policy prescriptions discussed above; it is also misleading because it assumes a clear causal relationship between doctrine and technology when none exists.

Military theories about using air power of the type provided by the helicopter were basically extant before the helicopter was invented. Yet this did not mean they caused that particular technological innovation to occur, or even that their existence was a necessary precondition. It is better to imagine doctrines and technology as objectives and means; that is, the former were an expression of what the military believed was an important objective and the latter provided one way to achieve those aims. Both objectives and means developed along parallel paths until they were joined to make a specific doctrine when imaginative military thinkers saw the helicopter's potential to provide new ways of achieving existing objectives. In the joining, knowledge of the technical characteristics of rotary-wing flight set the parameters for the doctrines about this new weapon; in exchange, ideas on how these might be employed shaped the exact development and application of that technology, since the closeness of the technical community and the military meant that the technology makers had many potential doctrines in mind when they developed technological innovations. During the twentieth century and especially after 1945, military decision makers placed great faith in technology, as indeed did society generally. They were aware of its potential and, more than ever before, sought technological applications to increase military capability.[18] Specific technological failures and delays rarely jeopardized the basic military concept with which they were linked; rather it was assumed that these difficulties would be overcome. Equally, doctrinal innovations were often made without an exact understanding of the limitations of existing technology. The unlikely appearance and rudimentary technology of early military helicopters only emphasizes the degree that faith in technology influenced its relationship with doctrine.

Gray, in answering his question, "Does theory lead technology?" maintained that the opposite was true because, even when a doctrine maker "is moved to advocate a new path in weapons development [he does so] because he perceives inadequacies in existing technology."[19] Yet such inadequacies were not perceived (and in that sense did not even exist) until a new doctrine began to appear, thereby revealing problems with current technology. The argument here is not that doctrinal innovation did in fact lead to technological innovation but rather that, as Gray himself concluded, the situation was too complex to make such distinctions. What is important was the reason for this complexity. Essentially the acts of technological or doctrinal innovation cannot be divorced from the people and the organizations through which they occurred. When it was knowledge, technology had relevance only to doctrine making through the forward planning of these people. Scientific knowledge about helicopters was like the multitude of ideas from which decision makers drew when formulating actual doctrines, and its most important effect was to provide a number of potential technological solutions to equipment requirements that a particular doctrine might have. The actual doctrines and technologies did not at this stage exist except in the minds of doctrine makers and weapons makers and, as a result, there was a seamless interlinking of the two in which neither one can be said to precede the other.

Commodities

The next stage in the relationship between technology and doctrine occurred when manufacturing or engineering organizations created a commodity: a proposed weapon that was not yet accepted by the military but which the company or design bureau wished to market and sell.[20] In the case of helicopters, commodities tended to develop directly from doctrine when the latter was in the process of gaining official sanction, because the decision to acquire technology (initiating the transition from knowledge to commodity) was made at the same senior level. Companies were generally happy to fulfill the requirements set down by doctrine not least because, by the 1960s, only the United States had more than one major helicopter manufacturer. Without the element of competition, the most profitable approach was not in changing doctrine but in meeting it as closely as possible. Some commodities have been created without this direct doctrinal link, thus requiring companies to establish a market for them; this was also necessary where a doctrine suddenly changed, rendering an existing commodity irrelevant. These markets could be established through a bureaucratic political process by pressuring the army to introduce doctrines that required those commodities. A clear example of the relationship between commodities and doctrine formulation comes from the United States where Bell Helicopter Textron, in response to the army's embryonic need for the RAH–66 Comanche, proposed a tilt-rotor aircraft possessing higher speed than a normal helicopter but with higher costs and greater development risks. What was critical here was that Bell was not proposing to build a better helicopter than its competitors in line with what the military wanted, but to ensure that the army's requirements were such that Bell was the only company that could succeed. However, Bell was unable to convince decision makers of its position, and there were few other cases of similar attempts to shape doctrine to meet commodities.

Commodities had a less profound but still important impact on the implementation of doctrine. The existence of a privately developed attack helicopter—the Cobra—made it easier for the U.S. Army to move immediately into this area of activity during the Vietnam War. Its existence and ready availability also helped to force the U.S. Army into doctrines that depended on that particular type of helicopter. Extant commodities thus narrowed the choice of technological solutions to doctrinal requirements where operational need made a quick purchase imperative.

Weapons

A particular commodity became a weapon when, having successfully negotiated the complex weapons acquisition process, it was accepted by the military as the technological partner to doctrine. The term encompasses entire weapons systems, individual armaments and even modifications to existing equipment. While knowledge affected doctrine more at the level of ideas and commodities were linked to official sanctioning of doctrine, weapons figured mainly in decisions about implementing doctrine. Although weapons came into service as a result of doctrinal decisions, once they began to be used the differences between imagined capability, which was so important in earlier decisions, and real capability became apparent. These differences then affected doctrine and the acceptance of innovation either positively or negatively depending on whether expectations about the technology were confirmed or not. For example, the AH–64 Apache convinced many who had otherwise been skeptical or undecided about deep-attack doctrines that these missions were both viable and potentially decisive. In contrast the inadequate British helicopters of the 1960s held back the cultural acceptance of airmobility as much as they impeded its theoretical development. Despite the fact that weapons, although affected considerably by the organizational and political nature of the acquisition process, were the result of stated doctrinal requirements, this relationship was not fixed. Because the introduction of new weapons demonstrated more clearly, by both observation and extrapolation, the potential capabilities that certain technologies might possess, it also caused changes in the basic relationship discussed above between knowledge and doctrines in development, thus affecting innovation at that level as well as influencing the implementation of existing doctrine.

Clearly, then, the relationship between doctrines and weapons was dependent upon the perceptions of the helicopter as a weapon. Whereas helicopter advocates were sufficiently impressed by potential technology to develop a doctrine based on it, those less convinced of the merit of helicopters required some tangible proof of the capabilities of these weapons. Indeed, even helicopter advocates did not recognize all the possibilities inherent in technology, and thus they benefited from the actual deployment and subsequent testing of helicopters. To use Millett's phrase, the "psychic satisfaction" that the military derived from technology was crucial in determining doctrinal acceptance—and this satisfaction was at its most potent when helicopters, with all their actual and symbolic power, were actually delivered to units.[21]

MILITARY PLANNING[22]

Strategy and Operational Planning

Strategy was vitally important in defining the basic circumstances, geographic locations, and enemies and allies that all shaped the development of helicopter doctrines. For example, West Germany's strategy of forward defense meant that the *Heeresflieger* never had to plan for operations outside Europe, nor was it able to develop avowedly offensive doctrines. Thus changes in strategy that altered the basic aims and concentration of a country's armed forces were a powerful motivation for innovation in helicopter doctrine. The Soviet Union's strategic concentration on conventional warfare in the 1980s and the United States' reemphasis of its role in Europe in the early 1970s are good examples. However, such changes were hardly the main determinant of the exact development of helicopter doctrines, nor were they necessary for such innovation. Rather, by creating an atmosphere in which armies asked new questions about their capabilities and sought new answers to their strategic problems, they set the scene in which other factors could work more directly.

A more important factor was a country's prevailing operational plan. Answering the questions posed by strategy, it set the basic goals for an army and assigned responsibilities for achieving them to the individual arms. Operational doctrine thus helped to establish the roles and missions in which the helicopter's particular military capabilities were to be used, providing a context in which specific doctrines could be developed. Because of the clear link between helicopter doctrine and operational planning, the latter had more impact than strategy in this regard, and it is notable that changes in helicopter doctrine very often coincided with reforms to operational planning. Take, for example, the way in which helicopter mobility, one of its prime characteristics, was treated differently in the two operational doctrines of the U.S. Army in the 1970s and 1980s. In the first of these, Active Defense, maneuver was seen as valuable, but its main function was to bring to bear sufficient firepower to defeat attacking enemy forces by attrition. In this context, attack helicopters were seen as highly mobile firing platforms for ATGMs, which were in themselves far more important than the helicopters. In the 1980s, with the rise of Airland Battle, American operational art began stressing that mobility could achieve results in its own right. Thus maneuver by air became a distinct role for helicopters, and the change saw the emergence of new doctrines for using helicopters in independent missions and behind enemy lines.

The Relevance of Military Planning

The current trend among analysts of decisionmaking is to challenge the superficial assumption that rational, objective military reasons shaped doctrine and to concentrate on other determinants. Organizational structure and bureaucratic politics have received more attention, even in works in which the theories behind these factors are never clearly articulated. One exception to this trend towards alternative explanations was the revisionist view of British armored warfare theories before 1939 in

Larson's *The British Army and the Theory of Armored Warfare*.[23] While it treated some of the social, economic and political factors normally associated with the topic rather too lightly, its concentration on purely military factors was a useful reminder of the role they can play in doctrine making. Central to every decision was an assessment of the combat value of helicopters in relation to some military problem. For this reason military factors cannot be ignored, no matter what the organizational and bureaucratic processes.

Most helicopter doctrines were subsets of an overarching operational concept. It is hard to judge whether the use within the doctrines of language descended from the concept indicated that the military conditions that were responsible for operational plans affected helicopter doctrines in a similar manner, or whether the hierarchical nature of the armed forces simply required various subdoctrines to be expressed as if they came directly from the superior doctrines.[24] Strategic and operational plans were reflections of current and predicted military conditions. In the initial mental creation of doctrine, it was the problems arising from these conditions that affected decisions about helicopters, and some creators of helicopter doctrines were influenced more by their own understanding of prevailing military conditions than by the officially authorized version. Yet subordinate doctrines had to be expressed or justified in terms of the prevailing superior military thinking if they were to receive official sanction and be accepted. Once again it is impossible to determine finally which was the cause and which the effect: it is sufficient to see that helicopter doctrines reflected operational doctrines in their language, even if the helicopter doctrines were in some ways revisionary or resistant to the concepts within the higher doctrines.

Threat Analysis

Threat analysis—a summary of perceptions, assumptions and intelligence about enemy capabilities and intentions—also affected the development of helicopter doctrines. This analysis, although more highly organized than at any other time in human history, was and remains imprecise and subjective. Indeed, intelligence assessments reflected the perceptions and assumptions of the analyst as much as the actual capabilities of the enemy.[25] The impact of threat analysis on decisionmaking has also been studied under the misleading title of "action-reaction," in which doctrinal or technological innovations by one side were construed as being direct reactions to similar acts by the other. As Stephen M. Meyer pointed out, such language "inappropriately constrains" any discussion of the topic because of the loss of control over decisionmaking that is implied by the word "reaction." [26] It is incorrect to imagine that helicopters were developed by a country's armed forces just because another was doing the same. Certainly the demonstration of the helicopter's effectiveness by one army led others to evaluate it—for example, the Soviet Union paid close attention to the United States' use of helicopters in Korea. However, a reaction to enemy initiatives was only one of many factors at work and was unlikely to have as much influence as factors internal to a particular country. It can be suggested, though, that innovation was more likely to occur when there was the security of knowing that, if proven wrong, then at least the enemy would be in a similar position. As well, the existence of a large number of enemy helicopters provided advocates in other countries with a powerful justification for innovation. In such cases, it was public and political perceptions of the importance of the enemy's capabilities that mattered rather more than military analysis of its force structure.

More important to the military were perceptions that the enemy possessed some capability that could only be countered effectively by using helicopters. If it was a threat with which existing force could not cope, then this was an overwhelming reason for innovation. A prime example was the rapid development in NATO of anti-tank helicopters. Prior to the late 1960s, helicopters had mainly been devoted to observation and transport missions; even though they had been tested as ATGM platforms, this task was not at all common. However, with the realization that nuclear weapons would not be used initially to answer the threat posed by the preponderance of Warsaw Pact armored forces, fast-moving and rapidly redeployable helicopters offered a valuable countermeasure. As a result, between 1965 and 1975 all the major Western armies either acquired or decided to acquire anti-tank helicopters.

EXPERIENCE

Experience was at the heart of all military decisionmaking, from specific helicopter doctrines to strategy; it also played a major role in shaping the way in which enemy capabilities were assessed and interpreted. The impact of personal experience helped to ensure that, despite the efforts of some policy and decision makers to construe their actions as rational, they in fact remained subjective. There were three types of experience, each of which will be dealt with in turn: actual combat experience, tests and trials, and received experience.

Actual Combat Experience

Perhaps the most significant example of day-to-day combat experience with helicopters was that gained by the majority of senior and middle-rank U.S. Army officers during a decade of fighting in Indochina. The most important effect on doctrine was that, although some doubts remained as to its viability in more intense forms of warfare, the helicopter was no longer an exotic piece of equipment seen perhaps once or twice a year at a major exercise. It was now a familiar part of army life. American officers were better placed to evaluate the helicopter; their judgment was more considered because it was not swayed (either for or against) by the helicopter's novelty. Thus, while experience helped advocates to argue more convincingly for helicopters, obviously it also reduced the extent of innovation required: as will be seen below in the discussion of organizational process, the larger the innovation, the more likely it is to be resisted. At the same time, these evaluations and arguments were colored by the fact that helicopters, as used in Vietnam, were closely associated with the defeat. Along with other factors, this association helped to change the emphasis of helicopter use away from infantry-based airmobile operations to anti-tank action.

Indicating the limits of direct experience, the Soviets' involvement in Afghanistan seemed to have little impact. While the design of the Mi–28 Havoc clearly benefited from lessons about helicopter vulnerability and there were some changes in air-landing practices, on the whole the experience was too much associated with counterinsurgency operations to affect the main concepts of the USSR's helicopter doctrine. For the Soviets, the Great Patriotic War probably held experiences of more value, even though helicopters had not actually been used. Afghanistan occurred when many doctrines were strongly established and thus were less likely to change. In contrast, Soviet experience in the later 1960s with exercises such as *Dnepr* came at a time when helicopter doctrine was in the process of development and thus more susceptible to influence. The war in Afghanistan did, however, influence the implementation of doctrine on air-land cooperation, by providing evidence of continuing failures in this regard as well as being a testing ground for new organizations and procedures.

Experience was only accessible through interpretation. Therefore, it was influenced and indeed dependent on the perspectives from which it was viewed and the preconceptions of its interpreters. Interpretation was particularly important where experience was gained in counterinsurgency campaigns and yet was being used to develop doctrines for more conventional warfare or was being translated into modern terms from earlier wars. Even where the lessons of experience were apparently clear cut, subjective assumption remained crucial. On the surface, the First World War convincingly demonstrated the ineffectiveness of horsed cavalry. Yet the mind-set of most cavalry officers was such that they interpreted the evidence against them as selectively as necessary to maintain their position. The development of helicopter doctrines was not affected to this degree, but nevertheless the subjective interpretation of the lessons of history and of trials was more influential on doctrine than any objective proof.[27]

What was the basis of this interpretation? It is often asserted that armed forces prepare for the next war as if it were going to be no different from the last, an argument usually supported by reference to almost every European country before 1914 and the Western allies before 1939. Yet none of these armies imagined that subsequent wars would be a simple replay of previous battles. As Millett commented, "Much [new] doctrine comes from a desire not to fight the last war again."[28] Military planners were aware that changes had to be made and often had a very clear vision—even if not the correct one—of what the future might hold. The problems arose not from overdependence on historical experience but when the vision of future war was based on what planners wanted to happen, rather than what was most likely to occur. Having established this vision—usually a rejection of all that had been wrong with the last war or a statement of what best suited their army's particular capabilities—they then constructed specific lessons to be learned from the previous war in terms of what they were expecting, that is wanting, to happen the next time. Leading American historian Jay Luvaas noted that an army tended "to select, combine and read its own peculiar lesson" from its experiences mainly because existing doctrines imposed "blinders on the vision of military observers."[29] The same is equally true of proposed or potential doctrines. Those seeking to prove or reject these doctrinal ideas reconstructed actual combat experience in light of their present concerns, blinkered by their preconceptions about these ideas. Thus, far from objectively determining what doctrines actually arose, combat experience provided evidence to support the imaginings and assumptions of decision makers about the use of helicopters in the next war.

Tests and Trials

Where direct wartime experience was not available to support doctrinal decisions, armies conducted trials. Since the Second World War and the introduction of operational research, decision makers have come to rely more and more heavily on these supposedly accurate and scientific simulations of combat; those dealing with helicopters were no exception. Trials were, however, quite different from combat. They were purposive, in that particular outcomes were clearly in mind when the trials were arranged and held. This in turn influenced the results that were achieved. Trials also had a specific and usually political context, as with the many tests conducted in the United States to determine possible increases or reductions in spending on Army Aviation. They were controlled, allowing ease of measurement at the expense of realism and thus were most authoritative for very specific procedural or tactical issues and much less accurate for general doctrinal concepts. Finally, they were replayable, ensuring that the "right" answer, the one that best suited the controllers of the trial, could be achieved, as occurred in tests of the OH–58D AHIP in the 1980s.

In an age where many decision makers were eager to demonstrate to outside arbiters their reliance on apparently objective testing and were themselves convinced of the validity of such testing, trials attempted to give scientific authority to decisions, proving they were "correct." Thus they were often used to supplement or to confirm what had been suggested by combat experience. However, trials did not provide objective answers. Rather, they usually supplied inconclusive evidence that, like actual experience, was then interpreted in an appropriate manner and used either to support or oppose helicopters. As Edward Katzenbach, in looking at the persistence of horsed cavalry in the twentieth century commented, "Nothing is more difficult to test than a weapon's effectiveness."[30] The reason, as Paret concluded, was that such tests were "affected by value judgements and imprecise knowledge . . . and subjective factors [were] introduced into the process."[31]

These views are supported by an examination of various helicopter tests in a number of countries. France's *4e division aéromobile* was not tested until after it had been established amid much publicity and approbation, thus suggesting that the actual results of exercise *Damoclés* were irrelevant to the general concept for the division's employment. Moreover, it would have been most surprising if the test's results had not confirmed the division's worth, since so many of the participants had a vested interest in such success. In all such trials, the more that their conclusions were invested with authority, the less objective they became, and they obtained a significance out of proportion with the actual results of the test. The most notable example was the Ansbach trial, which produced impressive statistics favoring anti-tank helicopters. The evidence from Ansbach was used relentlessly in the campaign to introduce these helicopters into the U.S. Army. It also became commonplace for Soviet writers to use the hypothetical exchange ratio obtained at Ansbach between helicopters and tanks to support their own conclusions—so much so that on occasion Western writers imagined that these figures had come from experiments that the Soviets themselves had conducted.[32] In the end, all that happened was that the data became entirely separate from its experimental foundation and was important only because of the value invested in it by those using it.

Received Experience

Trials and the evidence from actual combat were used not just by participants but also by observers in other countries, thereby becoming received experience. The doctrines and plans of allied countries also fit this category. Yet was another's experience as important as one's own exposure to helicopters? Its most obvious effect was to spark new ideas and to initiate a debate about existing practices. The French ATGM tests in the late 1950s and the fighting in Vietnam were very important in bringing the helicopter into focus in other countries, but the resolution of the debates over helicopters in these countries owed more to internal experience—a view supported by the diversity of doctrinal developments. Received evidence was much more obviously interpreted, often being used to support both sides of a doctrinal argument. It also appeared too remote from the decision-making process, especially for sanctioning and implementation and, as a result, decision makers tended to put their faith in their own experiences, which appeared to require no intervening adaptation.

ORGANIZATIONAL PROCESS[33]

Although this work is not based on organizational theory and does not seek to examine it in detail, chapters 1 to 4 were necessarily guided by the notion that structured groups or organizations tended to behave in certain ways and that this behavior affected decisionmaking by them or in relation to them. It is emphasized that this view of events relies not on the motives of individuals pursuing specific goals, but on the imperatives for action inherent in the structures in which those people were located.

It has been assumed by most commentators that military organizations have typically been conservative.[34] Yet despite the conservatism apparently dictated by organization theory, Posen discovered in studying European military doctrine in the 1920s and 1930s that the "stagnant doctrines" that theory predicted were not what he found.[35] Posen resolved this dissonance by defining innovation in a rather limited way, thus protecting the general assumption. Yet innovation and conservatism are not incompatible behaviors within an organization. They can be resolved by understanding the distribution of power and responsibility for decisions within the structure.

Conservatism

Conservatism stemmed from the fact that organizations generally sought to fulfill established goals according to existing priorities, to operate in a coherent fashion by using "rehearsed SOPs [standard operating procedures]" that regulated their diverse component groups, and to avoid uncertainty by working within "standard scenarios." Essentially they developed and used routine approaches to problems.[36] These routines delayed or even prevented the introduction of new weapons or concepts, particularly affecting the implementation of new ideas. Because they were not usually the ones responsible for creating the ideas, those people charged with implementing them were influenced more by existing doctrines. Fear of uncertainty, particularly in preparing for something as dangerous as combat, was a powerful influence—yet innovation required the organization to actively embrace uncertainty.[37] It was not that individual judgments were being made against it, but rather that the organization tended to continue doing what had been done before. An example of structural conservatism can be seen in the RAF, which concentrated upon strategic and then operational nuclear bombing, air superiority and air defense. These roles and missions defined the goals and priorities of the organization; thus routines developed around them and not around the provision of helicopter support to the army. The RAF was only responsible for flying and managing helicopters, not for working out why or how they should be used. Thus the RAF was "waiting for tasks to arise"[38] and was incapable of real innovation.

One effect of conservatism was that initial major innovations—for example the introduction of helicopters for anti-tank action—were by far the hardest. Once a new doctrine was sanctioned and began to be accepted, then even quite significant changes to it were much less restricted by organizational caution. Once initiated, helicopter doctrines developed their own routine within an organization and thus their own momentum, and it was much harder to reverse an innovative decision after it had become established. Another organizational imperative working in favor of helicopter advocates appeared when a large number of helicopters had been acquired for actual combat purposes. When the war ended, rather than abandon a large number of these aircraft and disrupt established routines of military activity, it was often easier to find a new task for them. This created a climate more favorable to doctrinal innovation—as with France and the United States' introduction of the anti-tank helicopter after the Algerian and Vietnam conflicts.[39]

The most significant organizational barrier to easy innovation with helicopters was the routine acceptance of the primacy of armored warfare, which had become a way of life for armies in the latter half of the twentieth century. The tank was not a part of the broader social fabric in the way that the horse had been some decades before, but it had become a crucial symbol of modern warfare and was the focus of land combat developments.[40] Armies were not foolish to assume that armored warfare was paramount, but the problem was that ideas such as those about light airmobile infantry, which did not fit comfortably within the conventional wisdom about armored warfare, faced a much greater battle for acceptance.[41] Nor was it surprising that the helicopter developments relating to anti-tank combat found the most favor, since they were clearly relevant to this paramount concern.

Another vital organizational imperative came from the fact that armies were concerned almost exclusively with ground combat. Therefore, aerial weapons and doctrines concerning airborne activity were not necessarily seen as fundamental to this primary function. Where aircraft were a means to prepare for ground combat, as with paratroop operations, then armies were more accepting. However, armies were more than normally cautious about the widespread deployment of armed helicopters—weapons that actually fought in the air—because of the apparent incongruity between flying and army activity. This caution was most often expressed as a fear that helicopter crews were not real soldiers and that they were unable to understand land combat. It also appeared in the apparently minor but intensely symbolic tension over the different uniforms usually worn by flying and nonflying soldiers. (One U.S. Army divisional commander would not allow the senior aviation officer to enter the division's headquarters until he had changed from his flying suit into standard battledress.) Armies had often set up their own flying arms so as to have more control over the use of air power for close support in land battles. Yet, while there was a political incentive for armies to be innovative with helicopters to gain that control, at the same time there were pressures against accepting or sanctioning decisions made within helicopter groups because of their anomalous status and their presumed lack of understanding of ground combat.

Innovation

In *Sources of Military Doctrine*, Posen argued that the RAF's insistence on the radical and innovative doctrine of strategic bombing before the Second World War was mainly due to that group's struggle within the armed forces to establish its independence and secure its future survival.[42] Posen believed that this doctrine could not be termed innovation. Yet there is strong evidence that the desire for change was a feature of organizational behavior.[43] There was a similar struggle for existence when the various aviation branches and corps that were responsible for military helicopters emerged from within existing structures. In Britain, the AAC expanded rapidly after its creation, moving from aerial observation and liaison to utility, reconnaissance and armed action. Notably, its interest in airmobility was for many years expressed only in terms of expanding the corps by taking over the RAF support helicopters. In the U.S. Army, changes to attack helicopter doctrine promoting the new idea of deep helicopter raids into enemy territory coincided with the creation of an independent Army Aviation branch.

Where a concentrated group did not exist, innovation was much harder. Britain's division of ownership of and responsibility for helicopters between the RAF and the AAC created a structure that inhibited concerted change by reducing the organizational power of helicopter advocates and by preventing a buildup of intellectual momentum among all those associated with helicopters. Since the U.S. Army, up until 1984, divided its aviation resources among other branches, its experience might appear to be an exception. However there was always a de facto leader: in the 1950s and 1960s it was the Infantry Branch; in the 1970s Armor. This integrated structure of ownership in the United States nevertheless affected the character of helicopter doctrine, with airmobility and then anti-tank combat tending to dominate while, respectively, Infantry and Armor had the most control.

A determined trend towards innovation also occurred in those cases where an air force exercised very tight control over aviation resources. There was also a political component to the struggle by armies for better access to close air support. However, the organization of armed forces was such that armies had little choice but to turn to groups within their own structure to provide such support. Moreover, while the helicopter's value may have recommended itself to armies in any case, it is interesting to ponder whether helicopters would have become so important had they not normally been the only suitable aircraft available outside the air force. It is instructive to remember that the U.S. Army was eager to develop fixed-wing aircraft for its close air support requirements, and it was USAF opposition and then wartime necessity that saw the appearance of the dedicated attack helicopter.

Imperialism

These two organizational behaviors can be resolved when it is understood that they had the same motive, which Allison called "imperialism."[44] While the term "imperialism" has a connotation of intent that might make it more useful in a bureaucratic political analysis, it is reasonable to assume that there was a structural dynamic that caused groups to pursue additional power in relation to other organizations with which they interacted. Clearly, one important method of achieving this was innovation. At the same time, however, imperialism can also refer to the conservation of power by an established group. While caused by the same tendency, these two differing organizational behaviors emerged because of the differential distribution of decision-making authority. It was uneven, with one or two groups dominating and thus forcing others to be innovative to maintain their separate identity. Because the latter were already in positions of power, their approach was usually more conservative.

The distribution of power had different effects on the various aspects of doctrine. The power to create the new ideas that might then become official doctrine lay with the helicopter advocates in the branches or corps, or officers assigned from there to central doctrine-making bodies. The power to sanction these ideas, however, lay with senior officers. The power to implement them effectively lay with the operational commanders who had helicopters in their units. Because of the structural weakness of helicopter organizations (lack of status, few senior officers) both the senior officers and the operational commanders were rarely aviators and usually had little direct interest in helicopters. Therefore, the full acceptance and implementation of helicopter doctrines was impeded by a structure in which those who had the most knowledge about the doctrines were unable to exercise full decision-making power in relation to them.[45] Clearly, this structure was a safeguard against the adoption of unrealistic ideas—but it also ensured that organizational imperatives for conservatism outweighed those for innovation.

BUREAUCRATIC POLITICS

Essentially the term "bureaucratic politics" describes the activities of individuals within the organizations just examined. The ultimate goal of the bureaucratic politic game, as it has come to be known, was the acquisition or retention of power for the organization to which the players were loyal, especially in relation to other organizations. In the military context, such power was reflected in the resources available to the organization and the recognition given to the value of the organization's contribution to the overall armed forces mission. A number of analysts have suggested that there is an association between bureaucratic politics and organizational imperatives in relation to decisionmaking. Some have even blurred the distinction between them.[46] However, the reasons why these factors influenced decisions were not the same. While organizations could appear to be political actors,[47] this phenomenon resulted from the process by which individuals identified with and embodied them. Organizational imperatives acted on whatever reason people had for making a particular decision. Groups were not imperialistic because their leaders decided they should be, but rather, according to organization theory, they simply were expansionist. In contrast, when acting in the bureaucratic process, people actively decided to pursue the acquisition of more power for their organization. Many studies have argued that structural and bureaucratic imperatives coincided, but it is a moot point as to whether that was because of an actual synchronization of individual and organizational behavior or because analysts have constructed the latter in the former's terms.

A more appropriate linkage of the two can be found in their effects on decisionmaking, especially in relation to the issue of innovation. In the history of helicopter developments, both organizational and bureaucratic political factors tended to work in favor of existing doctrines. Although both types also contained some force for change, their conservative elements were, on the whole, stronger. Bureaucratic political factors were not a priori favorable to conservatism—for example, the development of helicopters in Britain and America owed much to the battles between army and air force. However, as with organizational imperatives, the distribution of political power tended to preserve existing groups and doctrines.

The importance of politics first of all depended on whether or not a country was at war. As one eighteenth century commentator on military reform succinctly concluded, "A danger sensed by all muffles *the voice of intrigue*, and the innovation appears a smaller evil that must be accepted to avoid a greater. Conditions are different when a reform is to be instituted in times of peace [emphasis added]."[48] It was the same for helicopters. The USAF and the army were suddenly able to resolve the differences that had plagued them for twenty years while fighting in Vietnam. However, in times of peace bureaucratic aims were much less closely aligned with the ultimate goal of military effectiveness. After Vietnam, the collaboration between army and air force soon decreased. Military effectiveness was not as readily classifiable and various definitions arose, each favorable to a specific interest group and each competing with one another. When not at war, military structures were also less flexible because of the importance of such issues as career development, orderly training and recruitment and so on. These increased the nonmilitary interests that guided actors in the bureaucratic game.

A recent study by David Kozak and John Keagle, entitled *Bureaucratic Politics and National Security*, summarized the activities of the game in "classical statements," drawn from previous works on the subject. Some of these simply defined the subject matter. First, the delegation of power from the highest levels of authority ensured that lower-level bureaucrats and bureaucracies become involved in decisionmaking. Thus initial responsibility for helicopter doctrines was delegated, in the most part, to the acknowledged experts within the military hierarchy, but its final approval depended on those delegating their authority. Second, politics, meaning the pursuit of power and the representation of interest groups, was a fundamental part of bureaucratic activity.

Of the other statements, five are especially relevant to the study of helicopter doctrine. First, the agents involved were "driven by [the] highly particularised and parochial views, interests and values of their organizations," and second, they were continually competing for "stakes and prizes." Third, agents represented and acted for interest groups and constituencies within their organization. Fourth, "Policy made in an arena of bureaucratic politics is characterized by bargaining, accommodation and compromise"; and finally, "Proposals for organization change and reform are politically motivated."[49] The activities categorized by these five statements and their impact on doctrine making will be assessed in turn.

Motivation

Perhaps the most important question about bureaucratic politics concerns the motives of those seeking change. If bureaucratic players were acting purely for political reasons—to benefit their vested interests—then military decisionmaking was flawed because its ultimate aim did not coincide with the reasons for establishing the process in the first place. Instead of being the end result, military power would be merely the means for obtaining bureaucratic and personal power. However, it is far more convincing to suppose that, at the heart of most attempts to play the political game, there lies a genuine military motivation to improve or best utilize a country's military power. Even if these motivations were based on erroneous military views, there is no evidence that they were not truly believed by the people who acted on them. The various RAF officers who opposed more helicopter development did not do so simply because it meant more money and influence for their service, but rather because they believed that Britain's national defense would have been better served by buying fixed-wing aircraft instead. Nevertheless, this conclusion does nothing to diminish the importance of political influence on helicopter doctrine. Even though bureaucratic politics is best seen as a process through which military decisions passed, this does not imply that, at the end of the supposedly rational process, military factors won through and politics did not affect the eventual decisions. On the contrary, the means by which decisions were made were as important as the final product.

Loyalty

Although helicopter advocates had a basic interest in the activities of the main organization to which they belonged—usually the army—their main loyalty was to helicopters, helicopter doctrines and the group surrounding them. The root cause of this loyalty was a belief, based on their own military analysis, that helicopters were an essential part of modern warfare and that the main organization would be less capable without them. In the same manner armor, infantry or other enthusiasts supported their particular aspect of the army's business. While these beliefs were essentially military, their practical application was political. The value of the helicopter and changes in doctrines concerning it were judged in terms of their effect on increasing or decreasing the bureaucratic power of these organizations.

For example, the paratroop officers who led the push for airmobility had a vested interest in airborne operations of any kind, but those involving glider or parachute landings were clearly becoming impractical in modern warfare. Therefore, they had to develop new methods—such as using helicopters—or else their power and influence within the army would have declined. The loyalties of other participants in the doctrine-making process were usually not associated with helicopters. Once again, innovative decisions requiring the involvement of officers and commanders from outside the helicopter group and that favored the latter were much more difficult. Since all but the most minor doctrinal adjustments could not be kept within the ranks of the helicopter organization, there was an inherent barrier to the easy introduction of new helicopter doctrine.

Stakes and Prizes

During the last four-and-a-half decades, the stated needs of military organizations always expanded to meet and overmatch any increase in the supply of resources so that there was never enough money to satisfy all of their component groups. Moreover, for much of the period there was a very real financial shortage. Thus military groups competed with each other for those funds, which were then translated into more personnel and more equipment. In West Germany in the late 1960s, helicopter advocates were competing for money with those favoring new tanks, multiple rocket systems and Tornado aircraft.

In this often intense competition for a limited amount of money, the weaker players were much less likely to succeed. Political power was mainly based on the existing level of money and resources, and so a vicious circle tended to develop that was usually broken only by outside factors favoring innovation. New ideas that threatened the existing order of priorities for the allocation of funds were regarded as highly dangerous by groups that were favored by that order, and pressure against innovation resulted. Although a short-term increase in the funds for helicopters might have indicated some approval, ultimately bureaucratic success was measured by improvements in the power of helicopter organizations. Of the Western nations, where bureaucratic politics was most prevalent, only the U.S. Army Aviation Branch achieved sufficient stature to compete effectively in the process without the assistance of other players.

While recurrent funding was often important to cementing a doctrinal innovation in place, it was not as vital as the initial allocation of money. Airmobility in Britain is likely to be underfunded—as far as helicopter advocates are concerned—for many years. Yet at least sufficient funds have been assigned to allow some implementation of doctrine. In the 1970s—at a time of severe budgetary pressure—the fear that airmobility would consume large amounts of resources prevented even a modest first commitment to it. It was as if such a decision was seen as the thin end of the wedge, opening a requirement for spending that was possibly acceptable in limited circumstances but that would develop a momentum of its own and lead to reductions in existing groups' budgets. In contrast, once at least some resources had been acquired, later decisions to reduce funding did not affect doctrine as much. For example, despite uncertainties in the procurement of the Apache and AHIP helicopters, U.S. helicopter doctrine remained relatively stable.

Also at stake was the prize of being highly valued within the armed forces, with the status of helicopters and helicopter operations within the army defined by the importance of aviation relative to other arms. All countries' helicopter advocates argued for many years that helicopters were capable of more than just the supporting role that they played at first; instead, it was suggested, that when organized properly they were able to maneuver on the battlefield in the way that infantry and armor units could. Maneuver and support differed less in absolute value—no one would argue that artillery, despite its status as a supporting arm, was not essential—but rather maneuver units initiated combat, set its pace and determined the actions of the supporting arms. This wartime status translated into power in peacetime. Hence maneuver status, while militarily reasonable given the helicopters' superior mobility, was an important goal in the political process of doctrine making. Unsurprisingly, the existing maneuver arms sought to protect their exclusive hold on it.

Although usually closely linked, the goals of status and resources were not necessarily in accord. In America after the Vietnam War, senior aviation officers pushed firmly for helicopters to have a support role—despite encouragement from some superiors that they be treated as a maneuver force—so that Army Aviation would not pose a major threat to other arms' funding. In this way, the large expenditure on helicopters that did occur was justifiable to senior members of the Armor, Infantry and other branches in that they would get the benefit of being supported by helicopters.

Bargaining

As demonstrated by the previous example, bargaining and compromise were integral to the political process that affected helicopter doctrine. Compromise was essential to successful innovation. The radical ideas that often emerged from within helicopter organizations were most unlikely to succeed. Rather than abandon them entirely, more senior officers, who were the key players in the bureaucratic game of gaining official sanction for doctrine, modified them to improve their chances of acceptance. A clear example arose from the development of the latest airmobile unit in Britain. Despite the fact that its formation was described as a great innovation, 24 Airmobile Brigade was barely more than an air transported brigade similar to that which was planned in the 1960s. A more powerful unit along the lines of the West German airmechanized brigades was impossible because the other arms would have prevented the massive reorganization and redirection of funds needed to create such a different and expensive formation. Accepting these realities, supporters of airmobility compromised; at least a base had been established from which further developments might occur.

Representation

Other goals for bureaucratic actors came from their duty as representatives of constituencies. In peacetime, one of the most important concerns of individual officers was career opportunity. In the U.S. Army in the late 1970s, the creation of combat aviation brigades was very important in increasing the total number of officer positions, especially at the level of colonel, as well as bringing aviation career paths into line with other branches. The new arrangement was militarily sound, served political ends by increasing the number of senior officers in Army Aviation, but also fulfilled the personal goals of members of the aviation arm. Although it was the only obvious example of such influences on doctrine, generally the effect of decisions on officers' careers can, in times of peace, be at least as relevant as their effect on potential enemies.[50]

The Applicability of the Bureaucratic Model

As noted in chapter 2, there is a fundamental problem with applying the bureaucratic model to the Soviet Union's decisionmaking. Not only is there a lack of evidence that specific decisions were influenced by particular interest groups, especially in relation to helicopters, but this theory of doctrine making has been created almost entirely in reference to the United States, a society very different culturally and politically from the old Soviet Union.[51] Some detailed studies have found factors such as empire building, bargaining and competition at work in Soviet defense decisionmaking, while other Soviet analysts have rejected the application of bureaucratic politics to the USSR because of the concentration of power in the hands of a few people.[52] The use of bureaucratic politics to explore Soviet decisionmaking was, according to one analyst, "dependent upon an assumption of the dominance of the sub-groups within a single one."[53] While keen observers could easily identify such subgroups within the Soviet structure, it was much harder to quantify their degree of dominance, and it was likely that sufficient centralization of power existed in the Soviet Union to render bureaucratic politics less important than in a pluralistic Western society.[54]

Yet the Soviet Union was not as centralist as Marxist-Leninist theory might suggest, and in analyzing issues of strategic policy bureaucratic politics seemed to be more relevant. Not only did interest groups, such as the Communist Party, the KGB and the army, who were close to these decisions actually have considerable power but the decisions were so important that they directly affected these groups. Thus each had an interest in the final result and had the capacity to play the bureaucratic game. Instead of just one rigid hierarchy, there were a relatively small number, all of which competed and interacted over crucial decisions but only where the decision affected more than one of them.

But decisions about helicopters did not involve more than one hierarchy. Even within the army, only the ground forces were involved. Although the VVS was the nominal owner and operator, its task was not to give intellectual leadership in helicopter doctrine, but to provide as efficient a service as possible. If the VVS had a particular interest, it was in long-range aviation and not in close support of ground operations. Moreover, there is some doubt as to whether the ground forces and VVS can even be considered as separate hierarchies: Meyer concluded that "inter-service rivalry . . . is effectively squelched by a General Staff with very strong institutional power."[55] Thus there was little intersection of competing interests within the army over helicopters in the same manner as in Western armed forces. Therefore, bureaucratic politics as a tool for analyzing the Soviet Union, while it must be considered, has much less relevance to decisionmaking about military helicopter doctrine than for other issues.

This notion of proximity and power can also be usefully applied to the role of bureaucratic politics in the development of helicopters in Western countries. Undoubtedly a number of groups would have liked to affect the doctrine-making process—notably politicians, civilian bureaucrats and the helicopter manufacturers—but were on the whole restricted to making an impact on weapons acquisition, a process that usually followed the settlement of many doctrinal issues. Their interest in and connection to decisions about equipment was much more direct: the companies actually made the weapons, the civilians and politicians controlled the money that bought them. To return to an earlier example, Bell's failure to achieve its desired goal to change army doctrine and create a market for the tilt-rotor LHX despite an intense media and political lobbying effort came about because it simply was not able to exert sufficient political influence on the military decision makers.

The reason for this was the level of autonomy given to the military on matters of doctrine, especially where they related to conventional weapons used mainly in tactical situations. Although the approval of civilian bureaucrats was required, the authority and apparent expertise of serving officers was usually sufficient to win arguments on purely military matters. Even in the exceptional case of MacNamara's intervention in the debate over airmobility, it must be remembered that most of the impetus for that action came from within the military itself. Of course, doctrine was usually linked intimately to the introduction of new weapons and thus to funding and technical issues. In this way the division of power became blurred, particularly when weapons were first acquired, but even so doctrine was principally the reserve of military decision makers with other participants concentrating on equipment. Thus, generally speaking, bureaucratic politics was most crucial as a process in which competing arms and services within the military fought over the allocation of power and resources within their own organization.

CONCLUSION

Successful Innovations

Judging whether helicopter innovations were successful or not depends largely on the criteria that are used. With hindsight it might be tempting to assess whether the decision to innovate was correct or not. Were, for example, the Americans right to form an airmobile division in the early 1960s? It appears to be a successful innovation, mainly because of the division's successful employment in Vietnam; but what if some other conflict had occurred to which the airmobile unit was entirely unsuited? As discussed in the introduction to this book, such assessments about military effectiveness, which were themselves a part of the process of innovation, are fraught with risks. Ultimately, success is better judged in terms of whether an innovation remained a novelty or not. If innovations became established within the military, then they were no longer innovations and had, from the advocates' viewpoint, succeeded completely.

The groups of factors discussed previously were all capable of helping innovations to proceed given the right circumstances. Operational and strategic circumstances, along with a perceived enemy threat, could provide a compelling force for innovation; or experience could be favorable, as with the Ansbach and Vietnam trials of anti-tank helicopters. Politics could propel innovation where the helicopter advocates were either in a position of strength or in alliance with the strong. Organizational routines developed around helicopters also helped them to succeed as an innovation. Finally, where it was perceived to offer a substantial capability and was itself a convincing weapon, technology provided a cogent reason for the introduction of helicopters.

Yet, ultimately, successful military innovation consisted in attaining familiarity and acceptance within the army as a whole, and the groups of factors operated in the helicopter's favor by encouraging these characteristics. Once something was familiar and accepted within the military, then serious questions were rarely raised about its value and place in modern warfare. Armored vehicles and the concept of combined arms are two examples of twentieth century innovation that have reached this state. Moreover, once an innovative idea or weapon was accepted, those who were most closely associated with it and thus best placed to exploit all of its advantages were given much more control over the innovation. Decisions about how to employ the innovation were no longer made by those least qualified to make them. Acceptance was, once again, a highly subjective quality, residing in the minds of armed forces' members and not in official statements or force structures. It was engendered through the authority that an innovation possessed. Authority was partly generated by the hierarchy of command in the armed forces so that innovations that were approved at higher levels had, to some extent, to be accepted; moreover, official sanction also made innovations more convincing of themselves because generally the views of senior army leaders were respected.

Yet authority was also generated within individuals when they formed opinions about an innovation by factors external to the chain of command. The quality of experiences was important. If the arguments in favor of innovation were based on combat experiences that the individuals had shared, then the innovation held great authority. Indeed, because combat experience was real life and associated with bravery and sacrifice, it affected participants very deeply, thus becoming a particularly potent source of authority. Alternatively, the immediacy of the threat and a particularly strong view of the political situation within the armed forces made the technologically unconvincing British anti-tank helicopters of the late 1960s appear acceptable. In contrast to these rather frail machines, the Apache attack helicopter was, perhaps, the ultimate symbol of the authority that weapons themselves could generate. Familiarity with an innovation, which itself tended to lead to acceptance, could to some extent develop only with time. It is not surprising that the soldiers of the 1980s, with four decades of helicopter employment behind them, saw this particular innovation rather differently than their predecessors in the late 1940s. More specifically, familiarity developed through the use of helicopters in combat or in exercises, thus exposing non-aviators to them and removing some of their novelty.

Yet every innovation was, at some stage, a challenge to familiar and accepted thinking, and the most normal response was for the innovation to be defined in terms that were already known and understood in an attempt to retain control over the new weapon or idea. As a result, new technologies were often promoted simply as replacements for existing equipment, performing the same established missions. As many analysts have pointed out,[56] such innovation was not necessarily effective and—as in France in 1940—tended to be positively dangerous; the changes certainly did not display the "intellectual mastery" so important to the process of successful innovation.[57] The need for familiarity also led to a strong compulsion to use existing weapons and doctrines to solve strategic and operational planning problems rather than looking at potentially innovative solutions.

However, some scholars have concluded that an innovation's "success depends on *harmonising* new equipment with existing material and personnel [emphasis added]."[58] This is the fundamental paradox of successful innovation. It will be more accepted and succeed more quickly if it is relatively familiar, so advocates might deliberately define their innovative proposal in terms of existing concepts and ideas. Yet, in doing so, the greater possibilities of the innovation tend to be discounted or even eliminated. As far as helicopters were concerned, some advocates altered their proposals to make them more acceptable and others did not. In either case, the great majority of military officers saw helicopters in terms of the two existing components of air-land combat—close air support aircraft and airborne operations. Yet, from this established base of acceptance, helicopter innovation continued changing what was familiar from the inside. The development of the helicopter over the last forty-five years was, in effect, the insurgency of doctrinal innovation against established practice: conservatism was defeated by the enemy within.

Decisionmaking: An Interactive Process

While in one way demonstrating the complexity of the decision-making process, the summary of factors in Figure 6.1 also suggests that some separation or distinction can be drawn between the factors. Evangelista, confronted by a similar array of determinants and seeking to avoid a conclusion that " 'everything' plays a role in such decisions," linked specific factors (such as bureaucratic politics or organizational conservatism) to stages of the innovation process. He argued, for example, that military planning only became important in America when a new weapon or idea

Figure 6.1
Summary of the Decision-making Process

Mechanics of Decision-Making

Levels of Decision	Time Factor
development	complexity
sanction	caution
implementation	technology

Individuals and Organisations	Subjectivity	Dynamic and Interactive

Factors

Technology	- knowledge - product - weapons	Organisation Process	- conservatism - routine - innovation - imperialism - distribution of power
Military Planning	- strategy - operational doctrine - threat analysis	Bureaucratic Politics	- delegation - war or peace - political motivation - political behaviour - competition for prizes - bargaining - representation - limited applicability
Experience	- combat experience - tests and trials - received experience		

had already gathered considerable technological, organizational and political momentum, thus suggesting that only a very strong military argument against adopting the innovation would succeed.[59] While there is no doubt that such distinctions improve the utility of theory in policy making, they ignore the point that everything *did* play a role in decisionmaking. While some factors were more directly influential at certain times, all were part of an overall context in which the entire decision-making process was conducted, and it was the interaction of factors that was crucial in determining exactly how doctrines emerged.

To conclude that there was interaction does not simply mean that a decision was caused by more than one factor or even that a concatenation of factors was responsible. Interaction meant that no single factor affected the outcome of decision making except as part of the joint influ-

Figure 6.2
The Concept of Interaction

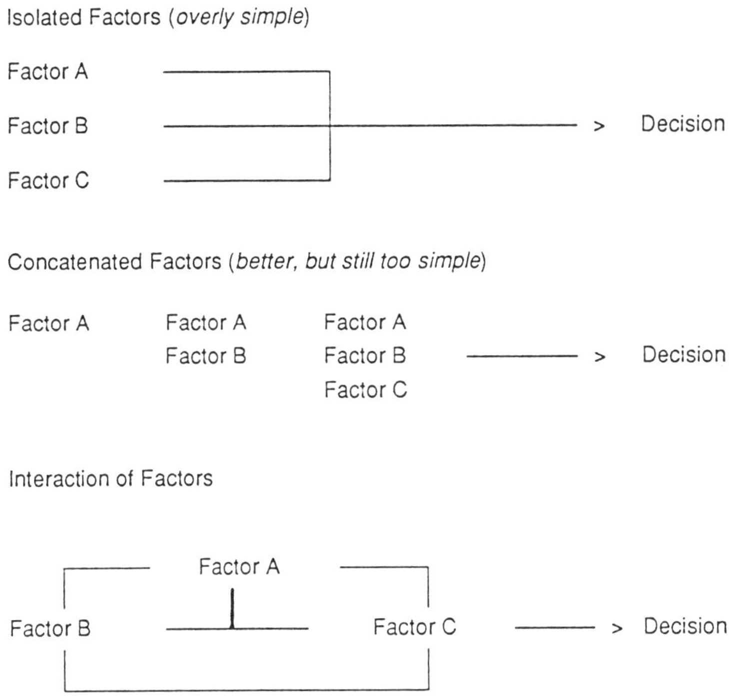

ence of all relevant inputs to that decision. The development of this concept is represented in Figure 6.2.

The last of these three diagrams in the most appropriate, reflecting that each factor operated within the context established by other possible determinants of doctrine. For example, when U.S. Army aviators were interpreting their experiences at the end of the Vietnam War to find support for helicopters, they were doing so with the knowledge of the USAF's campaign to prevent the army from acquiring a new helicopter. Yet the inverse was also occurring. In seeking to win the bureaucratic political battle against the USAF, the army was remembering its experiences of poor close air support by the air force in the Second World War and at the start of the Korean and Vietnam conflicts. It would also be simplistic to imagine either that individuals were in full control of decisionmaking or that they were mere spectators while the interplay of organizational and technological factors determined doctrines. The weakness of the rational actor model was not that it emphasized individual decision makers but that it assumed their actions were rational and objective. In the search for a more detailed understanding of decisionmaking and the accompanying critical assessment of this model, individuals have sometimes been ignored. Yet their role was crucial: not because some great men of history were able to impose their will on events, but rather because factors affected decisions only after the objective realities behind them had been subjected to interpretation by decision makers. Moreover, while it must be remembered that these individuals were the product of many of the forces that also affected doctrine, they were an essential focal point at which the interaction of factors took place. It would seem much more appropriate to stress the dual nature of decisionmaking. It was a drama in which the actors, reading from various scripts, constantly aware of other characters and defining their parts in relation to them, were continually forced to keep their balance on a stage that was itself always in motion.

NOTES

1. Allison, *Essence of Decision*, v–vi, 8–9 and 246–51.
2. James R. Kurth, "A Widening Gyre: The Logic of American Weapons Procurement," *Public Policy* 19.3 (1971): 374–80.
3. Perhaps the best example was Posen, *The Sources of Military Doctrine*, 221, 240–41.
4. As in Graham T. Allison, "Conceptual Models and the Cuban Missile Crisis," and R. G. Head, "Crisis Decisionmaking: Bureaucratic Politics and the Use of Force" in David C. Kozak and John M. Keagle (ed.), *Bureaucratic Politics and National Security* (Boulder, CO: Lynne Rienner, 1988) 54–90.
5. A. J. Alexander, *Decision-Making in Soviet Weapons Procurement*, Adelphi Paper No. 147–148 (London: IISS, 1978) 43.
6. Evangelista, *Innovation and the Arms Race*, 14–15.
7. See the revealing quotation from E. H. Carr in ibid., viii. Evangelista's arguments are summarized in the conclusion of this chapter.
8. Kurth, "A Widening Gyre: The Logic of American Weapons Procurement," 404.
9. John Baylis, "Defence Decision-making in Britain and the Determinants of Defence Policy," *JRUSI* 120.1 (1975): 47.
10. E. H. Carr, *What is History?* (Harmondsworth: Penguin, 1964), 89–90.
11. Morton H. Halperin, *Bureaucratic Politics and Foreign Policy* (Washington, DC: Brookings, 1974) 238.
12. Martin van Creveld, *Technology and War* (New York: Free Press, 1989) 146.
13. Michael Pearton, *The Knowledgeable State* (London: Burnett, 1982) 247–48.
14. "Commodities" is used in the sense of merchandise or saleable items. While not an entirely satisfactory word, it does imply a certain commercial interest in technology.
15. Van Creveld, *Technology and War*, 67ff.
16. Evangelista, *Innovation and the Arms Race*, 16; W. Harriet Critchley, "Does

Doctrine Precede Weaponry?" *International Journal* 33.3 (1978): 536–41; and William H. Kincade, "New Military Capabilities: Propellants and Implications" in Carl G. Jacobsen, *The Uncertain Course* (Oxford: Oxford University Press, 1987) 69.

17. Colin S. Gray, "Does Theory Lead Technology?" *International Journal* 33.3 (1978): 506–23; and Critchley, "Does Doctrine Precede Weaponry?" 524–56.

18. Kevin N. Lewis, "On the Appropriate Use of Technology," *Orbis* 27.3 (1983) 274.

19. Gray, "Does Theory Lead Technology?" 509.

20. Even the USSR reasonably fits this Western consumer model because, unlike other groups in the Soviet Union, the army had powers of choice over weapons—to accept, reject or demand changes: Carl G. Jacobsen, "Moscow's 'Military-industrial Complex': Its Nature and Impact" in Carl G. Jacobsen (ed.), *Strategic Power USA/USSR* (London: MacMillan, 1990) 305.

21. Millett, "Military Reform in America," 5.

22. See chapter 5 for a detailed examination of the operational issues involved in helicopter warfare.

23. Larson, *The British Army and the Theory of Armored Warfare 1918–1940*, 15–38.

24. Perhaps the most extreme example can be seen in the language used in U.S. Army manuals.

25. A study of Western analysis of the USSR's military capacity and intentions provides many examples of this phenomenon.

26. Stephen M. Meyer, "Soviet Strategic Programmes and the U.S. SDI," *Survival* 27.6 (1985): 284.

27. For more detailed discussion on the interpretation of experience, see Robert E. Harkavy, "The Lessons of Recent Wars: Towards a Comparative Analysis," in Robert E. Harkavy and Stephanie G. Neuman (ed.), *The Lessons of Recent Wars in the Third World*, Vol. 1 (Lexington, MA: D. C. Heath, 1985), 5–7.

28. Millett, "Military Reform in America," 9. See also Colonel General Mikhail A. Gareyev, "The Creative Nature of Soviet Military Science in the Great Patriotic War," *Voyenno-istoricheskiy zhurnal*, July 1985: 22ff (trans.) in *JPRS-UMA* 85–060 (1985): 18.

29. Jay Luvaas, "Lessons and Lessons Learned: A Historical Perspective," in Harkavy and Neuman, *The Lessons of Recent Wars in the Third World*, 60–65.

30. Edward L. Katzenbach, "The Horse Cavalry in the Twentieth Century: A Study in Policy Response," in Endicott and Stafford, *American Defense Policy*, 366.

31. Paret, *Innovation and Reform in Warfare*, 12.

32. Hussain et al., *The Future of the Military Helicopter*, 15.

33. In this and the following section the term "organization" is used to denote groups such as armies, air forces and aviation branches rather than specific military formations.

34. For example, Bernard Brodie, "Technological Change, Strategic Doctrine and Political Outcomes" in Klaus Knorr (ed.), *Historical Dimensions of National Security Problems* (Lawrence: University Press of Kansas, 1976) 300.

35. Posen, *Sources of Military Doctrine*, 222–27.

36. Allison, *Essence of Decision*, 78–87.

37. A. J. Alexander, *The Linkage between Technology, Doctrine, and Weapons Innovation: Experimentation for Use*, Rand Paper P–6621 (Santa Monica, CA: Rand Corp., 1981) 26.

38. Dowling, *RAF Helicopters—The First Twenty Years*, 20.

39. Also noted by Hussain et al., *The Future of the Military Helicopter*, 2.

40. Even the concept of combined arms was founded on attempts to improve the combat power of the tank, the central member of the combined arms team.

41. MacCormick, "The Dynamics of Doctrinal Change," 268, for further elaboration on the problems of innovation in the face of conventional wisdom. The best example of this can be seen in Britain in the mid–1970s.

42. Posen, *Sources of Military Doctrine*, 222–27.

43. See Paul Bracken, "Defense Organization and Management," *Orbis* 27.3 (1983): 253.

44. Allison, *Essence of Decision*, 93.

45. The exception that demonstrates the importance of this facet was the positive effect on helicopter innovation of a concentration in the U.S. Army of senior officers with aviation experience in the 1970s.

46. Halperin, *Bureaucratic Politics and Foreign Policy*, 28–62; and Posen, *Sources of Military Doctrine*, 222–28.

47. C. L. Jefferies, "Defence Decisionmaking in the Organizational-bureaucratic Context" in Endicott and Stafford, *American Defense Policy*, 237.

48. The commentator was a Hanoverian, von der Deckern, quoted in Paret, *Innovation and Reform in Warfare*, 7. See also the opening paragraph in Ian Bellany, *Inter-Service Rivalry and Military Innovation* (Lancaster: Centre for Study of Arms Control and International Security, 1987) 1.

49. David C. Kozak, "The Bureaucratic Politics Approach: The Evolution of a Paradigm" in Kozak and Keagle (ed.), *Bureaucratic Politics and National Security*, 5–10.

50. See also Paret, *Innovation and Reform in Warfare*, 7.

51. Roman Kolkowicz, "The Military" in H. G. Skilling and Frederic Griffiths, *Interest Groups in Soviet Politics* (Princeton, NJ: Princeton University Press, 1971) 132.

52. The differing views on this issue were well summarized in Stephen M. Meyer, "Soviet National Security Decisionmaking: What Do We Know and What Do We Understand?" in Juri Valenta and William Potter (ed.), *Soviet Decisionmaking for National Security* (London: George Allen and Unwin, 1984) 261–65.

53. Frederic Griffiths, "A Tendency Analysis of Soviet Policy-Making," in Skilling and Griffiths, *Interest Groups in Soviet Politics*, 335.

54. Gallagher and Spielmann, *Soviet Decision-Making for Defense*, 25.

55. Stephen M. Meyer, "Decision-making in Defence: The Soviet Case," in Jacobsen, *Strategic Power: USA/USSR*, 253.

56. See, for example, Katzenbach, "The Horse Cavalry in the Twentieth Century: A Study in Policy Response," 365; and Larson, *The British Army and the Theory of Armored Warfare*, 38.

57. Paret, *Innovation and Reform in Warfare*, 2.

58. Chris Harvie, "Technological Change and Military Power in Historical Perspective" in *New Conventional Weapons and East-West Security* Part I, Adelphi Paper No. 144 (London: IISS, 1978) 12.

59. Evangelista, *Innovation and the Arms Race*, 52, 227–28 and 239–40.

Appendix

Summary of Helicopters' Technical Characteristics

Giving a simple statistical summary of helicopter characteristics is all but impossible. Helicopter performance essentially depends on weight, which includes not only the number of passengers and weapons carried but also the amount of fuel; fuel in turn determines range. Speed also determines range because, if one flies very fast or very slowly, fuel is consumed less efficiently. Altitude is important since fuel use decreases as height above sea level increases; yet tactical considerations might require slow and fuel-consuming flight at NOE altitudes. Climate is also critical: the hotter it is, the less a helicopter can carry. Helicopter ranges, speeds and loads therefore are highly variable. For example, the Mi–24 Hind would probably be limited to about a two hundred kilometer range with a full combat load and could fly at no more than 250 kilometers per hour. Thus the following table must be used with caution. The statistics are essentially for comparison only.

Designations are those used by the military (where known) or, for the Soviet Union, the NATO reporting number and name. The maximum speeds, weights and cargo loads have been stated, and ranges given are with standard payload in level flight conditions, except where otherwise noted. The categories of Cargo/Passengers/Weapons are not inclusive: that is, the OH–58A can carry either 696 kilograms of cargo or 4 passengers. Passenger figures are standard and would be greatly increased in wartime—the British packed over one hundred personnel into a Chinook during the Falkland Islands Conflict. Weapons are given only for attack or anti-tank helicopters (most transport helicopters also carry machine guns for self-defense) and are the standard equipment: for example, the Apache could substitute more rockets for some of its Hellfire ATGMs if anti-tank action was not required.

Summary of Helicopters' Technical Characteristics

Designation	Maker	Speed	Range	Weight	Cargo/Passengers/Weapons
United States					
AH-1S Cobra	Bell	315 km/h	507 km	4,535 kg	8 TOW, 1 3x20mm C & 38 70mm rkts
AH-64A Apache	MDD	365 km/h	482 km	9,525 kg	16 Hellfire, 76 70mm rkts, 4 Stinger & 1 30mm C
CH-47A Chinook	Boeing	276 km/h	185 km[1]	12,882 kg	2,722 kg or 33 pax
CH-47D Chinook	Boeing	298 km/h	185 km[1]	19,178 kg	7,081 kg or 33 pax
OH-58A Kiowa	Bell	222 km/h	490 km	1,360 kg	696 kg or 4 pax
OH-58D AHIP	Bell	241 km/h	556 km	2,041 kg	4 Stinger or 4 Hellfire
UH-1H Iroquois	Bell	204 km/h	511 km	4,309 kg	1,899 kg or 11 pax
UH-60A Blackhawk	Sik.	296 km/h	600 km	7,708 kg	2,590 kg or 14 pax
Soviet Union					
Mi-2 Hoplite	Mil	210 km/h	580 km	3,700 kg	800 kg or 4 pax
Mi-6 Hook	Mil	300 km/h	620 km	42,500 kg	12,000 kg or 70 pax
Mi-8 Hip-C	Mil	250 km/h	500 km	12,000 kg	4,000 kg or 24 pax
Mi-8 Hip-E	Mil	250 km/h	500 km	12,000 kg	192 rkts, 4 Swatter & 1 12.7mm mg
Mi-24 Hind-D	Mil	320 km/h	740 km	10,000 kg	4 Swatter, 128 rkts & 4x12.7mm mg[2]
Mi-26 Halo	Mil	295 km/h	800 km	56,000 kg	20,000 kg or 85 pax
Mi-28 Havoc[3]	Mil	305 km/h	470 km	10,400 kg	16 Spiral ATGMs, rkts, SA-14s & 1 30mm(?) C
United Kingdom					
Chinook HC.1	Boeing	286 km/h	185 km[1]	17,463 kg	5,284 kg or 50 pax
Gazelle AH.1	W/A	310 km/h	709 km	2,202 kg	918 kg or 4 pax
Lynx AH.1	West.	259 km/h	630 km	4,763 kg	1,791 kg or 10 pax
Lynx AH.1 (TOW)	West.	259 km/h	630 km	4,763 kg	8 TOW (plus 8 extra rds)
Puma HC.1	W/A	274 km/h	590 km	6,700 kg	3,270 kg or 20 pax
Scout AH.1	West.	211 km/h	505 km	2,405 kg	800 kg or 4 pax or 4 SS-11
Germany					
BO 105P PAH-1	MBB	270 km/h	575 km	2,400 kg	6 HOT
BO 105M VBH	MBB	270 km/h	575 km	2,400 kg	268 kg or 3 pax
CH-53G	Sik.	315 km/h	413 km	19,050 kg	5,857 kg or 37 pax

France

SA.316 Alouette III	Aéro	210 km/h	500 km	2,100 kg	4 SS-11
SA 330 Puma	Aéro	274 km/h	590 km	6,700 kg	3,270 kg or 20 pax
SA 332 SuperPuma	Aéro	296 km/h	620 km	9,000 kg	4,525 kg or 20 pax
SA 342F Gazelle	Aéro	310 km/h	709 km	2,202 kg	918 kg or 4 pax
SA 342M Gazelle	Aéro	310 km/h	709 km	2,202 kg	4 HOT or 1 30mm C & 4 Mistral

Key: Hellfire, TOW, SS–11, Swatter and HOT are anti-tank guided missiles. Stinger, SA–14 and Mistral are air-to-air missiles. Rkts are air-to-ground rockets, either 2.75″ or 5″ in diameter. C are cannon; Mg are machine guns; "x" denotes multi barrel weapon; pax are passengers.

Aéro = Aérospatiale; MDH = McDonnell Douglas; Sik. = Sikorsky; West. = Westland; W/A = Westland/Aérospatiale

1. Max useful range is probably ca. 500 km; this figure corresponds to the specific payload given and is used for comparison purposes with other Chinook models.
2. Hind E and F carried the later model Spiral ATGM (up to 8) and had, respectively, two 23mm cannon or one 30mm cannon.
3. Details on Havoc are estimated.

Source: *Flight International, Jane's All the World's Aircraft* and various works cited previously.

Selected Bibliography

A NOTE ON SOURCES

General scholarly analyses of the development of military helicopters are scarce. Michael Hickey's *Out of the Sky* and John Galvin's *Air Assault* were more concerned with airborne operations of all types, while John Everett-Heath's books on British and Soviet helicopters concentrated more on design and development of the helicopters themselves. *The Future of the Military Helicopter*, produced by researchers at Britain's Royal United Services Institution, is a useful summary but must be read in the context of British military helicopter developments and its open argument for increased spending on army aviation in Britain. Indeed a major aim of the current work is to redress this lack of a general history of helicopter developments.

Given the contemporary nature of the subject and its concentration on military affairs, it is not surprising that there is also no substantial weight of official documents. Instead, research was focused on government publications; a wide range of military textbooks, manuals and journals; some trade magazines and newspapers; and direct contact with retired and serving military officers. There is some question about the extent to which manuals corresponded to actual practice and to which journal articles reflected official policy or were attempts to change that policy. But this work's concentration on the intellectual development of doctrine as much as actual implementation means this is not a serious concern.

This bibliography is very selective, listing either the most important, most representative or most accessible sources used. For this reason, full bibliographic details have been included with each reference in the text in case readers wish to read in more depth about military helicopters. However, it should be noted that the research relied on many more articles than are actually mentioned in the references and on a series of interviews which, for obvious reasons, remain confidential.

UNITED STATES—SELECTED SOURCES

Primary Sources

FM 90–4 *Air Assault Operations.* Washington, D.C.: Department of the Army, 1987.

FM 17–47 *Air Cavalry Combat Brigade.* Washington, D.C.: Department of the Army, 1982.

FM 90–4 *Airmobile Operations.* Washington, D.C.: Department of the Army, 1980.

FM 1–111 *Aviation Brigade.* Washington, D.C.: Department of the Army, 1986.

FM 1–100 *Combat Aviation Operations.* Washington, D.C.: Department of the Army, 1984.

FM 90–1 *Employment of Army Aviation in a High Threat Environment.* Washington, D.C.: Department of the Army, 1976.

FM 100–5 *Operations.* Washington, D.C.: Department of the Army, 1976.

FM 100–5 *Operations.* Washington, D.C.: Department of the Army, 1982.

FM 100–5 *Operations.* Washington, D.C.: Department of the Army, 1986.

Canedy, Colonel Charles E. "From ACCB to the 6th Cavalry Brigade (Air Combat)," *U.S. Army Aviation Digest* 21.5 (1975): 1–4.

Gavin, General James M. "Cavalry . . . and I Don't Mean Horses!" *Armor* 63.5–6 (1954): 18–22.

Howze, General Hamilton H. "The Case for the Helicopter," *Army* 29.3 (1979): 16–20.

———. "Speculations on the Future," *U.S. Army Aviation Digest* 20.6 (1974): 6–7.

Kinnard, Lt. General Harry W. O. "Airmobility Revisited Part I," *U.S. Army Aviation Digest* 26.6 (1980): 1–5, 26–27.

———. "Airmobility Revisited Part II," *U.S. Army Aviation Digest* 26.7 (1980): 8–14.

MacKenzie, Lt. Colonel John B. Jr. "Challenges for the Aviation Branch," *U.S. Army Aviation Digest* 30.12 (1984): 10–13.

McNair, Major General Carl H. Jr., and Captain John Reinsprecht. "Army Aviation Forces in the Airland Battle," *U.S. Army Aviation Digest* 27.7 (1981): 6–13.

Maddox, Major General Bobby J. "Army Aviation Branch Implementation," *U.S Army Aviation Digest* 29.8 (1983): 2–9.

Maddox, Major General William J. Jr. "Army Aviation Priorities," *U.S. Army Aviation Digest* 20.12 (1974): 4–5, 27–29.

———. "The Hinge," *U.S. Army Aviation Digest* 25.6 (1979): 2–4.

Merryman, Brigadier General James H. "Bringing Army Aviation Through the 70s . . . into the 80s," *U.S. Army Aviation Digest* 20.6 (1974): 2–5.

Merryman, Major General James H. "Air Cavalry Attack Brigade," *U.S. Army Aviation Digest* 25.11 (1979): 2–5.

"On Army Aviation," an interview with General Donn A. Starry, *U.S. Army Aviation Digest* 24.11 (1978): 2–6.

Ostovich, Brigadier General Rudolph III. "Airland Battle Part II: Views from Aviation Commanders," *U.S. Army Aviation Digest* 32.12 (1986): 31–39.

Otis, General Glen K. "Future Concepts and Capabilities in NATO's Central Region," *Journal of the Royal United Services Institute* 133.4 (1988): 17–19.

Parker, Major General Ellis D. "The Challenge of Winning," *U.S. Army Aviation Digest* 31.9 (1985): 2–9.

Richardson, Lt. General William R. "Airmobility in the 1980s," *U.S. Army Aviation Digest* 27.8 (1981): 2–5, 26.

RisCassi, Lt. General Robert W. "Army Aviation in the 1980s," *U.S. Army Aviation Digest* 32.1 (1986): 2–8.

Saint, Lt. General Crosbie E., and Colonel William H. Yates. "Attack Helicopter Operations in the Airland Battle: Close Operations," *Military Review* 68.6 (1988): 2–15.

———. "Attack Helicopter Operations in the Airland Battle: Deep Operations," *Military Review* 68.7 (1988): 2–9.

Smith, Major General John C. "AVNEC Report," *U.S. Army Aviation Digest* 24.7 (1978): 1.

Snyder, Major M., and Captain R. Hancock, "Aviation Employment Conference," *U.S. Army Aviation Digest* 30.1 (1984): 2–8.

Tolson, Lt. General John J. "The New Challenge," *U.S. Army Aviation Digest* 18.11 (1972): 2–3, 15–17.

Secondary Sources

Bergerson, Frederic A. *The Army Gets an Airforce*. Baltimore: Johns Hopkins University Press, 1980.

Butterworth, William E. *Flying Army*. New York: Doubleday, 1971.

Doughty, Robert A. *The Evolution of U.S. Army Tactical Doctrine 1946–76*. Ft. Leavenworth, KS: CSI, 1979.

Galvin, John R. *Air Assault: The Development of Airmobile Warfare*. New York: Hawthorn Books, 1969.

Romjue, John L. *From Active Defense to Airland Battle: The Development of Army Doctrine 1973–1982*. Ft. Monroe, VA: TRADOC, 1986.

Tolson, General John J. *Vietnam Studies: Airmobility 1961–1971*. Washington, D.C.: Department of the Army, 1973.

Weinert, Robert P. *A History of Army Aviation 1950–1962, Phase I: 1950–1954*. Ft. Monroe, VA: CONARC, 1971.

———. *A History of Army Aviation 1950–1962, Phase II: 1955–1962*. Ft. Monroe, VA: TRADOC, 1976.

SOVIET UNION—SELECTED SOURCES

Primary Sources

Belov, Colonel M. "Air Landing Forces," *Soviet Military Review*, January 1979: 22–23.

———. "Helicopters and Land Force Tactics," *Soviet Military Review*, December 1976: 22–24.

———. "Problems in Making Modern Armies Airmobile," *Krasnaya zvezda*, 7 March 1972: 3. Trans. in *JPRS-UMA* 802 (1972): 1–5.

Besschetnov, Colonel Ye. "A Friend Will Lend a Helping Hand," *Aviatsiya i kosmonavtika*, August 1986: 8ff. Trans. in *JPRS-UMA* 87–007 (1987): 15–21.

Folomeyev, Major General D. "Predicting the Combat Situation," *Aviatsiya i kosmonavtika*, August 1982: 4ff. Trans. in *JPRS-UMA* 1714 (1982): 10–14.

Grechko, Marshal A. A. *The Armed Forces of the Soviet Union*. Moscow: Progress Publishers, 1977.

Ivanov, Lt. Colonel G. and Lt. Colonel V. Karpov. "Firing against Abandoned Trenches: Why Helicopter Pilots Failed to Fulfill Their Mission," *Krasnaya zvezda*, 16 April 1982: 1. Trans. in *JPRS-UMA* 1698 (1982): 23–25.

Izgarchev, Colonel V. "The Helicopters Are Attacking," *Pravda*, 20 September 1979: 6. Trans. in *JPRS-UMA* 1480 (1979): 50–52.

Khorobrykh, Colonel A. "The Helicopter Gunship Attacks," *Krasnaya zvezda*, 6 June 1976: 4. Trans. in *JPRS-UMA* 1245 (1976): 51–54.

Konek, Major V. "They Provided Air Support," *Krasnaya zvezda*, 18 July 1984: 1. Trans. in *JPRS-UMA* 84–063, (1984): 35–37.

Nesterov, Colonel B. "The Quest for New Techniques," *Aviatsiya i kosmonavtika*, June 1983: 11. Trans. in *JPRS-UMA* 1798 (1983): 28–30.

Pismennyy, Colonel V. "Discrepancy in Guidance Documents Causes Lack of Coordination between Army Aviation and Motorized Rifle Troops," *Krasnaya zvezda*, 2 September 1989: 2. Trans. in *JPRS-UMA* 89–023, (1989): 8–9.

Radin, A. "Implementing the Concept of the Mi–24," *Krylya Rodiny*, May 1989: 24. Trans. in *JPRS-UMA* 89–024 (1989): 27–30.

Reznichenko, Lt. General Vasily G. (ed.). *Tactics*. Trans. 1984: Washington, D.C.: USAF, 1987.

———. *Tactics*. np: Moscow, 1987. Trans. in *JPRS-UMA* 88–008-L–1 (1988).

Savkin, Colonel V. Ye. *The Basic Principles of Operational Art and Tactics*. Trans. 1972: Washington, D.C.: USAF, 1974.

Sidorenko, Colonel A. A. *The Offensive*. Trans. 1970; Washington, D.C.: USAF, 1973.

Vorobyev, Major General Ivan N. "Tactics Today," *Soviet Military Review*, March 1985: 10–12.

Yefimov, Marshal A. N. "Soviet Military Art," *Voyenno-istoricheskiy zhurnal*, August 1986: 14ff. Trans. in *JPRS-UMA* 87–004 (1987): 21–28.

Yurkin, Major A. "Helicopters on Combat Course—The Main Thing is Effectiveness and Quality," *Krasnaya zvezda*, 28 May 1977: 1. Trans. in *JPRS-UMA* 1291 (1977): 83–84.

Collected Soviet Writings

Douglass, Joseph D. Jr., and Amoretta M. Hoeber (ed.). *Selected Readings from Military Thought 1963–73*. Washington, D.C.: USAF, 1982.

Kintner, William R., and Harriet Fast Scott (ed.). *The Nuclear Revolution in Soviet Military Affairs*. Norman, OK: University of Oklahoma Press, 1968.

Orenstein, Harry S. *Selected Translations from the Polish Military Press*, Ft. Leavenworth, KS: SASO, 1988.

Scott, Harriet Fast, and William F. Scott. *The Soviet Art of War*. Boulder, CO: Westview, 1982.

Scott, William F. (ed.). *Selected Soviet Military Writings 1970–1975*. Washington, D.C.: USAF, 1977.

Secondary Sources

Bort, Major Robert E. "Air Assault Brigades: New Element in the Soviet Desant Force Structure," *Military Review* 63.10 (1983): 21–38.

Donnelly, Chris N. "The Operational Manoeuvre Group," *International Defense Review* 15.9 (1982): 1177–86.

———. *The Place of the Helicopter on the Modern Battlefield*. Pamphlet A51. RMA Sandhurst: SSRC, 1983.

Erickson, John, Lynn Hansen, and William Schneider. *Soviet Ground Forces*. Boulder, CO: Westview, 1986.

Everett-Heath, John. *Soviet Helicopters: Design, Development and Tactics*. London: Jane's, 1988.

Holcomb, Major James F. Jr., and Graham H. Turbiville, Jr. *Exploiting the Vertical Dimension: Continuing Development of the Soviet Desant Force Structure*. Ft. Leavenworth, KS: SASO, 1987.

Holcomb, Major James F. Jr. *Recent Developments in Soviet Helicopter Tactics*. Pamphlet B34. RMA Sandhurst: SSRC, 1988.

Isby, David C. *Weapons and Tactics of the Soviet Army*. London: Jane's, 1988.

Murphy, P. J. (ed.). *The Soviet Air Forces*. Jefferson, NC: McFarland, 1984.

Scott, Harriet Fast, and William F. Scott. *The Armed Forces of the USSR*. Boulder, CO: Westview, 1984.

———. *Soviet Military Doctrine*. Boulder, CO: Westview, 1988.

Urban, Mark. *War in Afghanistan*. London: Macmillan, 1988.

UNITED KINGDOM—SELECTED SOURCES

Primary Sources

Air Support of Land and Maritime Operations. London: RUSI, 1968.

"Blackbird" (pseud.). "The Value of Their Mobility," *British Army Review* 64 (1980): 5–12.

Bagnall, General Sir Nigel. "Airmobile Operations in Northern Army Group," *NATO's Sixteen Nations* 33.10 (1988): 75–79.

Day, Group Captain John R. "The Contribution of Support Helicopters to Airmobile Operations," *Journal of the Royal United Services Institution*, supplement 134.12 (1989): 18–20.

Design for Military Operations—The British Military Doctrine. Prepared under the direction of the Chief of the General Staff, British Army. London: HMSO, 1989.

Farndale, General Sir Martin. "The Role of Helicopters in the Central Region to the Year 2000," *Journal of the Royal United Services Institution*, supplement 134.12 (1989): 2–6.

———. "The Use of Helicopters on the European Battlefield," *Army Quarterly and Defence Journal* 119.4 (1989): 137–42.

Gilbert, Major General G.C.A. "An Air-Mobile Force for the Future," *Journal of the Royal United Services Institution* 119.12 (1974): 51–53.

Goodman, Major General J.D.W. "The British Army Air Corps," *Army Air Corps Journal* (1987): 9–10.

Grist, Brigadier R. D. "Airmobile Forces in Central Europe," *Journal of the Royal United Services Institution* 133.3 (1988): 41–48 .

Kennedy, Lt. Colonel A.I.G. "Airmobility—A Practitioner's View," *British Army Review* 85 (1987): 16–20.

Learmont, Lt. General Sir John. "Military Theory and the Helicopter," *Army Air Corps Journal* (1989): 7–11.

Ramsbotham, Lt. General Sir David. "Battlefield Helicopter Operations and Airmobility," *Army Quarterly and Defence Journal* 119.4 (1989): 143–48.

———. "British Developments," *Journal of the Royal United Services Institution*, supplement 134.12 (1989): 7–10.

The Role of the Helicopter in the Land Battle. London: RUSI, 1972.

Secondary Sources

Beaver, Paul. "British Army Helicopter Needs in the 1990s," *Defence Helicopter World* 5.2–3 (1986): 6–14.

———. *Today's Army Air Corps*. Wellingborough, Northants.: PSL, 1987.

Everett-Heath, John. *British Military Helicopters*. London: Arms and Armour, 1986.

Hickey, Michael. *Out of the Sky*. London: Mills and Boon, 1979.

WEST GERMANY AND FRANCE—SELECTED SOURCES

Primary Sources

Bernhardt, Generalmajor Georg. "Airmobile Formations: The Federal Republic of Germany," *International Defense Review* 23.4 (1990): 379–81.

———. "The West German View," *Journal of the Royal United Services Institution*, supplement 134.4 (1989): 14–17.

Cannet, General Maurice. *"L'aéromobilité: le sens d'une évolution,"* *Armées d'Aujourd'hui* 35 (1978): 44–45.

Csoboth, Brigadier General. "Air-to-air fighting with Helicopters I," *International Defense Review* 22.2 (1989): 152–54.

de Reviers de Mauny, Colonel Xavier. *"La carte maîtresse: l'aéromobilité,"* *Armées d'Aujourd'hui* 68 (1982): 28–29, 47.

Dreibing, Brigadier Hans E. "Views on Airmobility," *U.S. Army Aviation Digest* 21.3 (1975): 1–3, 10–11.

Imbot, General Renée. "The French Rapid Action Force," *NATO's Sixteen Nations* 28, spec. ed. 1 (1983): 34–39.

Préaud, General H. "Airmobile Formations: France," *International Defense Review* 23.4 (1990): 377–79.
von Senger and Etterlin, General Ferdinand M. "New Operational Dimensions," *Journal of the Royal United Services Institution* 128.6 (1983): 11–15.

Secondary Sources

Beaver, Paul. "The French ALAT in the 1990s," *Defence Helicopter World* 4.6–8 (1985): 8.
Charnbost, G., and K. G. Benz. "The Franco-German Combat Helicopter—Reconciling Contradictory Requirements," *International Defense Review* 17.5 (1984): 575–84.
Jackson, Paul. "ALAT Regroups," *Armed Forces* 5.3 (1986): 137–39.

GENERAL—SELECTED SOURCES

Air-land Warfare

Hussain, Farooq, Ian Kemp, and Philip McCarty. *The Future of the Military Helicopter*. London: Royal United Services Institution, 1986.
Simpkin, Richard E. *Anti-tank—An Airmechanised Response to Armoured Threats in the 90s*. Oxford: Brassey's, 1982.
———. *Race to the Swift*. London: Brassey's, 1985.
———. *Tank Warfare*. London: Brassey's, 1979.

Decisionmaking and Innovation

Allison, Graham T. *Essence of Decision*. Boston: Little, Brown, 1971.
Doughty, Robert A. *The Evolution of U.S. Army Tactical Doctrine 1946–76*. Ft. Leavenworth, KS: Combat Studies Institute, 1979.
———. *The Seeds of Disaster—the Development of French Army Doctrine 1919–1939*. Hamden, CT: Archon Books, 1985.
Evangelista, Matthew. *Innovation and the Arms Race*. Ithaca, N.Y.: Cornell University Press, 1988.
Kozak, David C., and John M. Keagle (ed.). *Bureaucratic Politics and National Security*. Boulder, CO: Lynne Riener, 1988.
Millett, Allan R. "Military Reform in America," *Air University Review* 36.6 (1985): 3.
Paret, Peter. *Innovation and Reform in Warfare*. Colorado: USAF Academy, 1966.
Posen, Barry R. *The Sources of Military Doctrine*. Ithaca, N.Y.: Cornell University Press, 1984.

Index

About the Author

MATTHEW ALLEN lives in Western Australia where he is an independent scholar specializing in history. He is currently researching the history of military conscription in Australia.